Gengsheng Lawrence Zeng
Medical Image Reconstruction

Also of interest

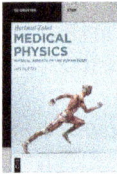

Medical Physics
Hartmut Zabel, 2023
Volume 1 Physical Aspects of the Human Body
ISBN 978-3-11-075691-3, e-ISBN (PDF) 978-3-11-075695-1

Medical Physics
Hartmut Zabel, 2023
Volume 2 Physical Aspects of Diagnostics
ISBN 978-3-11-075702-6, e-ISBN (PDF) 978-3-11-075709-5

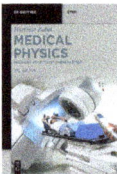

Medical Physics
Hartmut Zabel, 2023
Volume 3 Physical Aspects of Therapeutics
ISBN 978-3-11-116867-8, e-ISBN (PDF) 978-3-11-116873-9

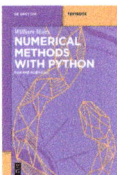

Numerical Methods with Python.
for the Sciences
William Miles, 2023
ISBN 978-3-11-077645-4, e-ISBN (PDF) 978-3-11-077664-5

Multi-level Mixed-Integer Optimization.
Parametric Programming Approach
Styliani Avraamidou, Efstratios Pistikopoulos, 2022
ISBN 978-3-11-076030-9, e-ISBN (PDF) 978-3-11-076031-6

Gengsheng Lawrence Zeng

Medical Image Reconstruction

From Analytical and Iterative Methods to Machine Learning

2nd Edition

DE GRUYTER

Author
Prof. Dr. Gengsheng Lawrence Zeng
Department of Computer Science
Utah Valley University
Orem, UT 84058
USA
larry.zeng@uvu.edu

Department of Radiology and Imaging Sciences
University of Utah
Salt Lake City, UT 84108
USA
larry.zeng@hsc.utah.edu

ISBN 978-3-11-105503-9
e-ISBN (PDF) 978-3-11-105540-4
e-ISBN (EPUB) 978-3-11-105570-1

Library of Congress Control Number: 2023933613

Bibliographic information published by the Deutsche Nationalbibliothek
The Deutsche Nationalbibliothek lists this publication in the Deutsche Nationalbibliografie;
detailed bibliographic data are available on the Internet at http://dnb.dnb.de.

© 2023 Walter de Gruyter GmbH, Berlin/Boston
Cover image: m_pavlov/iStock/Getty Images Plus
Typesetting: Integra Software Services Pvt. Ltd.
Printing and binding: CPI books GmbH, Leck

www.degruyter.com

In loving memory of my parents
Yuqing Lu and Yiduo Zeng

To my family
Ya Li
Andrew Fang
Kathy Fang
Megan Zeng

Preface

This tutorial text introduces the classical and modern image reconstruction technologies to the general audience. It covers the topics in two-dimensional parallel-beam and fan-beam imaging, and three-dimensional parallel ray, parallel plane, and cone-beam imaging. Both analytical and iterative methods are presented. The applications in X-ray CT (computed tomography), single-photon emission computed tomography, positron emission tomography, and magnetic resonance imaging are also discussed. Contemporary research results in exact region-of-interest reconstruction with truncated projections, Katsevich's cone-beam filtered backprojection (FBP) algorithm, and reconstruction with highly undersampled data are also included in this book.

Chapter 8 of the book is devoted to the techniques of using a fast analytical FBP algorithm to emulate an iterative reconstruction with an edge-preserving denoising constraint. These techniques are the author's most recent research results, and the author also produced CT images of some patients leading to the third place winner of the 2016 Low-Dose CT Grand Challenge.

This book is written in an easy-to-read style. The readers who intend to get into medical image reconstruction will gain the general knowledge of the field in a painless way. I hope you enjoy reading it as much as I enjoy writing (and drawing) it.

Larry Zeng, Ph.D. (larry.zeng@uvu.edu, larry.zeng@hsc.utah.edu)
Professor of Computer science, Utah Valley University, Orem, Utah, USA
Adjunct Professor of Radiology and Imaging science, University of Utah, Salt Lake City, Utah, USA

August 8, 2016, in Salt Lake City

https://doi.org/10.1515/9783111055404-202

Contents

1 Basic principles of tomography

This is an introductory chapter that presents the fundamental concepts in tomography. It first defines what the tomography is and shows how a tomographic image can be obtained from its measurements using two simple examples. The concept of projection is then explained. Next, the filtered backprojection (FBP) image reconstruction method is introduced using a point source example. Finally, the concept of backprojection is discussed.

1.1 Tomography

The Greek word *tomos* means a section, a slice, or a cut. *Tomography* is the process of imaging a cross section. For example, if you are given a watermelon and would like to see inside, the easiest way to do so is to cut it open (Figure 1.1). Clearly, this approach to obtain a cross-sectional image is not a good idea in medicine. Nobody wants to be cut open in order to see what is inside.

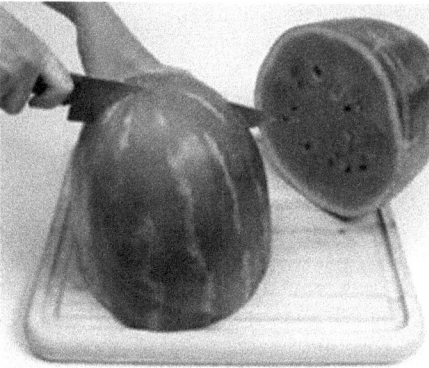

Fig. 1.1: Cutting open to see what is inside.

Let us look at another example. You are visiting a small park, which is closed for maintenance. You walk around the park and take a few pictures of it. After you get home, you can use your pictures to make a map of the park. To make your life easier, let us assume that there are two large trees in the park, and you take two pictures from the east and the south (Figure 1.2, left). Using these two pictures, you can map out where these two trees are (Figure 1.2, right). This can be done by positioning the pictures at the original orientations at which the pictures were taken, drawing a line from each tree, and finding the intersections. If you have enough pictures, it is not hard to find out where the trees are.

https://doi.org/10.1515/9783111055404-001

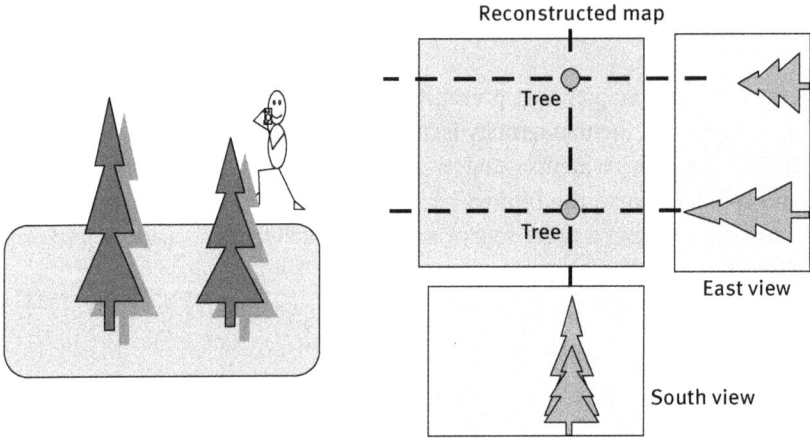

Fig. 1.2: Reconstruct a map from two pictures.

Tomography is a mathematical problem. Let us do a fun mathematical exercise here. We have a 2×2 matrix. We do not tell you what it is yet. Here are the hints: The sum of the first row is 5, the sum of the second row is 4, the sum of the first column is 7, and the sum of the second column is 2 (see Figure 1.3). Now, you figure out what this 2×2 matrix is.

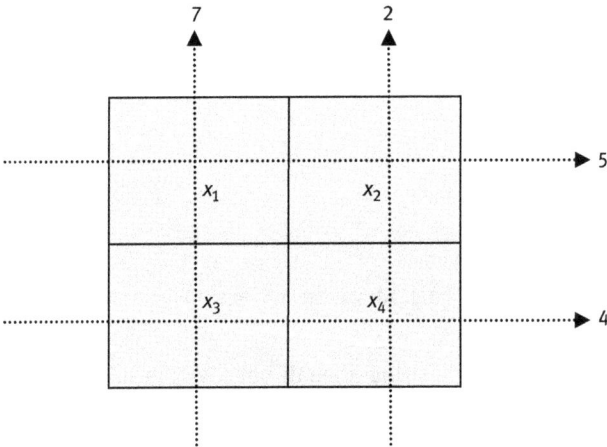

Fig. 1.3: A 2×2 matrix puzzle.

You can solve this puzzle by setting up a system of linear equations with the matrix entries as unknowns:

$$x_1 + x_2 = 5, \quad x_3 + x_4 = 4, \quad x_1 + x_3 = 7, \quad \text{and} \quad x_2 + x_4 = 2. \tag{1.1}$$

Solving these equations, you will get

$$x_1 = 3, \quad x_2 = 2, \quad x_3 = 4, \quad \text{and} \quad x_4 = 0. \tag{1.2}$$

Congratulations! You have just mathematically solved a tomography problem. Usually, a tomography problem is solved mathematically; hence, the term "computed tomography" (CT). The row sum or column sum in this example can be generalized as a *ray sum*, a *line integral*, or a *projection*. The procedure to produce a tomographic image from projections is called *image reconstruction*.

What if the tomography problem gets more complicated? If there are many more trees in the park, taking only two pictures may not provide us enough information to map out the park. If the matrix size is larger than 2×2, the row sum and column sum alone do not form enough equations to solve for the matrix entries.

We need more views! For the matrix identification case, we need to sum the matrix diagonally at various angles. In turn, more sophisticated mathematics is required to solve the tomography problem.

1.2 Projection

In order to understand the concept of projection (ray sum, line integral, or Radon transform), we will present more examples here.

In the first example, the object is a uniform disk on the x–y plane, the center of the disk is at the origin, and the (linear) density of the disk is ρ (see Figure 1.4). The projection (i.e., the line integral) of this object can be calculated as the chord length t times the linear density ρ, that is,

$$p(s) = \rho t = 2\rho \sqrt{R^2 - s^2} \text{ if } |s| < R;\ p(s) = 0 \text{ otherwise.} \tag{1.3}$$

In this particular example, the projection $p(s)$ is the same for any view angle θ, which is the orientation of the detector.

If the object is more complicated, the projection $p(s, \theta)$ is angle θ dependent (see Figure 1.5).

In the next example we use a point source on the y-axis to further illustrate the angle θ dependency of the projection $p(s, \theta)$ (see Figure 1.6). Here we pay attention to the location s of the spike on the one-dimensional detector, which can be evaluated as

$$s = r \sin \theta. \tag{1.4}$$

This is a sine function with respect to θ. If you display this point source projection data set $p(s, \theta)$ in the s–θ coordinate system (see Figure 1.6, right), you will see the trajectory of a sine wave. Because of this observation, people refer to the projection data set as a *sinogram*.

The fourth example is a discrete object of a 2×2 matrix, similar to that in Figure 1.3. The detector is also discrete with four detector bins (see Figure 1.7). Each matrix element

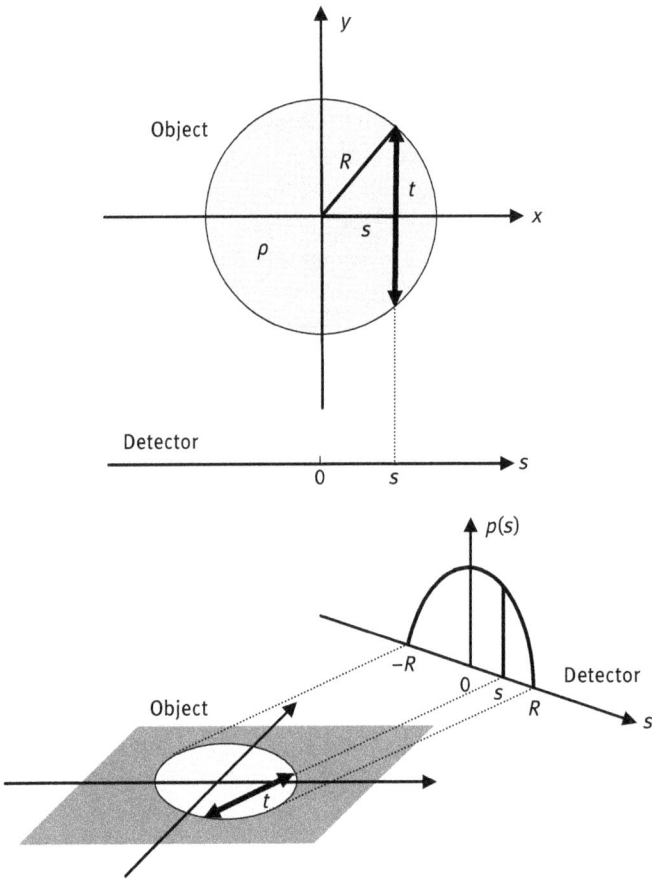

Fig. 1.4: The line integral across the disk is the length of a chord times the density.

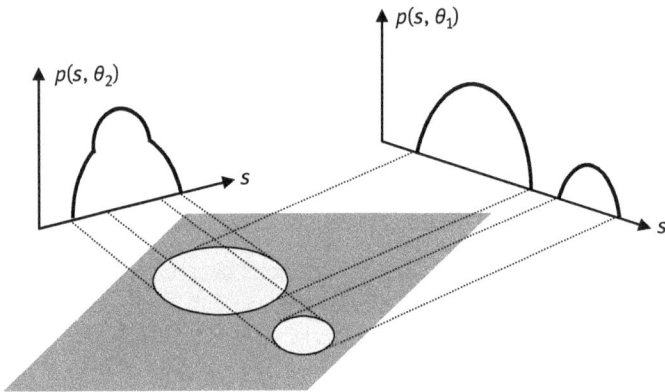

Fig. 1.5: The projections are usually different at a different view angle.

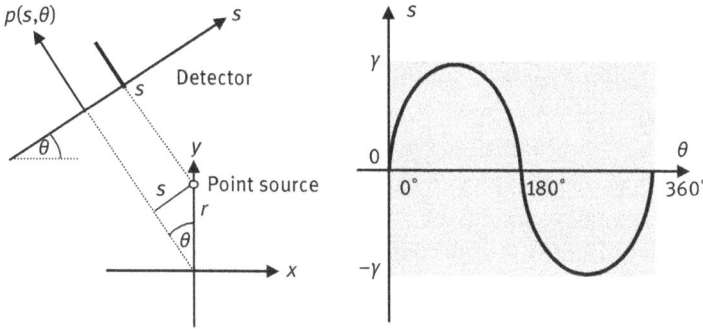

Fig. 1.6: A sinogram is a representation of the projections on the $s-\theta$ plane.

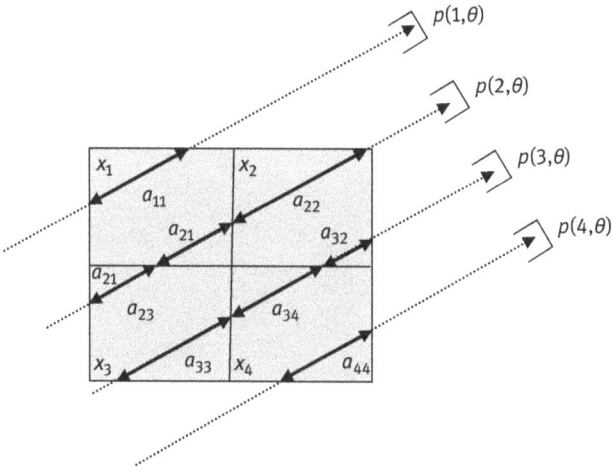

Fig. 1.7: The projections are weighted by the line length within each pixel.

represents a uniform pixel, and x_i ($i = 1, 2, 3, 4$) is the linear density in the ith pixel. Here we would like to find the line-integral $p(s, \theta)$ of the matrix at a view angle θ. The quantity a_{ij} ($i = 1, 2, 3, 4$ and $j = 1, 2, 3, 4$) is the segment length of the path toward the detector bin i within the pixel j, and $a_{ij} = 0$ if the jth pixel is not on the path to the ith detector bin. The projection $p(i, \theta)$ is calculated as

$$p(i, \theta) = a_{i1}x_1 + a_{i2}x_2 + a_{i3}x_3 + a_{i4}x_4 \qquad \text{for } i = 1, 2, 3, 4. \qquad (1.5)$$

1.3 Image reconstruction

In this section, we illustrate the common image reconstruction strategy by considering a point source. Let us consider an empty two-dimensional (2D) plane with an x–y coordinate system, and we place a small dot with a value, say 1, somewhere on this plane and not necessarily at the origin (see Figure 1.8). We now imagine that there is a detector (e.g., a camera) rotating around the origin, acquiring images of projections. At a particular angle θ, we denote the projection as $p(s, \theta)$, where s is the coordinate on the detector.

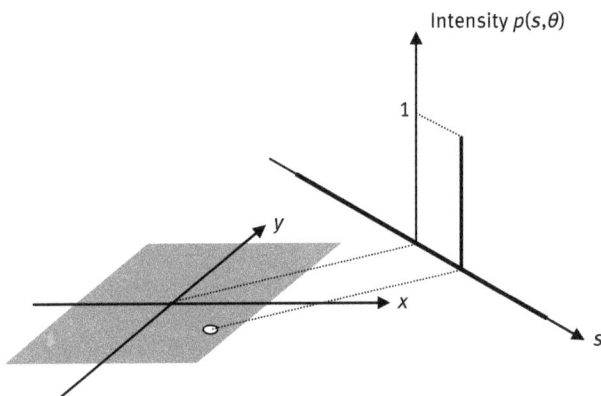

Fig. 1.8: Projection of a point source object.

The projection $p(s, \theta)$ is formed by drawing a line across the x–y plane, orthogonal to the detector, and meeting on the detector at location s. Then we evaluate the line integral along this line, and the integral value is $p(s, \theta)$. In our example, if the line does not touch the point source, $p(s, \theta)$ is zero. If the line passes through the point source, then $p(s, \theta)$ is 1.

Now we are going to reconstruct the image using the projections $p(s, \theta)$. Our strategy is similar to that in the tree-map example in Section 1.1, where we drew a line from each tree on the detector and found the location of the intersections. In image reconstruction, we not only need to find the location but also the intensity value of the object of interest.

As shown in Figure 1.9(a), a number of projections are taken from the point source at various view angles. We attempt to reconstruct the point source image in the following manner.

When you look at the projections $p(s, \theta)$ at one view θ, you see a spike of intensity 1. This spike is the sum of all activities along the projection path. To reconstruct the image, you must redistribute the activity in the spike back to its original path. The problem is that you do not know where you need to put more activity along the path and where you put less. Before you give up, you decide to put equal amounts of activity everywhere along the path, and the amount is the magnitude of the projection spike (see Figure 1.9b). If you do that for few more angles, you will have the situation as shown in Figure 1.9(c). Due to the superposition effect, there will be a tall spike in the x–y plane at the location of the point source.

What you have just done is a standard mathematical procedure called *backprojection*. If you backproject from all angles from 0° to 360°, you will produce an image similar to the one shown in Figure 1.9(d).

After backprojection, the image is still not quite the same as the original image but rather is a blurred version of it. To eliminate the blurring, we introduce negative "wings" around the spike in the projections before backprojection (see Figure 1.9e). The procedure of adding negative wings around the spike is called *filtering*. The use of the negative wings results in a clear image (see Figure 1.9f). This image reconstruction algorithm is very common and is referred to as an FBP algorithm.

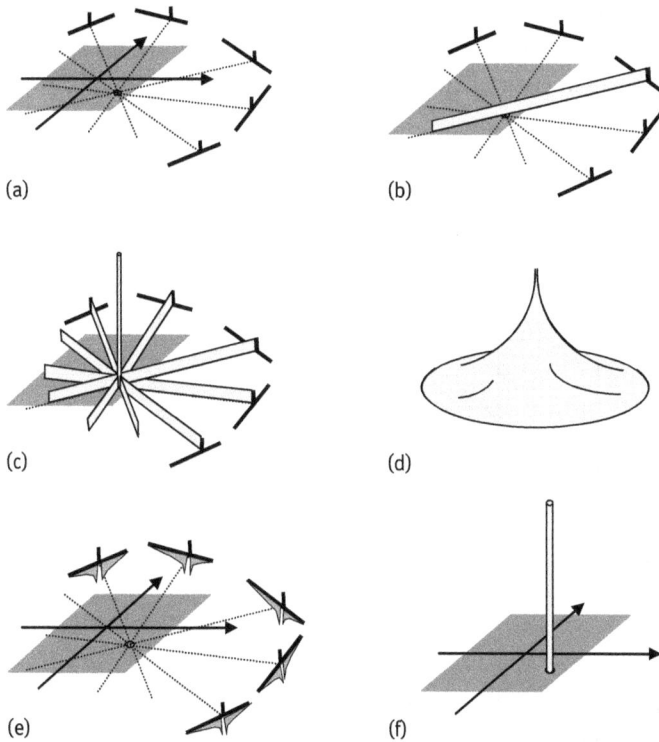

(a)

(b)

(c)

(d)

(e)

(f)

Fig. 1.9: Reconstruction of a point source image by backprojecting unfiltered and filtered data. (a) Project the point source; (b) backproject from one view; (c) backproject from a few views; (d) backproject from all views; (e) add negative wings; and (f) backproject modified data.

In this section, we use a point source to illustrate the usefulness of filtering and backprojection with many views in image reconstruction. We must point out that if the object is a point source, we only need two views to reconstruct the image, just like the map-making example in Section 1.1.

1.4 Backprojection

One must first define projection before backprojection can be defined. We must make it clear that backprojection is not the inverse of projection. Backprojection alone is not sufficient to reconstruct an image. After you backproject the data, you do not get the original image back. We will illustrate this point by a simple discrete 2×2 problem below (see Figure 1.10).

The original image is defined as $x_1 = 3$, $x_2 = 2$, $x_3 = 4$, and $x_4 = 0$. The associated projections are $p(1, 0°) = 7$, $p(2, 0°) = 2$, $p(1, 270°) = 5$, and $p(2, 270°) = 4$. The projections are formed one view at a time (see Figure 1.10). The backprojected image is also formed one view at a time. The final backprojected image is the summation of the backprojections from all views, as shown in Figure 1.11. Please note that the backprojected image is different from the original image.

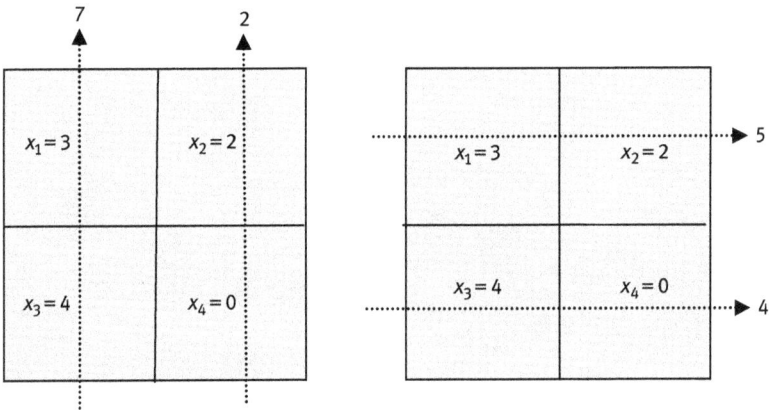

Fig. 1.10: View-by-view projection.

Even though the backprojected image is not the original image, they are closely related. Their relationship will be further discussed in Chapter 2.

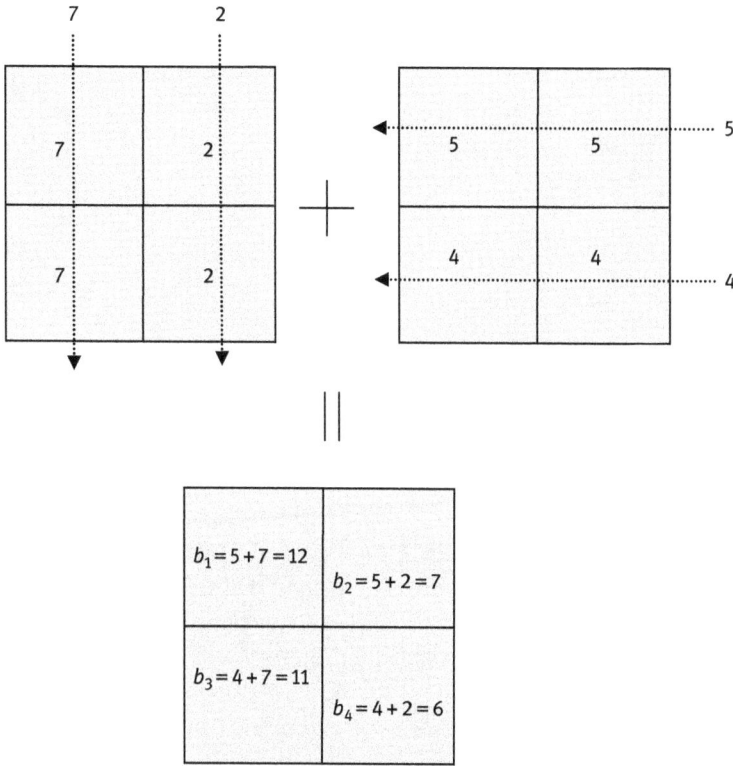

Fig. 1.11: View-by-view backprojection, then sum all backprojected images.

1.5 Mathematical expressions

In every chapter we dedicate a section especially to mathematical expressions. These mathematical expressions help the advanced readers to better grasp the main concepts which are discussed in the chapter. Mathematical expressions of projection and backprojection for 2D parallel-beam imaging are presented in this section. The Dirac δ function has an important role in analytic algorithm development; its definition and some properties are also covered in this section.

1.5.1 Projection

Let $f(x, y)$ be the density function in the x–y plane. The projection (ray sum, line integral, or Radon transform) $p(s, \theta)$ has many equivalent expressions such as

$$p(s, \theta) = \int\limits_{-\infty}^{\infty} \int\limits_{-\infty}^{\infty} f(x,y)\delta(x \cos \theta + y \sin \theta - s)dxdy, \qquad (1.6)$$

$$p(s, \theta) = \int\limits_{-\infty}^{\infty} \int\limits_{-\infty}^{\infty} f(x,y)\delta\left(\vec{x} \cdot \vec{\theta} - s\right)dxdy, \qquad (1.7)$$

$$p(s, \theta) = \int\limits_{-\infty}^{\infty} f(s \cos \theta - t \sin \theta, s \sin \theta + t \cos \theta)dt, \qquad (1.8)$$

$$p(s, \theta) = \int\limits_{-\infty}^{\infty} f\left(s\vec{\theta} + t\vec{\theta}^{\perp}\right)dt, \qquad (1.9)$$

$$p(s, \theta) = \int\limits_{-\infty}^{\infty} f_\theta(s + t)dt, \qquad (1.10)$$

where $\vec{x} = (x, y)$, $\vec{\theta} = (\cos \theta, \sin \theta)$, $\vec{\theta}^{\perp} = (-\sin \theta, \cos \theta)$, δ is the Dirac delta function, and f_θ is the function f rotated by θ clockwise. We assume that the detector rotates counterclockwise around the object or that the object rotates clockwise while the detector stays still. The coordinate systems are shown in Figure 1.12.

1.5.2 Backprojection

Backprojection is the adjoint of projection. Here "adjoint" is a mathematical term. It refers to the conjugate transpose in linear algebra. For a real matrix A, its adjoint is simply the transposed matrix A^T. In the discrete case, as in Section 1.4, the projection is

$$P = AX, \qquad (1.11)$$

where X represents an image, but in a column form. For example, the 2×2 image is expressed as (see Figure 1.3 or 1.10)

$$X = [x_1, x_2, x_3, x_4]^T. \qquad (1.12)$$

The column matrix P represents the projections. If we use the example in Figure 1.10,

$$P = [p(1, 0°), \ p(2, 0°), \ p(1, 270°), p(2, 270°)]^T = [7, \ 2, \ 5, \ 4]^T. \qquad (1.13)$$

The matrix A is the projection operator. Its entries a_{ij} are defined in Figure 1.7. Using the example of Figure 1.10, the backprojection of P can be calculated using matrix multiplication as

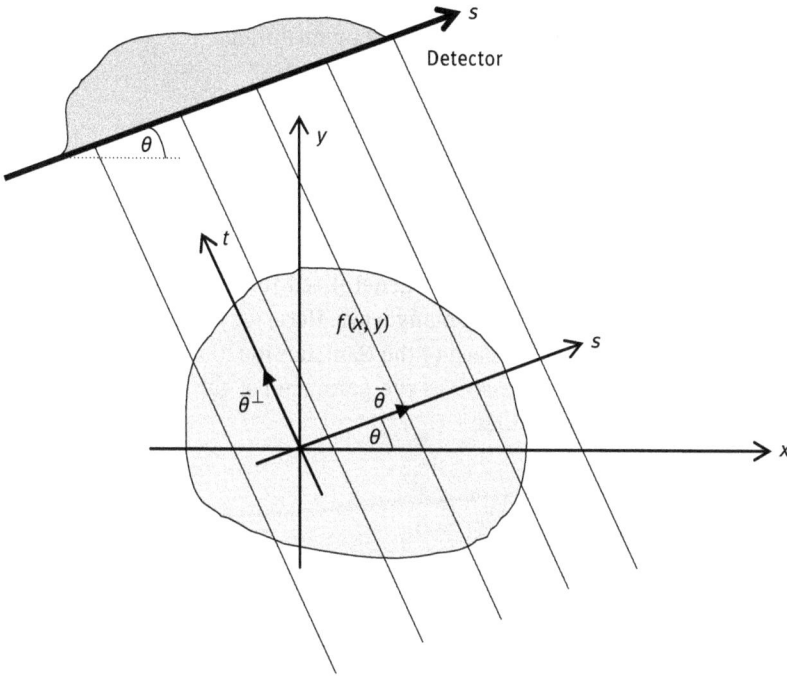

Fig. 1.12: Coordinate systems for 2D parallel-beam imaging.

$$B = A^T P = \begin{bmatrix} 1 & 0 & 1 & 0 \\ 0 & 1 & 0 & 1 \\ 1 & 1 & 0 & 0 \\ 0 & 0 & 1 & 1 \end{bmatrix}^T \begin{bmatrix} 7 \\ 2 \\ 5 \\ 4 \end{bmatrix} = \begin{bmatrix} 1 & 0 & 1 & 0 \\ 0 & 1 & 1 & 1 \\ 1 & 0 & 0 & 1 \\ 0 & 1 & 0 & 1 \end{bmatrix} \begin{bmatrix} 7 \\ 2 \\ 5 \\ 4 \end{bmatrix} = \begin{bmatrix} 12 \\ 7 \\ 11 \\ 6 \end{bmatrix}, \tag{1.14}$$

which is the same as the result obtained "graphically" in Figure 1.11.

For the continuous case, the backprojection image $b(x, y)$ can be expressed in the following equivalent ways:

$$b(x, y) = \int_0^\pi p(s, \theta)|_{s = x\cos\theta + y\sin\theta} d\theta, \tag{1.15}$$

$$b(x, y) = \int_0^\pi p(s, \theta)|_{s = \vec{x} \cdot \vec{\theta}} d\theta, \tag{1.16}$$

$$b(x, y) = \int_0^\pi p\left(\vec{x}, \vec{\theta}, \theta\right) d\theta, \tag{1.17}$$

$$b(x,y) = \frac{1}{2} \int_0^{2\pi} p(x\cos\theta + y\sin\theta, \theta)\,d\theta. \tag{1.18}$$

1.5.3 The Dirac δ-function

The Dirac δ-function is not a regular function that maps a value in the domain to a value in the range. The Dirac δ-function is a generalized function or a distribution function. The δ-function can be defined in many ways. Here, we use a series of Gaussian functions to define the δ-function. Each of the Gaussian functions (see Figure 1.13) has a unit area underneath its curve, and as the parameter n gets larger, the curve gets narrower and taller (see Figure 1.13):

$$\left(\frac{n}{\pi}\right)^{1/2} e^{-nx^2}. \tag{1.19}$$

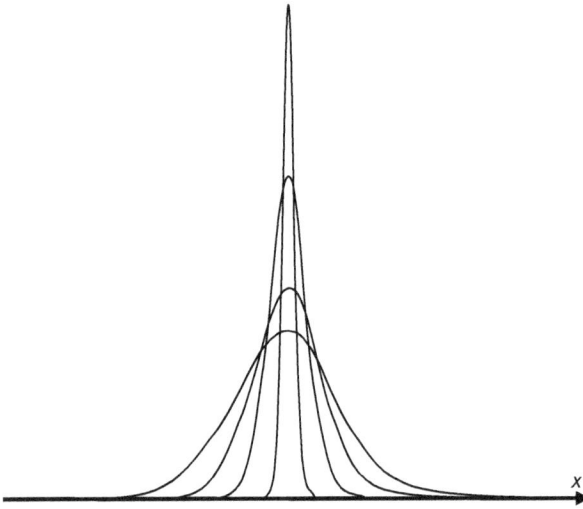

Fig. 1.13: Using a train of Gaussian functions to define the δ-function.

Let $f(x)$ be the smooth function that is differentiable everywhere with any order and $\lim_{x\to\infty} x^N f(x) = 0$ for all N. Then the δ-function is defined implicitly as

$$\lim_{x\to\infty} \int_{-\infty}^{\infty} \left(\frac{n}{\pi}\right)^{1/2} e^{-nx^2} f(x)\,dx = \int_{-\infty}^{\infty} \delta(x)f(x)\,dx = f(0). \tag{1.20}$$

The δ-function has some properties:

$$\int_{-\infty}^{\infty} \delta(x-a)f(x)dx = \int_{-\infty}^{\infty} \delta(x)f(x+a)dx = f(a),\tag{1.21}$$

$$\int_{-\infty}^{\infty} \delta(ax)f(x)dx = \frac{1}{|a|}f(0),\tag{1.22}$$

$$\int_{-\infty}^{\infty} \delta^{(n)}(x)f(x)dx = (-1)^n f^{(n)}(0) \; [\text{the } n\text{th - order derivative}],\tag{1.23}$$

$$\delta(g(x))f(x) = f(x) \sum_n \frac{1}{|g'(\lambda_n)|}\delta(x-\lambda_n),\tag{1.24}$$

where λ_ns are the zeros of $g(x)$.

In 2D and 3D cases, $\delta(\vec{x}) = \delta(x)\delta(y)$ and $\delta(\vec{x}) = \delta(x)\delta(y)\delta(z)$, respectively. In the last property, $|g'|$ will be replaced by $|\mathrm{grad}(g)| = \sqrt{(\partial g/\partial x)^2 + (\partial g/\partial y)^2}$ and $|\mathrm{grad}(g)| = \sqrt{(\partial g/\partial x)^2 + (\partial g/\partial y)^2 + (\partial g/\partial z)^2}$, respectively, in 2D and 3D cases.

In 2D imaging, we use a 2D δ-function $\delta(\vec{x} - \vec{x}_0)$ to represent a point source at location $\vec{x} = \vec{x}_0$. The Radon transform of $f(\vec{x}) = \delta(\vec{x} - \vec{x}_0) = \delta(x - x_0)\delta(y - y_0)$ is given as

$$p(s,\theta) = \int_{-\infty}^{\infty}\int_{-\infty}^{\infty} f(\vec{x})\delta\left(\vec{x}\cdot\vec{\theta} - s\right)d\vec{x},\tag{1.25}$$

$$p(s,\theta) = \int_{-\infty}^{\infty}\int_{-\infty}^{\infty} \delta(x,x_0)\delta(y-y_0)\delta(x\cos\theta + y\sin\theta - s)dxdy,\tag{1.26}$$

$$p(s,\theta) = \int_{-\infty}^{\infty} \delta(y-y_0)\left[\int_{-\infty}^{\infty} \delta(x,x_0)\delta(x\cos\theta + y\sin\theta - s)dx\right]dy,\tag{1.27}$$

$$p(s,\theta) = \int_{-\infty}^{\infty} \delta(y-y_0)\delta(x_0\cos\theta + y\sin\theta - s)dy,\tag{1.28}$$

$$p(s,\theta) = \delta(x_0\cos\theta + y_0\sin\theta - s),\tag{1.29}$$

which is a sinogram similar to that shown in Figure 1.6.

1.6 Worked examples

Example 1: If you see two separate trees on both views, can you uniquely reconstruct the map of trees (see Figure 1.14)? If not, you may need to take more pictures. If you are only allowed to take one more picture, at which direction should you take the picture?

Solution
Both of the two situations as shown in Figure 1.15 can satisfy the two views.
   If we take another picture at 45°, we are able to solve the ambiguity.

Example 2: Find the projections of a uniform disk. The center of the disk is not at the center of detector rotation.

Solution
We already know that if the center of the disk is at the center of detector rotation, the projection can be evaluated as

$$p(s, \theta) = 2\rho\sqrt{R^2 - s^2} \quad \text{if } |s| < R; \ p(s, \theta) = 0 \text{ otherwise.} \tag{1.30}$$

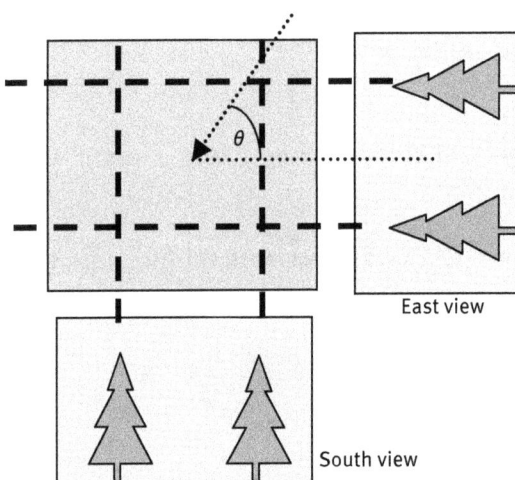

Fig. 1.14: Two trees can be seen on both views.

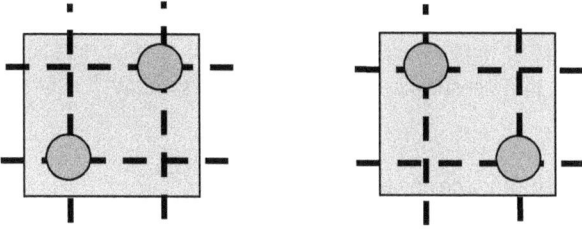

Fig. 1.15: Two potential solutions for the mapping problem.

Without loss of generality, we now assume that the center of the disk is on the positive x-axis with the coordinates $(r, 0)$ (Figure 1.16).

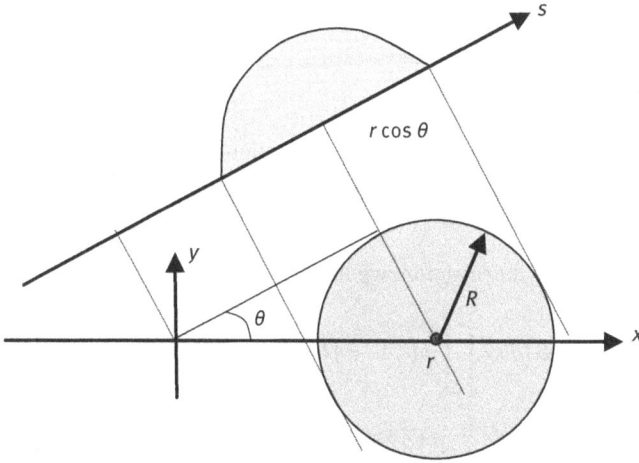

Fig. 1.16: Imaging of an off-centered disk.

For this new setup, we need to shift the projection data on the s-axis. The shifting distance is $r \cos \theta$. That is,

$$p(s, \theta) = 2\rho\sqrt{R^2(s - r \cos \theta)^2} \text{ if } |s = r \cos \theta| < R; \ p(s, \theta) = 0 \text{ otherwise.} \quad (1.31)$$

Example 3: Show that the parallel-beam data redundancy condition is $p(s, \theta) = p(-s, \theta + \pi)$.

Proof. Using the projection definition in Section 1.5, we have

$$p(-s, \theta, \pi) = \int_{-\infty}^{\infty} \int_{-\infty}^{\infty} f(x,y)\delta(x\cos(\theta+\pi)+y\sin(\theta+\pi)-(-s))dxdy$$

$$= \int_{-\infty}^{\infty} \int_{-\infty}^{\infty} f(x,y)\delta(-x\cos\theta-y\sin\theta+s)dxdy$$

$$= \int_{-\infty}^{\infty} \int_{-\infty}^{\infty} f(x,y)\delta(-(x\cos\theta+y\sin\theta-s))dxdy$$

$$= \int_{-\infty}^{\infty} \int_{-\infty}^{\infty} f(x,y)\delta(x\cos\theta+y\sin\theta-s)dxdy$$

$$= p(s, \theta). \tag{1.32}$$

[The δ-function is an even function.]

Example 4: Show that the point spread function of the projection/backprojection operator is $1/r$, where $r = \|\vec{x} - \vec{x}_0\|$ and the point source object is $f(\vec{x}) = \delta(\vec{x} - \vec{x}_0)$.

Proof. Using the definition of the backprojection, we have

$$b(\vec{x}) = \int_0^{\pi} p\left(\vec{x}\cdot\vec{\theta}, \theta\right)d\theta = \int_0^{\pi} \int_{-\infty}^{\infty} f\left(\left(\vec{x}\cdot\vec{\theta}\right)\vec{\theta}+t\vec{\theta}^{\perp}\right)dtd\theta. \tag{1.33}$$

We realize that the line integral $\int_{-\infty}^{\infty} f\left(\left(\vec{x}\cdot\vec{\theta}\right)\vec{\theta}+t\vec{\theta}^{\perp}\right)dt$ is along the line that passes through the point $\vec{x}$ and in the direction of $\vec{\theta}^{\perp}$ (see Figure 1.17), so

$$\int_{-\infty}^{\infty} f\left(\left(\vec{x}\cdot\vec{\theta}\right)\vec{\theta}+t\vec{\theta}^{\perp}\right)dt = \int_{-\infty}^{\infty} f\left(\vec{x}-\hat{t}\vec{\theta}^{\perp}\right)d\hat{t}. \tag{1.34}$$

Therefore, the projection/backprojection image can be obtained as

$$b(\vec{x}) = \int_0^{\pi} \int_{-\infty}^{\infty} f\left(\vec{x}-\hat{t}\vec{\theta}^{\perp}\right)d\hat{t}d\theta. \tag{1.35}$$

Let $\hat{\vec{x}} = \hat{t}\vec{\theta}^{\perp}$ with $\|\hat{\vec{x}}\| = |\hat{t}|$, $d\hat{\vec{x}} = |\hat{t}|d\hat{t}d\theta$. The above expression becomes

$$b(\vec{x}) = \int_{-\infty}^{\infty} \int_{-\infty}^{\infty} \frac{f\left(\vec{x}-\hat{\vec{x}}\right)}{\|\hat{\vec{x}}\|}d\hat{\vec{x}}. \tag{1.36}$$

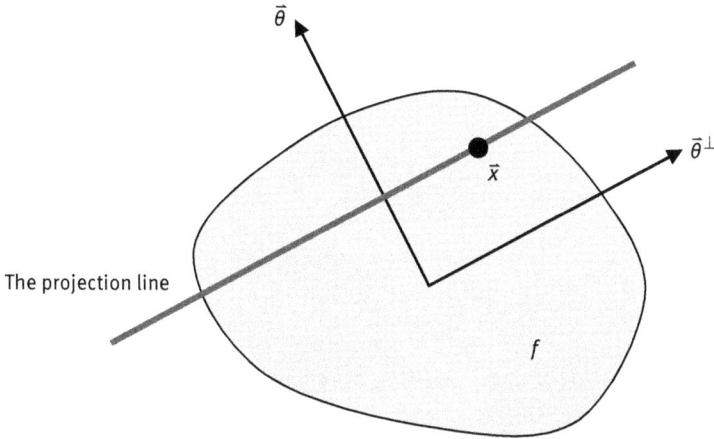

Fig. 1.17: The line integral is performed on a line passing through the backprojection point.

Let $f(\vec{x}) = \delta(\vec{x} - \vec{x}_0)$. The point spread function of the projection/backprojection operator is

$$b(\vec{x}) = \int\limits_{-\infty}^{\infty} \int\limits_{-\infty}^{\infty} \frac{\delta\left(\vec{x} - \vec{x}_0 - \widehat{\vec{x}}\right)}{||\widehat{\vec{x}}||} d\widehat{\vec{x}} = \frac{1}{||\vec{x} - \vec{x}_0||} = \frac{1}{r}. \qquad (1.37)$$

Example 5: Evaluate $\displaystyle\int\limits_{-\infty}^{\infty} \delta\left(e^{2(x-3)(x+4)} - 1\right) f(x)\,dx.$

Solution
Let

$$g(x) = e^{2(x-3)(x+4)} - 1. \qquad (1.38)$$

Solving $g(x) = e^{2(x-3)(x+4)} - 1 = 0$, we obtain the zeros of $g(x)$ as

$$\lambda_1 = 3 \quad \text{and} \quad \lambda_2 = -4. \qquad (1.39)$$

The derivative of $g(x)$ is

$$g'(x) = 2[(x-3) + (x+4)]e^{2(x-3)(x+4)} = 2(2x+1)e^{2(x-3)(x+4)}. \qquad (1.40)$$

Thus, at the two zeros of $g(x)$, we have

$$g'(3) = 14 \quad \text{and} \quad g'(-4) = -14. \qquad (1.41)$$

Hence,

$$
\int_{-\infty}^{\infty} \delta\left(e^{2(x-3)(x+4)} - 1\right)f(x)dx = \int_{-\infty}^{\infty} \frac{\delta(x-3)}{|14|}f(x)dx + \int_{-\infty}^{\infty} \frac{\delta(x+4)}{|14|}f(x)dx
$$
$$
= \frac{f(3) + f(-4)}{14}
$$

(1.42)

1.7 Summary

- Tomography is a process of taking projection data and converting the data into cross-sectional images. Projection data from multiple views are required.
- A projection is a line integral (or ray sum, Radon transform) of an object. Projection data are acquired with detectors. Objects overlap on the detectors.
- Backprojection is a superposition procedure and it sums the data from all projection views. Backprojection evenly distributes the projection domain data back along the same lines from which the line integrals were formed.
- Image reconstruction is a mathematical procedure that dissolves the overlapping effect in the projection data and recreates the original image with nonoverlapped objects. A mathematical procedure is called an algorithm.
- Dirac's δ-function usually acts as a point source in algorithm development.
- The readers are expected to understand two main concepts in this chapter: projection and backprojection.

Problems

Problem 1.1 If a 2D object is a point source, it is sufficient to use the projection data from two different detector views to obtain an exact reconstruction. Let us consider a 2D object that consists of three point sources which are not on a same straight line (i.e., they are not colinear). Determine the smallest number of detector views so that sufficient projection data are available to obtain an exact reconstruction.

Problem 1.2 It is known that the Radon transform of a shifted point source $\delta(x-x_0, y - y_0)$ is $\delta(x_0 \cos \theta + y_0 \sin \theta - s)$. This result can be extended to a general object $f(x, y)$. If $p(s, \theta)$ is the Radon transform of the unshifted object $f(x, y)$, determine the Radon transform of the shifted object $f(x-x_0, y - y_0)$.

Problem 1.3 Use the definition of the δ-function to prove that the following two definitions of the Radon transform are equivalent:

$$p(s,\theta) = \int\limits_{-\infty}^{\infty} \int\limits_{-\infty}^{\infty} f(x,y)\delta(x\cos\theta + y\sin\theta - s)dxdy,$$

$$p(s,\theta) = \int\limits_{-\infty}^{\infty} f(s\cos\theta - t\sin\theta, s\sin\theta + t\cos\theta)dt.$$

Problem 1.4   The backprojection in the Cartesian coordinate system is defined as

$$b(x,y) = \int\limits_{0}^{\pi} p(x\cos\theta + y\sin\theta, \theta)d\theta.$$

Give an equivalent expression $b_{\text{polar}}(r, \varphi)$ of the backprojection in the polar coordinate system.

Bibliography

[1]   Barrett H, Swindell W (1988) Radiological Imaging, Academic, New York.
[2]   Bracewell RN (1955) Two-Dimensional Imaging, Prentice Hall, Englewood Cliffs, NJ.
[3]   Bushberg JT, Seibert JA, Leidoldr EM, Boone JM (2002) The Essential Physics of Medical Imaging, 2nd ed., Lippincott Williams and Wilkins, Philadelphia.
[4]   Carlton RR, Adler AM (2001) Principles of Radiographic Imaging: An Art and a Science, 3rd ed., Delmar, Albany, NY.
[5]   Cho ZH, Jones JP, Signh M (1993) Foundations of Medical Imaging, Wiley, New York.
[6]   Cormack AM (1963) Representation of a function by its line integrals, with some radiological applications. J Appl Phys 34:2722–2727.
[7]   Herman G (1981) Image Reconstruction from Projections: The Fundamentals of Computerized Tomography, Academic Press, New York.
[8]   Kak AC, Staney M (1987) Principles of Computerized Tomography, IEEE Press, Piscataway, NJ.
[9]   Macovski A (1983) Medical Imaging Systems, Prentice Hall, Englewood Cliffs, NJ.
[10] Natterer F (1986) The Mathematics of Computerized Tomography, Wiley-Teubner, New York.
[11] Prince JL, Links JM (2006) Medical Imaging Signals and System, Pearson Prentice Hall, Upper Saddle River, NJ.
[12] Shung KK, Smith MB, Tsui BMW (1992) Principles of Medical Imaging, Academic Press, San Diego, CA.
[13] Zeng GL (2001) Image reconstruction – a tutorial. Comput Med Imaging Graph 25:97–103.

2 Parallel-beam image reconstruction

This chapter introduces the central slice theorem, which is the function of tomography. This theorem relates the two-dimensional (2D) image with its one-dimensional (1D) projections in the Fourier domain. From this theorem, many image reconstruction algorithms are derived. Among these algorithms, the filtered backprojection (FBP) algorithm is the most popular one, which consists of a ramp-filtering step and a backprojection step. The filtering can be implemented as a multiplication in the Fourier domain or as a convolution in the spatial domain. A special topic of region-of-interest (ROI) reconstruction with truncated projections is included in this chapter.

2.1 Fourier transform

The concept of the Fourier transform is based on the fact that it is possible to form a function $p(s)$ as a *weighted summation* of a series of sine and cosine terms of various frequencies, ω, with a weighting function $P(\omega)$. You can use a prism to decompose the sunlight into a spectrum of different colors; you can also reproduce the original light by recombining the spectrum of different colors (see Figure 2.1).

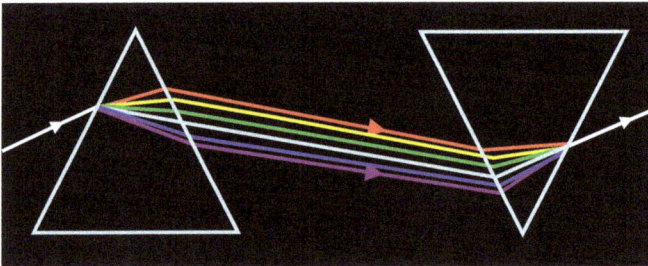

Fig. 2.1: The white light can be decomposed into color lights, which can be converted back to the original white light.

The weighting function $P(\omega)$ for each frequency ω is called the Fourier transform of $p(s)$. One can easily use mathematical formulas to find $P(\omega)$ from $p(s)$ and to recover $p(s)$ from $P(\omega)$. If you know one function (either $p(s)$ or $P(\omega)$), you know the other. In this pair, one function is denoted by a lowercase letter with a variable s, and the other function is denoted by an uppercase letter with a variable ω.

A Fourier transform pair $p(s)$ and $P(\omega)$ is shown in Figure 2.2, where $P(\omega)$ is the Fourier transform of $p(s)$. The function $P(\omega)$ tells us that the triangle function $p(s)$ has rich low-frequency components because the central lobe has a large amplitude. As the frequency gets higher (i.e., as $|\omega|$ gets larger), the amplitude of the lobes becomes smaller. We also see some notches in $P(\omega)$; those notched frequencies are not in the triangle

https://doi.org/10.1515/9783111055404-002

function $p(s)$. Using the Fourier transform can help us understand some hidden mathematical relationships, which are not easy to see without the Fourier transform.

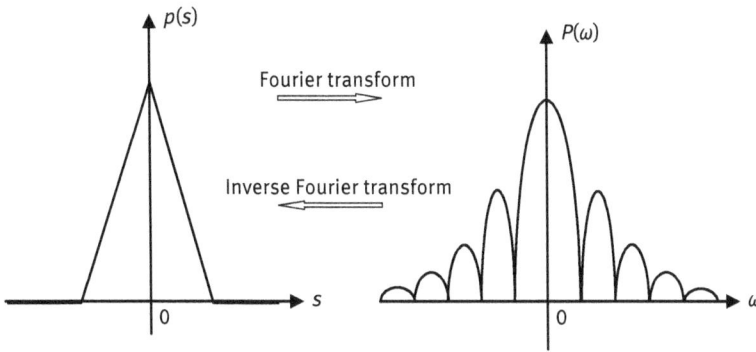

Fig. 2.2: A Fourier transform pair.

One can also find the Fourier transform for a function with two or more variables. We will denote the Fourier transform of the function $f(x, y)$ as $F(\omega_x, \omega_y)$, where ω_x is the frequency in the x direction and ω_y is the frequency in the y direction.

2.2 Central slice theorem

Central slice theorem is the foundation of tomography. It has other names: projection slice theorem and Fourier slice theorem. The central slice theorem in 2D states that the 1D Fourier transform $P(\omega)$ of the projection $p(s)$ of a 2D function $f(x, y)$ is equal to a slice (i.e., a 1D profile) through the origin of the 2D Fourier transform $F(\omega_x, \omega_y)$ of that function which is parallel to the detector (see Figure 2.3).

If we rotate the detector around the object at least for 180°, the corresponding "central slice" in the 2D Fourier transform $F(\omega_x, \omega_y)$ will rotate synchronously and will cover the entire 2D Fourier space, that is, the ω_x–ω_y plane (see Figure 2.4). In other words, by rotating the detector 180°, the entire 2D Fourier transform $F(\omega_x, \omega_y)$ is "measured." Once $F(\omega_x, \omega_y)$ is available, the original 2D function $f(x, y)$ can be readily obtained by a mathematical procedure called the 2D inverse Fourier transform.

Backprojecting projection data at one view is equivalent to adding a "central slice" of $F(\omega_x, \omega_y)$ in the ω_x–ω_y plane (i.e., the Fourier space). Backprojection over 180° fully reconstructs the 2D Fourier transform $F(\omega_x, \omega_y)$. Due to the property of the Fourier pair, the original function $f(x, y)$ can be readily found from $F(\omega_x, \omega_y)$.

In Chapter 1, we learned that backprojection alone does not get you the original image back; instead, a blurred image is obtained. Is there a contradiction here? No. Let us take another look of the ω_x–ω_y plane in Figure 2.4. After we add the "central slices" to the ω_x–ω_y plane, we see higher density of the "central slices" at the origin of the ω_x–ω_y

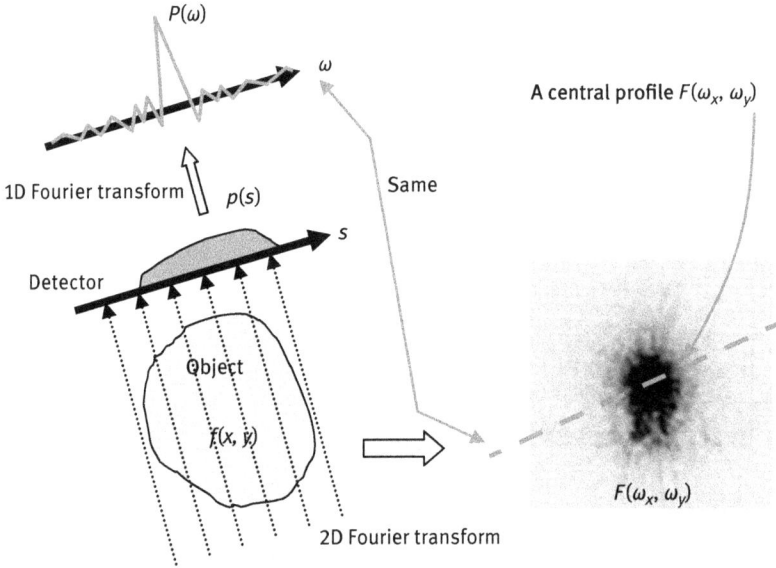

Fig. 2.3: Illustration of the 2D central slice theorem.

plane and lower density at the regions away from the origin. The central region of the Fourier space represents low frequencies. Overweighting with low-frequency components blurs the image.

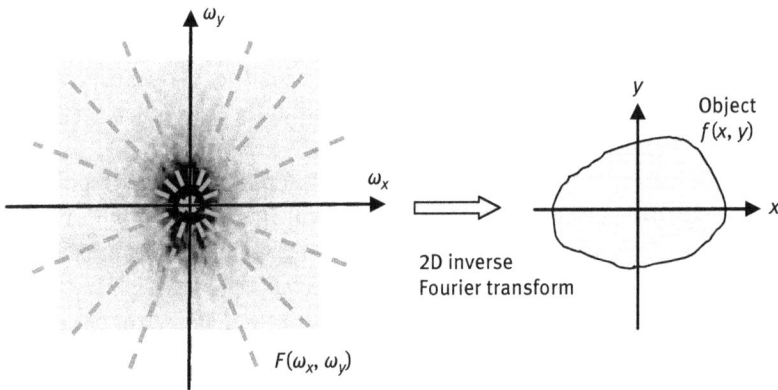

Fig. 2.4: Each view adds a line in the Fourier space. The 2D inverse Fourier transform reconstructs the original image.

To counter this blurring effect, we must compensate for the nonuniformity in the Fourier space. The nonuniform density in the Fourier space is proportional to $1/\sqrt{\omega_x^2 + \omega_y^2}$. There are two ways to do the compensation.

One way is to multiply the ω_x–ω_y space Fourier "image" by $\sqrt{\omega_x^2 + \omega_y^2}$, and the resultant Fourier space "image" is $F(\omega_x, \omega_y)$. The 2D inverse Fourier transform of $F(\omega_x, \omega_y)$ gives the exact original image $f(x, y)$.

The other way is to multiply the 1D Fourier transform $P(\omega, \theta)$ of the projection data $p(s, \theta)$ by $|\omega|$. We then take the 1D inverse Fourier transform of $|\omega| P(\omega, \theta)$. After this special treatment (i.e., filtering) of the projection data, the treated (i.e., filtered) projection data are backprojected, and the exact original image $f(x, y)$ is obtained. Note that in the previous discussion (see Figure 2.3), we ignored the second variable in $P(\omega, \theta)$ and $p(s, \theta)$ on purpose to allow the reader to pay attention to the first variable.

These two methods (i.e., multiplying the Fourier transformed backprojected image by $\sqrt{\omega_x^2 + \omega_y^2}$ and multiplying the Fourier transformed projections by $|\omega|$) will be further discussed later in this book. The second method, which is referred to as the FBP algorithm, is more popular than the first. The function $|\omega|$ is called the *ramp filter* in tomography, named after its appearance (see Figure 2.5). It is the second method that added negative "wings" around the spikes in the projections in Chapter 1.

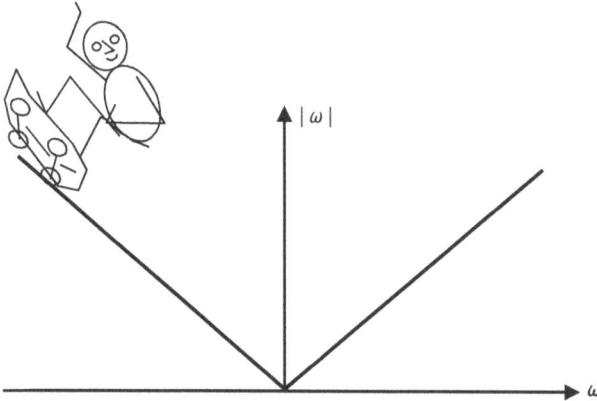

Fig. 2.5: The 1D ramp filter transfer function.

2.3 Reconstruction algorithms

Backprojection accomplishes most of the work in image reconstruction; it converts the projection data at various views into an image, which is almost what we wanted except for the blurring effect. The blurring effect is caused by the $1/|\omega|$ nonuniform weighting, with $|\omega| = \sqrt{\omega_x^2 + \omega_y^2}$, in the 2D Fourier space (i.e., the $\omega_x \omega_y$ plane). Compensation for $1/|\omega|$ function can be realized by filtering the projection data $p(s, \theta)$ or equivalently its 1D Fourier transform $P(\omega, \theta)$ with a ramp filter $|\omega|$. If the filtered projection data are backprojected, the exact image can be obtained.

2.3.1 Method 1

Precisely, this FBP algorithm can be implemented according to the steps shown in Figure 2.6, in which the ramp filtering is implemented as follows:

(i) Find the 1D Fourier transform of $p(s, \theta)$ with respect to the first variable s, obtaining $P(\omega, \theta)$.

(ii) Multiply $P(\omega, \theta)$ with a ramp filter $|\omega|$, obtaining $Q(\omega, \theta)$.

(iii) Find the 1D inverse Fourier transform of $Q(\omega, \theta)$ with respect to the first variable ω, obtaining $q(s, \theta)$.

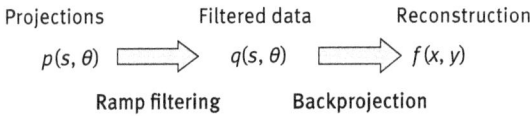

Projections   Filtered data   Reconstruction

$p(s, \theta)$ $\Longrightarrow$ $q(s, \theta)$ $\Longrightarrow$ $f(x, y)$

   Ramp filtering   Backprojection

Fig. 2.6: The procedure of the filtered backprojection (FBP) algorithm.

2.3.2 Method 2

There are more ways than one to do ramp filtering. In fact, we can perform ramp filtering without using the Fourier transform at all. According to the Fourier transform theory, multiplication in one domain (say, the ω domain) corresponds to convolution in the other domain (say, the s domain) (see Figure 2.7).

Thus steps (i), (ii), and (iii) in Method 1 are equivalent to a mathematical procedure called *convolution*. The ramp-filtered data $q(s, \theta)$ can be obtained by convolution as follows:

$$q(s, \theta) = p(s, \theta) * h(s), \tag{2.1}$$

where "$*$" denotes the convolution operation, which is an integral with respect to the variable s. Here $h(s)$ is the convolution kernel and is the 1D inverse Fourier transform of $H(\omega) = |\omega|$.

Here we give an example of convolution so that you can understand what convolution can do for you. If the convolution kernel $h(s)$ is not symmetric, you first flip it left–right, making it $h(-s)$. Second, you imagine that your function $p(s)$ can be decomposed into vertical spikes (or Dirac's delta functions). Third, replace each spike by $h(-s)$, that is, the $s = 0$ position of $h(-s)$ is at the spike's position and scale $h(-s)$ by the amplitude (which can be negative) of the spike. Finally, sum up all the shifted and scaled versions of $h(-s)$, obtaining $q(s)$ (see Figure 2.8).

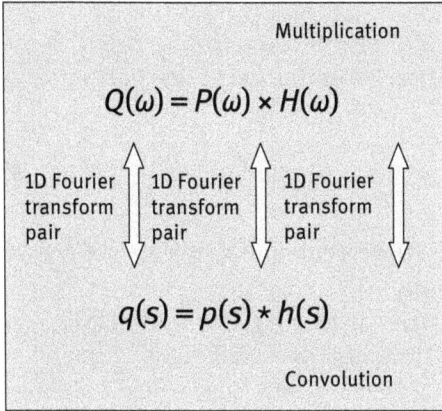

Fig. 2.7: An important property of Fourier transform: multiplication in one domain is equivalent to convolution in the other domain.

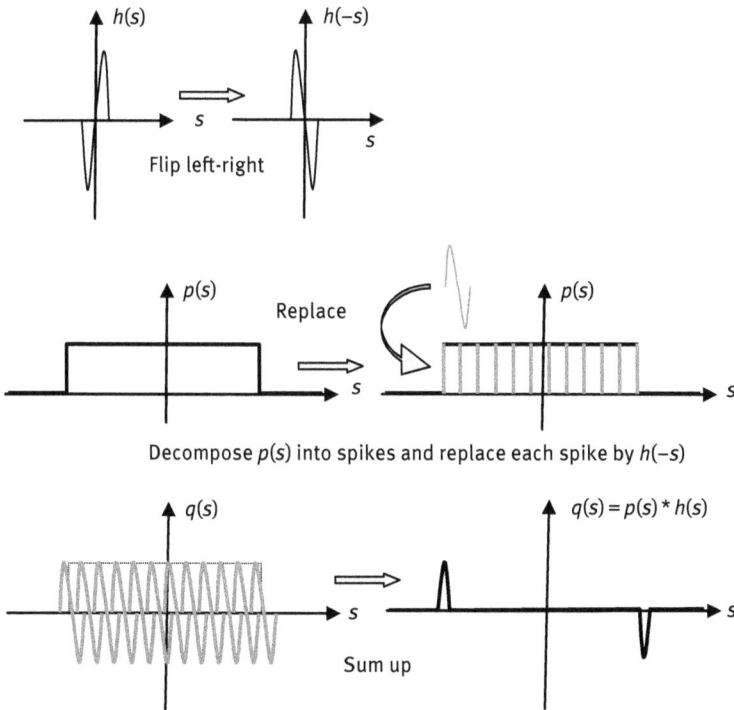

Fig. 2.8: An illustration of the procedure of convolution.

2.3.3 Method 3

There is a third way to implement ramp filtering. Let us factor the ramp filter into two parts:

$$H(\omega) = |\omega| = i2\pi\omega \times \frac{1}{i2\pi}\,\mathrm{sgn}(\omega), \tag{2.2}$$

where $i = \sqrt{-1}$, $\mathrm{sgn}(\omega) = 1$ if $\omega > 0$, $\mathrm{sgn}(\omega) = -1$ if $\omega < 0$, and $\mathrm{sgn}(\omega) = 0$ if $\omega = 0$. Here we will use two properties of the Fourier transform.

Fact 1: Multiplication by $i2\pi\omega$ in the Fourier domain (i.e., the ω domain) corresponds to the derivative with respect to s in the spatial domain (i.e., the s domain).

Fact 2: The inverse Fourier transform of $-i\,\mathrm{sgn}(\omega)$ is $1/(\pi s)$. Convolution with $1/(\pi s)$ is called the Hilbert transform.

Using the relationship shown in Figure 2.7, the ramp filtering can be realized as

$$q(s, \theta) = \frac{dp(s, \theta)}{ds} * \frac{-1}{2\pi^2 s}, \tag{2.3}$$

which is a combination of the derivative and the Hilbert transform.

2.3.4 Method 4

If we switch the order of ramp filtering and backprojection, we can get another way to reconstruct the image – *backprojection then filtering*. After backprojection, we get a blurred image $b(x, y)$, which is 2D. We need to apply a 2D ramp filter to it. One way to do it is as follows:
(i)   Find the 2D Fourier transform of $b(x, y)$, obtaining $B(\omega_x, \omega_y)$.
(ii) Multiply $B(\omega_x, \omega_y)$ with a ramp filter $|\omega| = \sqrt{\omega_x^2 + \omega_y^2}$, obtaining $F(\omega_x, \omega_y)$.
(iii) Find the 2D inverse Fourier transform of $F(\omega_x, \omega_y)$, obtaining $f(x, y)$.

2.3.5 Method 5

Method 3 consists of three components: the derivative, the Hilbert transform, and the backprojection. If we switch the order, we obtain yet another reconstruction algorithm:
(i)   Find the derivative of the projection data $p(s, \theta)$ with respect to s, obtaining $dp(s, \theta)/ds$.
(ii) Backproject $dp(s, \theta)/ds$ over 180°.
(iii) Perform the line-by-line Hilbert transform, in the direction parallel to the detector at the 90° position.

2.3.6 Method 6

By switching the order of the first two steps in Method 5, we can obtain still another method to reconstruct an image – the finite Hilbert transform of the derivative of the backprojection:

(i) Backproject $p(s, \theta)$ to obtain two blurred images $b_x(x, y)$ and $b_y(x, y)$, where $-\sin \theta$ is used as the weighting function in creating $b_x(x, y)$ and $\cos \theta$ is used as the weighting function in creating $b_y(x, y)$.

(ii) Perform partial derivatives and obtain $b(x,y) = \partial b_x(x, y)/\partial x + \partial b_y(x, y)/\partial y$.

(iii) Perform the line-by-line Hilbert transform, in the direction parallel to the detector at the 90° position.

One distinctive feature of Method 6 is that no filtering is applied to the raw projection data. This feature is important when the list-mode projection data are acquired, for example, in nuclear medicine. In the list-mode data, every event is stored independently and usually there is no easy way to find a neighboring event with which to take a derivative. This algorithm can be easily applied to other imaging geometries and to some situations where the projection data are truncated (i.e., not fully measured).

You must have noticed that we are playing the order changing game to create more and more reconstruction algorithms. We simply change the order of ramp filtering and backprojection, or change of the order of performing derivative, the Hilbert transform, and backprojection. If we keep playing this game, we can make a list of some possible image reconstruction algorithms (see Figure 2.9). Each algorithm has its advantages and disadvantages.

The table in Figure 2.9 does not exhaust all possibilities. For example, the Hilbert transform can be implemented as convolution in the spatial domain or as multiplication in the Fourier domain. The Hilbert transform has another form, which is not convolution but is an integral over a finite interval. The finite Hilbert transform has an important application in handling truncated projection data.

You may see that the backprojection is used in all algorithms. This does not have to be the case. You do not have to use a spatial domain backprojector in a reconstruction algorithm. You can use the central slice theorem by assigning the Fourier domain projection data $P(\omega, \theta)$ at the proper ω_x–ω_y locations. This is the Fourier domain implementation of the backprojection. However, this Fourier domain backprojection has limited applications because large interpolation errors could be introduced in the polar coordinate (ω, θ) system to the Cartesian $(\omega_x$–$\omega_y)$ system transformation.

Method	Step 1	Step 2	Step 3
1	1D ramp filter with Fourier transform	Backprojection	
2	1D ramp filter with convolution	Backprojection	
4	Backprojection	2D Ramp filter with Fourier transform	
	Backprojection	2D Ramp filter with 2D convolution	
3	Derivative	Hilbert transform	Backprojection
5	Derivative	Backprojection	Hilbert transform
6	Backprojection	Derivative	Hilbert transform
	Hilbert transform	Derivative	Backprojection
	Hilbert transform	Backprojection	Derivative
	Backprojection	Hilbert transform	Derivative

Fig. 2.9: A list of parallel-beam analytical image reconstruction algorithms.

2.4 A computer simulation

In Figure 2.10, we show an example of the FBP algorithm in action. The original image $f(x, y)$ is shown in the lower right corner; it consists of a large disk and four small disks. The projection data $p(s, \theta)$ are generated analytically using computer software. The projections of the two outer small disks trace two sine curves in the sinogram, which is an s–θ coordinate display of the projection data.

After applying the ramp filter to $p(s, \theta)$, the filtered data $q(s, \theta)$ looks sharper because the ramp filter, which is a high-pass filter, suppresses the low-frequency components and enhances the high-frequency components. One can see the darker edges around the projection of the disks. These darker edges are the negative "wings" that we discussed in Chapter 1.

The image labeled (A) is the backprojection of $q(s, \theta)$ at the first angle. The filtered data $q(s, \theta)$ at the first angle are simply copied across the entire image. This action is sometimes described as even distribution of a value along the projection path. Progressing through images (A)–(G), as data from more and more angles are backprojected, the image takes shape and gets closer and closer to the original image. Remember that backprojection consists of two actions: smear back and superposition. The image is reconstructed when the backprojection is performed over 180°. The backprojection can also be performed over 360° (then divide the image value by 2) because the data are redundant by observing the fact that $p(s, \theta) = p(-s, \theta + \pi)$ and $q(s, \theta) = q(-s, \theta + \pi)$.

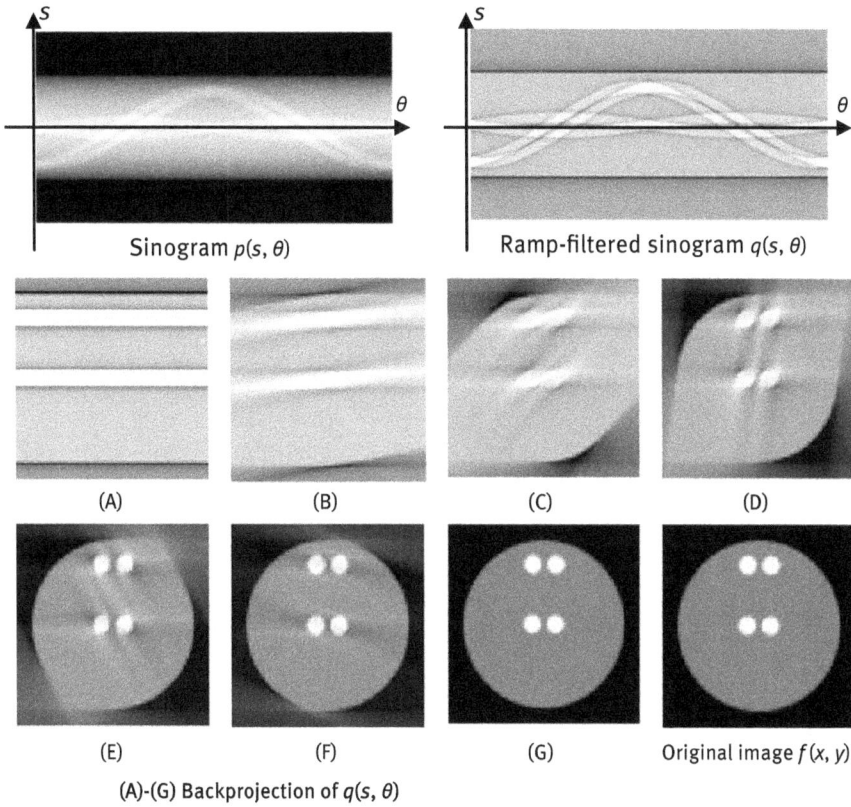

Sinogram $p(s, \theta)$

Ramp-filtered sinogram $q(s, \theta)$

(A)   (B)   (C)   (D)

(E)   (F)   (G)   Original image $f(x, y)$

(A)-(G) Backprojection of $q(s, \theta)$

Fig. 2.10: An FBP algorithm in action: filtering and view-by-view backprojection.

2.5 ROI reconstruction with truncated projections

The situation of the ROI reconstruction with truncated projections is illustrated in Figure 2.11, where the detector is not large enough to cover the entire object but is large enough to cover the ROI. In tomographic theory, this ROI reconstruction problem is called the interior problem. Only the circular FOV (field-of-view) is fully measured. For a general interior problem, only an approximate solution can be obtained. Exact reconstructions exist only for some special cases.

One solvable ROI reconstruction problem is shown in Figure 2.12, and the reconstruction is provided by the "derivative–backprojection–Hilbert transform" algorithm. The ROI is the dark shaded smaller region within the circular FOV. Both the derivative operation and the backprojection operation are local. That is, they only use the locally available data to find the results. After the derivative and backprojection, the data within the FOV are exact.

The next step in the reconstruction algorithm is to perform the line-by-line 1D Hilbert transform along the direction indicated as the vertical direction in Figure 2.12. The usual Hilbert transform is in the form of convolution with a convolution kernel $1/s$ that does not vanish. Therefore, the Hilbert transform requires data along the entire filtering line.

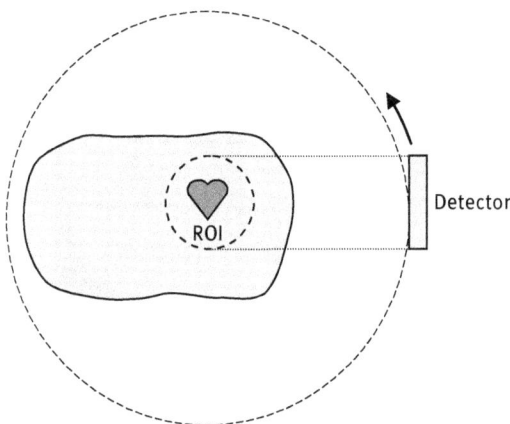

Fig. 2.11: The detector is only large enough to cover the ROI, but not large enough to cover the entire object.

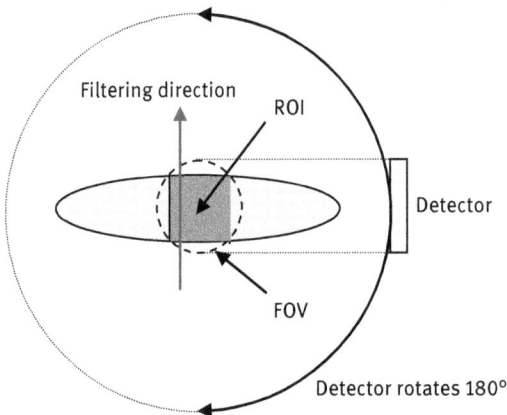

Fig. 2.12: The derivative–backprojection–Hilbert transform algorithm is able to exactly reconstruct the ROI.

Thanks to a finite inverse Hilbert transform formula, we are able to perform the inverse Hilbert transform with finite data. If a function $f(s)$ is nonzero on the interval $[a, b]$ and if the Hilbert transform of $f(s)$ is $g(s)$, then the finite inverse Hilbert transform formula only needs $g(s)$ on $[a, b]$ to recover $f(s)$.

Using this finite inverse Hilbert transform formula, we are now able to evaluate the 1D inverse Hilbert transform along the filtering line to reconstruct the image because the required exact data are available in the FOV. This finite inverse Hilbert transform formula is not in the form of convolution; therefore, it is not efficient to perform this step using the Fourier transform methods.

It has been shown that an exact ROI reconstruction can be obtained for a less restricted case shown in Figure 2.13. For this case, the ROI has the same size as the FOV. Here only one end of the filtering line is required to be outside the object. The proof of exact reconstruction for this case is based on analytic continuation. The drawback of this method is that we have not found a closed-form formula to reconstruct the ROI image.

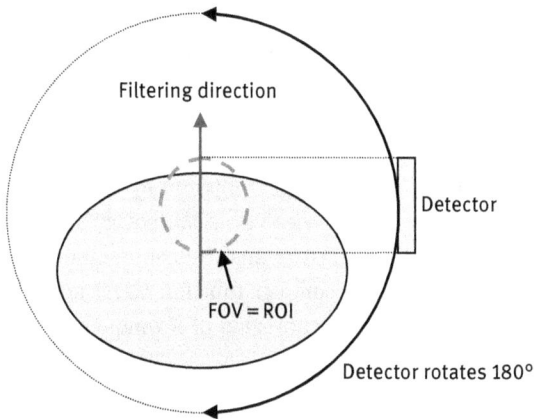

Fig. 2.13: The measured data are sufficient to exactly reconstruct the FOV. A part of the FOV is outside the object.

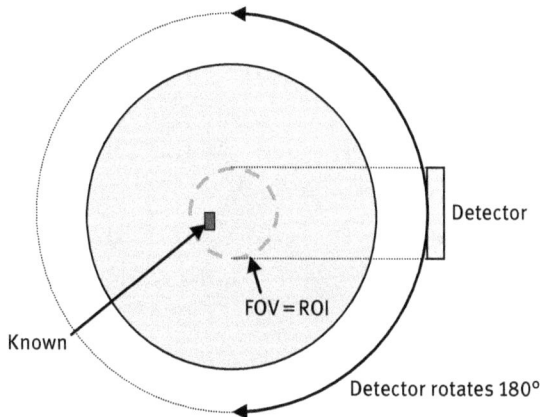

Fig. 2.14: If a small region in the FOV is known, the FOV can be exactly reconstructed.

An immediate extension of this analytic continuation method can be applied to the general interior problem, where a small region of the image within the FOV is exactly known in advance (see Figure 2.14). The known region in the image can be very small.

We now explain what we mean by analytic continuation. The analytic continuation is a subject in a mathematical branch called complex analysis, which studies functions with complex variables.

If x is a real variable, $h(x) = 2x + i(3 + 5x)$, with $i = \sqrt{-1}$, is not a function with a complex variable. However, $h(z) = 2z$, with $z = x + iy$ and both x and y are real, is a function with a complex variable and is defined in a region (called domain) in the complex plane.

A complex function $h(z)$ is said to be analytic in a complex region R, if and only if the function $h(z)$ is differentiable at every point in R. If a complex function is analytic on a region R, it is *infinitely* differentiable in R. Therefore, you can have a power expansion of an analytic function $h(z)$ anywhere in R. You can imagine that an analytic function is very smooth.

Let $h_1(z)$ and $h_2(z)$ be analytic functions in regions R_1 and R_2, respectively, and suppose that the intersection $R_1 \cap R_2$ is not empty and that $h_1(z) = h_2(z)$ on $R_1 \cap R_2$. Then $h_2(z)$ is called an analytic continuation of $h_1(z)$ to R_2, and vice versa. Moreover, the analytic continuation of $h_1(z)$ to R_2 is unique. This uniqueness of analytic continuation is a rather amazing and extremely powerful statement. If a complex function $h(z)$ is analytic in a region R, knowing the value of $h(z)$ in a very small subregion of R uniquely determines the value of the function $h(z)$ at every other point in R.

If our 2D image $f(x, y)$ is known in a small region Ω_{known}, we can somehow determine the image $f(x, y)$ in a larger region Ω. Note that $f(x, y)$ is not an analytic function. We do not perform analytic continuation on $f(x, y)$.

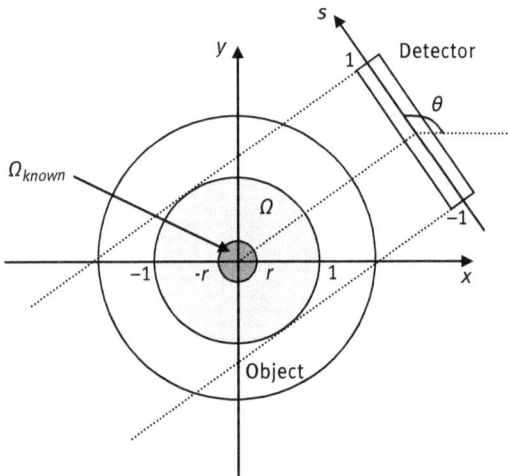

Fig. 2.15: The known image values on $[-r, r]$ together with the truncated projections can be used to determine image values in $(-1, 1)$.

First, we draw a line, say, the x-axis, passing through the small region Ω_{known}, as shown in Figure 2.15. Without loss of generality, we only consider the image values on one line: $y = 0$. We can do the same for other lines that pass through the known region Ω_{known}.

It is given that the image $f(x, 0)$ is known when $-r \le x \le r$. In the following, we are going to define a function $h(x)$, convert it into a complex function $h(z)$, and apply analytic continuation to $h(z)$ so that the image $f(x, 0)$ can be estimated in a larger interval $(-1, 1)$. In this particular imaging problem, the detector is smaller than the object. Projection data are only available for $-1 \le s \le 1$. We can use the FBP algorithm to reconstruct the image as

$$f(x,0) = \frac{1}{4\pi} \int_0^{2\pi} d\theta \int_{-\infty}^{\infty} \frac{1}{s - x\cos\theta} \frac{\partial p(s,\theta)}{\partial s} ds. \qquad (2.4)$$

This expression can be written in two terms: the first term with measured data and the second term with unmeasured data:

$$f(x,0) = \frac{1}{4\pi} \int_0^{2\pi} d\theta \int_{-1}^{1} \frac{1}{s - x\cos\theta} \frac{\partial p(s,\theta)}{\partial s} ds + \frac{1}{4\pi} \int_0^{2\pi} d\theta \int_{|s|>1} \frac{1}{s - x\cos\theta} \frac{\partial p(s,\theta)}{\partial s} ds. \qquad (2.5)$$

The first term can be evaluated using the measured data on $[-1, 1]$ and it contains the sharp edges of the image f. However, the second term is very smooth (almost a constant) but cannot be evaluated because the data are not available. Let us denote the second term as $h(x)$. We have

$$h(x) = \frac{1}{4\pi} \int_0^{2\pi} d\theta \int_{|s|>1} \frac{1}{s - x\cos\theta} \frac{\partial p(s,\theta)}{\partial s} ds, \quad |x| < 1. \qquad (2.6)$$

In the above expression, $|x| < 1$ and $|s| > 1$ make $1/(s - x\cos\theta)$ differentiable of any order. We further assume that $\partial p / \partial s$ is continuous, which is usually satisfied in practice. We claim that if we replace the real variable x by a complex variable z, then $h(z)$ is an analytic function within $|z| < 1$.

The analytic function $h(z)$ is known as $[-r, r]$, because $h(x)$ can also be expressed as

$$h(x) = f(x,0) - \frac{1}{4\pi} \int_0^{2\pi} d\theta \int_{-1}^{1} \frac{1}{s - x\cos\theta} \frac{\partial p(s,\theta)}{\partial s} ds, \qquad (2.7)$$

where $f(x, 0)$ is given as $[-r, r]$, and the second term in the above expression can be evaluated with the measured data. By using analytic continuation, $h(z)$ is uniquely determined in $|z| < 1$. Thus, $h(x)$ is determined in $(-1, 1)$. Once $h(x)$ is found, the image value in $(-1, 1)$ can be obtained as

$$f(x,0) = \frac{1}{4\pi} \int_0^{2\pi} d\theta \int_{-1}^{1} \frac{1}{s - x\cos\theta} \frac{\partial p(s,\,\theta)}{\partial s} ds + h(x), \quad |x| < 1. \tag{2.8}$$

In the above discussion, we know that we should have enough data to determine the image in region Ω. However, doing analytic continuation is easier said than done. Mathematically, the right way to do analytic continuation is by doing the Taylor expansions, but this approach is not practical. Currently iterative approaches are used for this purpose.

Another practical problem is the stability issue. In some cases we can prove mathematically that the limited data can uniquely determine an image, but the image reconstruction method can be extremely ill-conditioned and unstable. The limited angle imaging problem could be such an ill-conditioned problem. In a limited angle imaging problem, the ROI does not have full 180° angular measurements. It may only have, for example, 10° angular measurements.

2.6 Mathematical expressions

This section presents the mathematical expressions of the 1D Fourier transform pair, the Hilbert transform pair, and two versions of the inverse finite Hilbert transform. The proof of the central slice theorem is given. This section also derives the FBP and backprojection-then-filtering algorithms. The spatial domain filtering methods, such as the convolution backprojection algorithm and the Radon inversion formula, are also included.

2.6.1 The Fourier transform and convolution

The 1D Fourier transform is defined as

$$P(\omega) = \int_{-\infty}^{\infty} p(s)e^{-2\pi i s\omega} ds, \tag{2.9}$$

and the 1D inverse Fourier transform is

$$p(s) = \int_{-\infty}^{\infty} P(\omega)e^{2\pi i s\omega} d\omega. \tag{2.10}$$

The convolution of two functions f and g is defined as

$$(f * g)(t) = \int_{-\infty}^{\infty} f(\tau)g(t-\tau)d\tau = \int_{-\infty}^{\infty} f(t-\tau)g(\tau)d\tau. \tag{2.11}$$

If we denote the Fourier transform operator as **F**, the convolution theorem can be stated as

$$\mathbf{F}(f * g) = \mathbf{F}(f) \times \mathbf{F}(g). \tag{2.12}$$

2.6.2 The Hilbert transform and the finite Hilbert transform

The Hilbert transform is less well known than the Fourier transform. Unlike the Fourier transform that converts a real function into a complex function, the Hilbert transform converts a real function into another real function.

In this section, we use **H** to denote the Hilbert transform operator. If you apply the Hilbert transform twice, you get the original function back, except for a sign change, that is

$$\mathbf{H}(\mathbf{H}f) = \mathbf{H}^2 f = -f. \tag{2.13}$$

In other words, the inverse Hilbert transform is the negative value of the Hilbert transform.

A real function and its Hilbert transform are orthogonal. If $g = \mathbf{H}f$, then

$$\int_{-\infty}^{\infty} f(t)g(t)dt = 0. \tag{2.14}$$

For example, the Hilbert transform of $\cos(t)$ is $\sin(t)$. The Hilbert transform of $\sin(t)$ is $-\cos(t)$.

The Hilbert transform can be defined in the Fourier domain. Let the Fourier transform of a real function $f(t)$ be $F(\omega)$, the Hilbert transform of $f(t)$ be $g(t)$, and the Fourier transform of $g(t)$ be $G(\omega)$, then

$$G(\omega) = -i\,\text{sgn}(\omega)F(\omega), \tag{2.15}$$

with $i = \sqrt{-1}$ and $\text{sgn}(\omega)$ being the signum:

$$\text{sgn}(\omega) = \begin{cases} 1, & \omega > 0 \\ 0, & \omega = 0 \\ -1, & \omega < 0. \end{cases} \tag{2.16}$$

Since the magnitude of $-i\,\text{sgn}(\omega)$ is 1, the Hilbert transform is an application of an all-pass filter with a $\pm 90°$ phase shift.

Equivalently, the Hilbert transform can also be expressed as convolution with the convolution kernel

$$h(t) = \frac{1}{\pi t} \tag{2.17}$$

as

$$g(t) = h(t) * f(t) = \text{p.v.} \int_{-\infty}^{\infty} f(\tau) \frac{1}{\pi(t-\tau)} d\tau \tag{2.18}$$

where "p.v." means principal value.

We have already seen that the ramp filter can be decomposed into the Hilbert transform and the derivative. It does not matter whether you perform the derivative first and then perform the Hilbert transform or you perform the Hilbert transform first and then perform the derivative. In fact, we have

$$\mathbf{H}\big(f'(t)\big) = \frac{d(\mathbf{H}(f(t)))}{dt}. \tag{2.19}$$

One can also change the order of the Hilbert transform and the 180° backprojection. To understand this property better, we will use the central slice theorem and think in the Fourier domain. Let us consider the backprojection problem

$$b(x,y) = \int_{-\pi/2}^{\pi/2} \mathbf{H}p(s,\theta)|_{s=x\cos\theta+y\sin\theta} d\theta, \tag{2.20}$$

where the Hilbert transform is with respect to the variable s. Let the Fourier transform of $p(s,\theta)$ be $P(\omega,\theta)$. The central slice theorem applied to this particular problem is shown in Figure 2.16, where $P(\omega,\theta)$ is multiplied by either i or $-i$ before putting it in the 2D Fourier domain.

An equivalent way to do this backprojection is to backproject $p(s,\theta)$, then multiply the entire left half (ω_x, ω_y) plane by i and multiply the entire right half (ω_x, ω_y) plane by $-i$. This multiplication action can be achieved by performing line-by-line 1D Hilbert transform in the x direction. Thus, we have

$$\mathbf{H}_x \int_{-\pi/2}^{\pi/2} p(s,\theta)|_{s=x\cos\theta+y\sin\theta} d\theta = \int_{-\pi/2}^{\pi/2} \mathbf{H}_s p(s,\theta)|_{s=x\cos\theta+y\sin\theta} d\theta. \tag{2.21}$$

In the following, we will briefly introduce the concept of the finite Hilbert transform, which plays an important role in the ROI reconstruction with truncated projection data.

Without loss of generality, we assume that a real function $f(t)$ is supported in $(-1, 1)$, that is, $f(t) = 0$ if $|t| \geq 1$. Then the Hilbert transform of $f(t)$ is given as

$$g(t) = \text{p.v.} \int_{-1}^{1} f(\tau) \frac{1}{\pi(t-\tau)} d\tau. \tag{2.22}$$

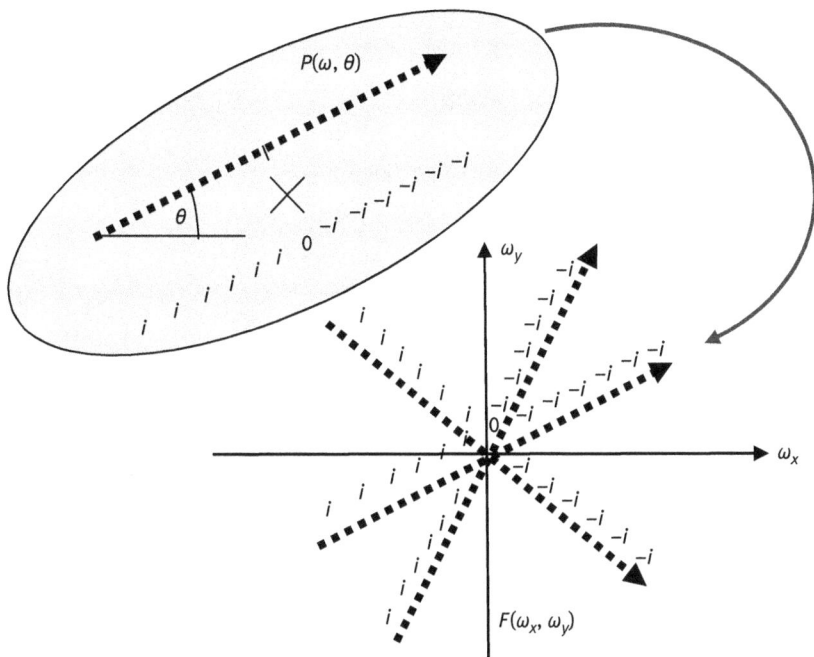

Fig. 2.16: Backprojection of H$p(s, \theta)$ over $(-\pi/2, \pi/2)$ is equivalent to backprojection of $p(s, \theta)$ over $(-\pi/2, \pi/2)$, then multiply i in the left half (ω_x, ω_y) plane and multiply $-i$ in the right half (ω_x, ω_y) plane.

Even though $f(t)$ is supported in a finite interval, $g(t)$ may not have a finite support. There are some formulas that are able to recover $f(t)$ using $g(t)$ only on $[-1,1]$. For example

$$f(t) = \frac{1}{\pi}\sqrt{\frac{t-1}{t+1}}\text{p.v.} \int_{-1}^{1} \sqrt{\frac{\tau+1}{\tau-1}}\frac{g(\tau)}{t-\tau}\, d\tau, \tag{2.23}$$

or

$$f(t) = \frac{1}{\pi\sqrt{1-t^2}}\left[\int_{-1}^{1} f(\tau)d\tau + \text{p.v.} \int_{-1}^{1} \frac{\sqrt{1-\tau^2}}{\tau-t} g(\tau)d\tau\right]. \tag{2.24}$$

2.6.3 Proof of the central slice theorem

The central slice theorem is given as

$$P(\omega, \theta) = F(\omega\cos\theta, \omega\sin\theta). \tag{2.25}$$

Proof. We start with the definition of the 1D Fourier transform:

$$P(\omega) = \int_{-\infty}^{\infty} p(s)e^{-2\pi i s\omega}\,ds, \tag{2.26}$$

then use the definition of $p(s, \theta)$ (see Section 1.5), obtaining

$$P(\omega, \theta) = \int_{-\infty}^{\infty}\left[\int_{-\infty}^{\infty}\int_{-\infty}^{\infty} f(x,y)\delta(x\cos\theta + y\sin\theta - s)\,dxdy\right]e^{-2\pi i s\omega}\,ds. \tag{2.27}$$

Changing the order of integrals yields

$$P(\omega, \theta) = \int_{-\infty}^{\infty}\int_{-\infty}^{\infty} f(x,y)\left[\int_{-\infty}^{\infty}\delta(x\cos\theta + y\sin\theta - s)e^{-2\pi i s\omega}\,ds\right]dxdy. \tag{2.28}$$

Using the property of the δ function, the inner integral over s can be readily obtained, and we have

$$P(\omega, \theta) = \int_{-\infty}^{\infty}\int_{-\infty}^{\infty} f(x,y)e^{-2\pi i(x\cos\theta + y\sin\theta)\omega}\,dxdy, \tag{2.29}$$

that is,

$$P(\omega, \theta) = \int_{-\infty}^{\infty}\int_{-\infty}^{\infty} f(x,y)e^{-2\pi i(xu+yv)}\big|_{u=\omega\cos\theta, v=\omega\sin\theta}\,dxdy. \tag{2.30}$$

Finally, using the definition of the 2D Fourier transform yields

$$P(\omega, \theta) = F(\omega_x, \omega_y)\big|_{\omega_x=\omega\cos\theta, \omega_y=\omega\sin\theta}. \tag{2.31}$$

In the polar coordinate system, the central slice theorem can be expressed as

$$P(\omega, \theta) = F_{\text{polar}}(\omega, \theta). \tag{2.32}$$

2.6.4 Derivation of the FBP algorithm

We start with the 2D inverse Fourier transform in polar coordinates

$$f(x,y) = \int_{0}^{2\pi}\int_{0}^{\infty} F_{\text{polar}}(\omega, \theta)e^{2\pi i\omega(x\cos\theta + y\sin\theta)}\omega\,d\omega d\theta. \tag{2.33}$$

Because $F_{\text{polar}}(\omega, \theta) = F_{\text{polar}}(-\omega, \theta+\pi)$, we have

$$f(x,y) = \int_0^\pi \int_{-\infty}^\infty F_{\text{polar}}(\omega, \theta)|\omega|e^{2\pi i\omega(x\cos\theta + y\sin\theta)}\,d\omega d\theta. \tag{2.34}$$

By using the central slice theorem, we can replace F by P:

$$f(x,y) = \int_0^\pi \int_{-\infty}^\infty P(\omega, \theta)|\omega|e^{2\pi i\omega(x\cos\theta + y\sin\theta)}\,d\omega d\theta. \tag{2.35}$$

We recognize that $|\omega|$ is the ramp filter. Let $Q(\omega, \theta) = |\omega|P(\omega, \theta)$, then

$$f(x,y) = \int_0^\pi \int_{-\infty}^\infty Q(\omega, \theta)e^{2\pi i\omega(x\cos\theta + y\sin\theta)}\,d\omega d\theta. \tag{2.36}$$

Using the definition of the 1D inverse Fourier transform and denoting the inverse Fourier transform of Q as q, we have

$$f(x,y) = \int_0^\pi q(x\cos\theta + y\sin\theta, \theta)d\theta, \tag{2.37}$$

or

$$f(x,y) = \int_0^\pi q(s, \theta)|_{s=x\cos\theta + y\sin\theta}\,d\theta. \tag{2.38}$$

This is the backprojection of $q(s, \theta)$ (see Section 1.5).

2.6.5 Expression of the convolution backprojection algorithm

Let the convolution kernel $h(s)$ be the inverse Fourier transform of the ramp filter $|\omega|$; that is, $h(s) = \int_{-\infty}^\infty |\omega|\, e^{2\pi i\omega s}\,ds$. It is readily obtained from Section 2.6.4 that

$$f(x,y) = \int_0^\pi [h(s) * p(s, \theta)]|_{s=x\cos\theta + y\sin\theta}\,d\theta$$

$$= \int_0^\pi \int_{-\infty}^\infty h(x\cos\theta + y\sin\theta - s)p(s, \theta)\,ds d\theta. \tag{2.39}$$

2.6.6 Expression of the Radon inversion formula

The derivative, the Hilbert transform, and the backprojection algorithm are also referred to as the Radon inversion formula, which can be obtained by factoring the ramp filter $|\omega|$ into the derivative part and the Hilbert transform part:

$$|\omega| = (2\pi i\omega) \times \left[\frac{1}{2\pi}(-i\operatorname{sgn}(\omega))\right]. \tag{2.40}$$

The inverse Fourier transform of $-i\operatorname{sgn}(\omega)$ is $1/(\pi s)$ and the inverse Fourier transform of $2\pi i\omega$ is the derivative operator. Thus,

$$q(s,\theta) = \frac{\partial p(s,\theta)}{\partial s} * \frac{1}{2\pi^2 s}, \tag{2.41}$$

and

$$f(x,y) = \int_0^\pi \int_{-\infty}^\infty \frac{\partial p(s,\theta)}{\partial s} \frac{1}{2\pi^2(x\cos\theta + y\sin\theta - s)}\,ds\,d\theta. \tag{2.42}$$

2.6.7 Derivation of the backprojection-then-filtering algorithm

Let us first look at the backprojection $b(x, y)$ of the original data $p(s, \theta)$ (without filtering). The definition of the backprojection is

$$b(x,y) = \int_0^\pi p(s,\theta)|_{s=x\cos\theta+y\sin\theta}\,d\theta. \tag{2.43}$$

Using the definition of the inverse Fourier transform, $p(s, \theta)$ can be represented with its Fourier transform

$$b(x,y) = \int_0^\pi \left[\int_{-\infty}^\infty P(\omega,\theta)e^{2\pi i\omega(x\cos\theta + y\sin\theta)}\,d\omega\right]d\theta. \tag{2.44}$$

Using the central slice theorem, we can replace P by F:

$$b(x,y) = \int_0^\pi \left[\int_{-\infty}^\infty F_{\text{polar}}(\omega,\theta)e^{2\pi i\omega(x\cos\theta + y\sin\theta)}\,d\omega\right]d\theta, \tag{2.45}$$

or

$$b(x,y) = \int\limits_{0}^{\pi} \int\limits_{-\infty}^{\infty} \frac{F_{polar}(\omega,\theta)}{|\omega|} e^{2\pi i \omega (x \cos\theta + y \sin\theta)} |\omega| d\omega d\theta. \tag{2.46}$$

This is the 2D inverse Fourier transform in polar coordinates. If we take the 2D Fourier transform (in the polar coordinate system) on both sides of the above equation, we have

$$B_{polar}(\omega,\theta) = \frac{F_{polar}(\omega,\theta)}{|\omega|}, \tag{2.47}$$

or in the Cartesian system,

$$B(\omega_x, \omega_y) = \frac{F(\omega_x, \omega_y)}{\sqrt{\omega_x^2 + \omega_y^2}}; \tag{2.48}$$

that is,

$$F(\omega_x, \omega_y) = \sqrt{\omega_x^2 + \omega_y^2} B(\omega_x, \omega_y). \tag{2.49}$$

The backprojection-then-filtering algorithm follows immediately.

2.6.8 Expression of the derivative–backprojection–Hilbert transform algorithm

This algorithm is Method 5 discussed in Section 2.3.5 and can be expressed as

$$f(x,y) = \mathbf{H}_x \int\limits_{0}^{\pi} \frac{\partial p(s,\theta)}{\partial s} \Big|_{s=x\cos\theta + y\sin\theta} d\theta. \tag{2.50}$$

The line-by-line 1D Hilbert transform in the x-direction is denoted as $\mathbf{H}_x$, which can be implemented either as a convolution with a kernel $1/(\pi x)$ or in the Fourier domain as a multiplication with a transfer function $i\, \text{sgn}(\omega_x)$.

Both derivative and backprojection are local operators. The x-direction Hilbert transform $\mathbf{H}_x$ is not a local operator; however, it only depends on the data on a finite range as discussed in Section 2.6.2. This algorithm can be used to reconstruct a finite ROI if the projections are truncated in the y direction (see Section 2.5).

2.6.9 Derivation of the backprojection–derivative–Hilbert transform algorithm

This algorithm is Method 6 discussed in Section 2.3.6, and it can be readily derived from Method 5 by observing that $\partial/\partial s$ is a directional derivative, which is equivalent to $-\sin\theta(\partial/\partial x)+\cos\theta(\partial/\partial y)$ Thus, we have

$$\int_{0}^{\pi} \left.\frac{\partial p(s,\theta)}{\partial s}\right|_{s=x\cos\theta+y\sin\theta} d\theta$$

$$= -\frac{\partial}{\partial x} \int_{0}^{\pi} \left. p(s,\theta)\sin\theta\right|_{s=x\cos\theta+y\sin\theta} d\theta + \frac{\partial}{\partial y} \int_{0}^{\pi} \left. p(s,\theta)\cos\theta\right|_{s=x\cos\theta+y\sin\theta} d\theta$$

$$= \frac{\partial b_x(x,y)}{\partial x} + \frac{\partial b_y(x,y)}{\partial y}, \tag{2.51}$$

where

$$b_x(x,y) = -\int_{0}^{\pi} \left. p(s,\theta)\sin\theta\right|_{s=x\cos\theta+y\sin\theta} d\theta,$$

$$b_y(x,y) = \int_{0}^{\pi} \left. p(s,\theta)\cos\theta\right|_{s=x\cos\theta+y\sin\theta} d\theta. \tag{2.52}$$

Equation (2.51) gives the first two steps in Method 6. The third step, the line-by-line x-direction Hilbert transform, is the same in both Methods 5 and 6.

2.7 Worked examples

Example 1: Write a short Matlab program to illustrate the importance of using suffi-cient view angles in a tomography problem. The Matlab function "*phantom*" generates a mathematical Shepp–Logan phantom. The function "*radon*" generates the projection data, that is, the Radon transform of the phantom. The function "*iradon*" performs the FBP reconstruction.

Solution

The Matlab code:

```
P = phantom(128); %Generate the Shepp-Logan phantom
                  %in a 128x128 array
angle = linspace(0,179,180); %Sampling angles
R = radon(P, angle); %Generate the Radon transform over 180°,
I1 = iradon(R, angle); %Inverse Radon transform, i.e., FBP
                       %reconstruction
I2 = iradon(R, angle,'linear','none'); %Backprojection
                                       %without ramp-filtering
subplot(1,3,1), imshow(P), title('Original')
subplot(1,3,2), imshow(I1), title('Filtered backprojection')
subplot(1,3,3), imshow(I2,[])
title('Unfiltered backprojection')
```

We get the following results (see Figure 2.17).

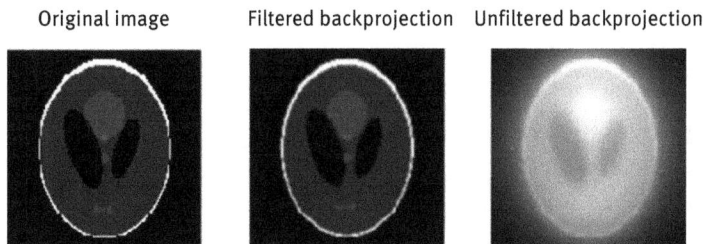

Original image   Filtered backprojection   Unfiltered backprojection

Fig. 2.17: The true image and the reconstructed images with the Matlab function "iradon."

If we change the line angle = linspace(0,179,180) to

$$\text{angle} = \text{linspace}(0, 179, 10)$$
$$\text{angle} = \text{linspace}(0, 179, 20)$$
$$\text{angle} = \text{linspace}(0, 179, 40)$$
$$\text{angle} = \text{linspace}(0, 179, 80)$$

respectively, we get the following results (see Figure 2.18).

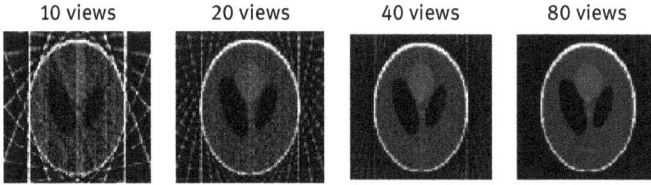

Fig. 2.18: Insufficient views cause artifacts.

Example 2: Run a simple Matlab code to show the noise effect. Apply three different window functions to the ramp filter to control noise.

Solution
The Matlab code:

```
P = phantom(128);
angle = linspace(0,179,180);
R = radon(P,angle);
R = 1e12*imnoise(1e-12*R,'Poisson'); %Add Poisson Noise
I1 = iradon(R,angle,'Ram-Lak');
subplot(1,3,1), imshow(I1,[]), title('w/ Ram-Lak filter')
I1 = iradon(R,angle,'Cosine');
subplot(1,3,2), imshow(I1,[]), title('w/ Cosine filter')
I1 = iradon (R,angle,'Hann');
subplot(1,3,3), imshow(I1,[]), title('w/ Hann filter')
```

The FBP reconstruction results with different window functions are shown in Figure 2.19.

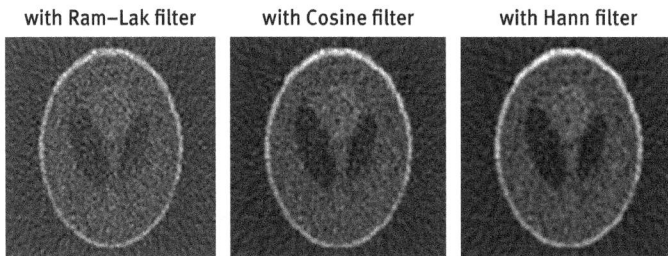

Fig. 2.19: Using different window functions to control noise.

Example 3: Find the discrete "Ramachandran–Lakshminarayanan" convolver, which is the inverse Fourier transform of the ramp filter. The cut-off frequency of the ramp filter is $\omega = 1/2$.

Solution
When the projection is to be filtered by the ramp filter, only a band-limited ramp filter is required, since the projection is band limited by 0.5. The band-limited ramp filter is defined in the Fourier domain as

$$\text{RAMP}(\omega) = \begin{cases} |\omega| & \text{if } |\omega| \le 0.5 \\ 0 & \text{if } |\omega| > 0.5 \end{cases}. \tag{2.53}$$

Let us first use the inverse Fourier transform to find the continuous version of the convolver $h(s)$:

$$h(s) = \int_{-1/2}^{1/2} |\omega| e^{2\pi i \omega s} d\omega = \frac{1}{2} \frac{\sin(\pi s)}{\pi s} - \frac{1}{4} \left[\frac{\sin\left(\frac{\pi s}{2}\right)}{\frac{\pi s}{2}} \right]^2. \tag{2.54}$$

The convolver $h(s)$ is band limited. According to the Nyquest sampling principle, the proper sampling rate is twice the highest frequency, which is 1 in our case. To convert $h(s)$ to the discrete form, let $s = n$ (integer) and we have

$$h(n) = \begin{cases} \frac{1}{4}, & n = 0 \\ 0, & n \text{ even} \\ \frac{-1}{n^2 \pi^2} & n \text{ odd} \end{cases}. \tag{2.55}$$

The continuous and discrete convolvers are displayed in Figure 2.20.

Example 4: How to obtain a discrete transfer function for the ramp filter in an FBP algorithm? In other words, what is the proper way to obtain a discrete version of eq. (2.52)?

Solution
It is a violation of the Nyquest sampling principle to directly sample eq. (2.52), because its "spectrum" $h(s)$ is not supported in a finite region of s. On the other hand, the discrete $h(n)$ can be used to perform exact ramp filtering.

It is impractical to find the discrete Fourier transform (DFT) of $h(n)$ to obtain a discrete version of eq. (2.52) because the number of samples of $h(n)$ is not finite. The good news is that the detector of a medical imaging system has a finite size, and only a finite central portion of the discrete kernel $h(n)$ is required to obtain exact convolution.

If the support of the projections has a length N, only the central N samples of the filtered projection are needed in backprojection. This requires that the minimal length of the convolution kernel be $2N - 1$. In order to make both functions to have the same

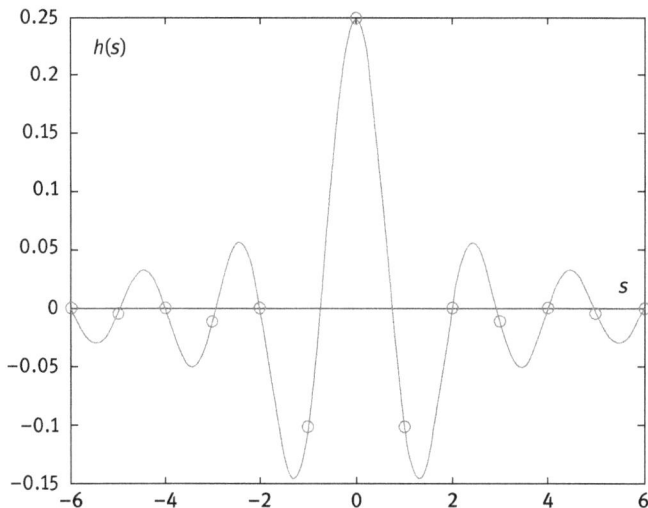

Fig. 2.20: The continuous and sampled "Ramachandran–Lakshminarayanan" convolution kernels.

length, one must pad the projections with $N - 1$ zeros. In general, a length M must be selected which is at least $2N - 1$ and is a power of 2. The projections are then padded with $M{-}N$ zeros. Both functions can now be treated as periodic with the same period M. We treat discrete signals to be periodic because we use the fast Fourier transform (FFT) method to indirectly compute convolution and the FFT method can only compute circular convolution.

Let the discrete version of eq. (2.52) be $H(k)$, $k = 0, 1, 2, \ldots, M - 1$, and RAMP($\omega$) in eq. (2.52) be $H_c(\omega)$, whose corresponding discrete convolution kernel is $h(n)$. Then $H(k)$ can be calculated as

$$
\begin{aligned}
H(k) &= \sum_{n=0}^{M-1} h_c(n) e^{-j2\pi\frac{nk}{M}} \\
&= \sum_{n=0}^{M-1} \int_{-0.5}^{0.5} H_c(\omega) e^{j2\pi\omega n} d\omega \, e^{-j\pi\frac{nk}{M}} \\
&= \int_{-0.5}^{0.5} H_c(\omega) \sum_{n=0}^{M-1} e^{j2\pi n(\omega - k/M)} d\omega \\
&= \int_{-0.5}^{0.5} H_c(\omega) \frac{1 - e^{j2\pi M(\omega - k/M)}}{1 - e^{j2\pi(\omega - k/M)}} d\omega \\
&= \int_{-0.5}^{0.5} H_c(\omega) \frac{\sin(\pi M(\omega - k/M))}{\sin(\pi(\omega - k/M))} e^{j\pi(M-1)(\omega - k/M)} d\omega.
\end{aligned}
\tag{2.56}
$$

Here, $\frac{\sin(\pi M(\omega - k/M))}{\sin(\pi(\omega - k/M))}$, similar to the sinc function, has a removable singularity at $\omega = k/M$, and it is a continuous function if one defines $\frac{\sin(\pi M(\omega - k/M))}{\sin(\pi(\omega - k/M))} = M$ at $\omega = k/M$. If $H_c(\omega)$ is an even function, eq. (2.56) reduces to

$$H(k) = \int_{-0.5}^{0.5} H_c(\omega) \frac{\sin\left(\pi M\left(\omega - \frac{k}{M}\right)\right)}{\sin\left(\pi\left(\omega - \frac{k}{M}\right)\right)} \cos\left(\pi(M-1)\left(\omega - \frac{k}{M}\right)\right) d\omega. \qquad (2.57)$$

It is learned from eq. (2.57) that in general

$$H(k) \neq H_c(k/M). \qquad (2.58)$$

When M is chosen as an integer at least $2N-1$, eq. (2.57) can be further written as

$$H(k) = \frac{1}{4} - \frac{2}{\pi^2} \sum_{\substack{n=1 \\ n \text{ odd}}}^{(M/2)-1} \frac{\cos(2\pi nk/M)}{n^2}. \qquad (2.59)$$

When using DFT or FFT to implement the ramp filter, eq. (2.59) is the correct formula to represent the ramp filter.

Errors may occur if the sampled continuous ramp filter is used. The errors are expressed as

$$\text{Error}\left(\omega = \frac{k}{M}\right) = H(k) - \frac{|k|}{M} \quad \text{for } \frac{|k|}{M} \leq 0.5. \qquad (2.60)$$

Some numerical values according to eq. (2.60) are listed in Tables 2.1 and 2.2 with $M = 2N$. It is observed from Tables 2.1 and 2.2 that if one uses the directly sampled ramp filter in DFT or FFT implementation, the DC gain needs to be corrected, while the errors for other frequency gains are small enough and can be ignored when the size M is large.

Tab. 2.1: Errors for $\omega = 0$, $M = 2N$.

N	k	H(k)	Sampled ramp filter value	Error
8	0	1.26×10^{-2}	0	1.26×10^{-2}
16	0	6.32×10^{-3}		6.32×10^{-3}
32	0	3.17×10^{-3}		3.17×10^{-3}
64	0	1.58×10^{-3}		1.58×10^{-3}
128	0	7.92×10^{-4}		7.92×10^{-4}
256	0	3.96×10^{-4}		3.96×10^{-4}
512	0	1.99×10^{-4}		1.99×10^{-4}
1,024	0	9.89×10^{-5}		9.89×10^{-5}
2,948	0	4.95×10^{-5}		4.95×10^{-5}

Tab. 2.2: Errors for $\omega = 1/16 = 0.0625$, $M = 2\,N$.

N	k	H(k)	Sampled ramp filter value	Error
8	1	6.11×10^{-2}	6.25×10^{-2}	-1.41×10^{-3}
16	2	6.28×10^{-3}		2.50×10^{-4}
32	4	6.25×10^{-3}		3.64×10^{-5}
64	8	6.25×10^{-3}		4.79×10^{-6}
128	16	6.25×10^{-3}		6.07×10^{-7}
256	32	6.25×10^{-3}		7.61×10^{-8}
512	64	6.25×10^{-3}		9.52×10^{-9}
1,024	128	6.25×10^{-3}		1.19×10^{-9}
2,048	256	6.25×10^{-3}		1.49×10^{-10}

Example 5: Use MATLAB to numerically evaluate the 1,024-point-induced ramp filter H by the kernel h defined in eq. (2.55). What is the DC gain of this induced ramp filter?

Solution
The MATLAB code is given as follows:

```
h1=zeros(1,1024);
for n=1:2:1023
    h1(n)=(-1/pi^2)/n^2;
end
h=[fliplr(h1) 1/4 h1];
h(end)=[];
H=fft(fftshift(h));
figure(1)
subplot(1,2,1),plot(-1024:1023,h),axis([-16 16 -0.12 0.25])
subplot(1,2,2),plot(-1024:1023,fftshift(real(H))),...
    axis([-1024 1023 0 0.5])
H(1)
```

The results are plotted below, and the value of DC gain is 9.8946×10^{-5}.

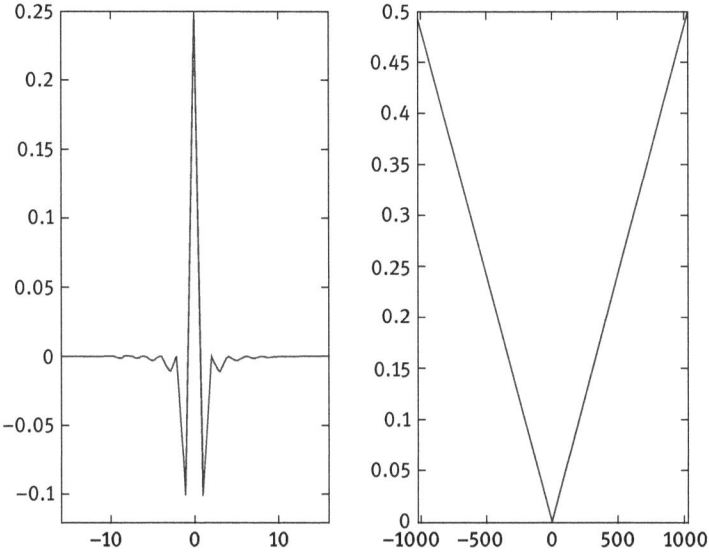

2.8 Summary

- The Fourier transform is a useful tool to express a function in the frequency domain. The inverse Fourier transform brings the frequency representation back to the original spatial domain expression.
- In the Fourier domain, the projection data and the original image are clearly related. This relationship is stated in the popular central slice theorem. The 1D Fourier transform of the projection data at one view is one line in the 2D Fourier transform of the original image. Once we have a sufficient number of projection view angles, their corresponding lines will cover the 2D Fourier plane.
- The backprojection pattern in the 2D Fourier domain indicates that the sampling density is proportional to $1/|\omega|$. Therefore, the backprojection itself can only provide a blurred reconstruction. An exact reconstruction requires a combination of ramp filtering and backprojection.
- The most popular image reconstruction algorithm is the FBP algorithm. The ramp filtering can be implemented as multiplication in the frequency domain or as convolution in the spatial domain.
- One has the freedom to switch the order of ramp filtering and backprojection.
- Ramp filtering can be further decomposed to the Hilbert transform and the derivative operation. Therefore, we have even more ways to reconstruct the image.
- Under certain conditions, the ROI can be exactly reconstructed with truncated projection data.
- The readers are expected to understand two main concepts in this chapter: the central slice theorem and the FBP algorithm.

Problems

? *Problem 2.1* Let a 2D object be defined as

$$f(x,y) = \cos(2\pi x) + \cos(2\pi y).$$

Find its 2D Fourier transform $F(\omega_x, \omega_y)$. Determine the Radon transform of this object $f(x, y)$ using the central slice theorem.

Problem 2.2 Let $f_1(x, y)$ and $f_2(x, y)$ be two 2D functions, and their Radon transforms are $p_1(s, \theta)$ and $p_2(s, \theta)$, respectively. If $f(x, y)$ is the 2D convolution of $f_1(x, y)$ and $f_2(x, y)$, use the central slice theorem to prove that the Radon transform of $f(x, y)$ is the convolution of $p_1(s, \theta)$ and $p_2(s, \theta)$ with respect to variable s.

Problem 2.3 Let the Radon transform of a 2D object $f(x, y)$ be $p(s, \theta)$, the 2D Fourier transform of $f(x, y)$ be $F(\omega_x, \omega_y)$, and the 1D Fourier transform of $p(s, \theta)$ with respect to the first variable s be $P(\omega, \theta)$. What is the physical meaning of the value $F(0, 0)$? What is the physical meaning of the value $P(0, \theta)$? Is it possible that $\int_{-\infty}^{\infty} g(s,\, 0^\circ) ds \neq \int_{-\infty}^{\infty} g(s,\, 90^\circ) ds$?

Problem 2.4 Prove that the Hilbert transform of the function

$$f(t) = \begin{cases} \sqrt{1-t^2}, & |t| < 1 \\ 0, & |t| \geq 1 \end{cases}$$

is

$$g(t) = \begin{cases} t, & |t| < 1 \\ t - \sqrt{t^2 - 1}\,\mathrm{sgn}(t), & |t| \geq 1. \end{cases}$$

Bibliography

[1] Bracewell RN (1956) Strip integration in radio astronomy. Aust J Phys 9:198–217.
[2] Bracewell RN (1986) The Fourier Transform and Its Applications, 2nd ed., McGraw-Hill, New York.
[3] Carrier GF, Krook M, Pearson CE (1983) Functions of a Complex Variable Theory and Technique, Hod, Ithaca.
[4] Clackdoyle R, Noo F, Guo J, Roberts J (2004) A quantitative reconstruction from truncated projections of compact support. Inverse Probl 20:1281–1291.
[5] Deans SR (1983) The Radon Transform and Some of Its Applications, John Wiley & Sons, New York.
[6] Defrise M, Noo F, Clackdoyle KH (2006) Truncated Hilbert transform and image reconstruction from limited tomographic data. Inverse Probl 22:1037–1053.
[7] Hahn SL (1996) Hilbert Transforms in Signal Processing, Artech House, Boston.
[8] Huang Q, Zeng GL, Gullberg GT (2007) An analytical inversion of the 180° exponential Radon transform with a numerically generated kernel. Int J Image Graphics 7:71–85.
[9] Kak AC, Slaney M (1998) Principles of Computerized Tomographic Imaging, IEEE Press, New York.
[10] Kanwal RP (1971) Linear Integral Equations Theory and Technique, Academic Press, New York.
[11] Natterer F (1986) The Mathematics of Computerized Tomography, Wiley, New York.
[12] Noo F, Clackdoyle R, Pack JD (2004) A two-step Hilbert transform method for 2D image reconstruction. Phys Med Biol 49:3903–3923.

[13] Sidky EY, Pan X (2005) Recovering a compactly supported function from knowledge of its Hilbert transform on a finite interval. IEEE Trans Signal Process Lett 12:97–100.

[14] Shepp L, Logan B (1974) The Fourier reconstruction of a head section. IEEE Trans Nucl Sci 21:21–43.

[15] Tricomi FG (1957) Integral Equations, Interscience, New York.

[16] Ye Y, Yu H, Wei Y, Wang G (2007) A general local reconstruction approach based on a truncated Hilbert transform. Int J Biom Imag 2007: 63634.

[17] Ye Y, Yu H, Wang G (2007) Exact interior reconstruction with cone-beam CT. Int J Biom Imag 2007: 10693.

[18] You J, Zeng GL (2006) Exact finite inverse Hilbert transforms. Inverse Probl 22:L7–L10.

[19] Yu H, Yang J, Jiang M, Wang G (2009) Interior SPECT – exact and stable ROI reconstruction from uniformly attenuated local projections. Commun Numer Methods Eng 25:693–710.

[20] Yu H, Ye Y, Wang G (2008) Interior reconstruction using the truncated Hilbert transform via singular value decomposition. J X-Ray Sci Tech 16:243–251.

[21] Zeng GL (2007) Image reconstruction via the finite Hilbert transform of the derivative of the backprojection. Med Phys 34:2837–2843.

[22] Zeng GL (2015) Re-visit of the ramp filter. IEEE Trans Nucl Sci 62:131–136.

[23] Zeng GL, You J, Huang Q, Gullberg GT (2007) Two finite inverse Hilbert transform formulae for local tomography. Int J Imag Syst Tech 17:219–223.

3 Fan-beam image reconstruction

The image reconstruction algorithms discussed in Chapter 2 are for parallel-beam imaging. If the data acquisition system produces projections that are not along parallel lines, the image reconstruction algorithms presented in Chapter 2 cannot be applied directly. This chapter uses the flat detector and curved detector fan-beam imaging geometries to illustrate how a parallel-beam reconstruction algorithm can be converted to user's imaging geometry for image reconstruction.

3.1 Fan-beam geometry and the point spread function

The fan-beam imaging geometry is common in X-ray computed tomography (CT), where the fan-beam focal point is the X-ray source. A fan-beam imaging geometry and a parallel-beam imaging geometry are compared in Figure 3.1.

For the parallel-beam geometry, we have a central slice theorem to derive reconstruction algorithms. We do not have an equivalent theorem for the fan-beam geometry. We will use a different strategy – converting the fan-beam imaging situation into the parallel-beam imaging situation and modifying the parallel-beam algorithms for fan-beam use.

In Chapter 2, when we discussed parallel-beam imaging problems, it was not mentioned, but we always assumed that the detector rotates around at a constant speed and has a uniform angular interval when data are taken. We make the same assumption here for the fan beam.

For the parallel-beam imaging geometry, this assumption results in a shift-invariant point spread function (PSF) for projection/backprojection. In other words, if you put a point source in the x–y plane (it does not matter where you put it), calculate the projections, and perform the backprojection, then you will always get the same star-like pattern (see Figure 3.2). This pattern is called the *point spread function* of the projection/backprojection operation.

In the parallel-beam case, when you find the backprojection at the point (x, y), you draw a line through this point and perpendicular to each detector. This line meets the detector at a point, say, s^*. Then add the value $p(s^*, \theta)$ to the location (x, y).

In the fan-beam case, when you find the backprojection at the point (x, y), you draw a line through this point and each focal-point location. This line has an angle, say, γ^*, with respect to the central ray of the detector. Then add the value $g(\gamma^*, \beta)$ to location (x, y).

It can be shown that if the fan-beam focal-point trajectory is a complete circle, the PSF is shift invariant (i.e., the pattern does not change when the location of the point source changes) and has the same PSF pattern as that for the parallel-beam case (see Figure 3.3).

https://doi.org/10.1515/9783111055404-003

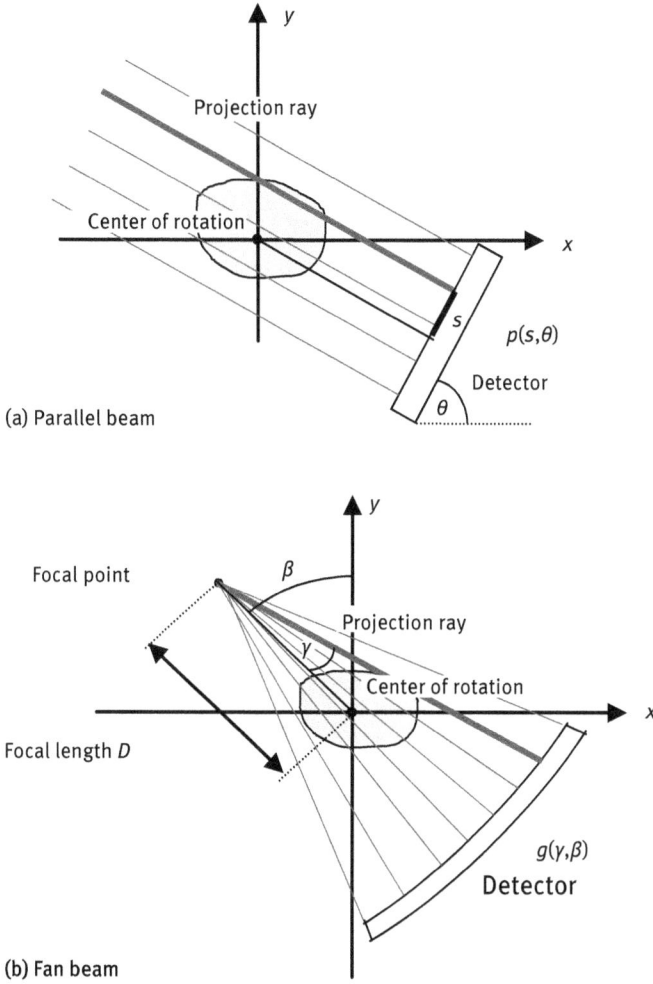

(a) Parallel beam

(b) Fan beam

Fig. 3.1: Comparison of the parallel-beam and the fan-beam imaging geometries.

This observation is important. It implies that if you project and backproject an image, you will get the same blurred version of that image, regardless of the use of parallel-beam or fan-beam geometry.

If the original image is $f(x, y)$ and if the backprojection of the projection data is $b(x, y)$, then the PSF can be shown to be $1/r$, where $r = \sqrt{x^2 + y^2}$. Then $f(x, y)$ and $b(x, y)$ are related by

$$b(x,y) = f(x,y) ** \frac{1}{r},$$ (3.1)

where "$**$" denotes 2D convolution. The Fourier transform of b is B, and the Fourier transform of f is F. Thus the above relationship in the Fourier domain becomes

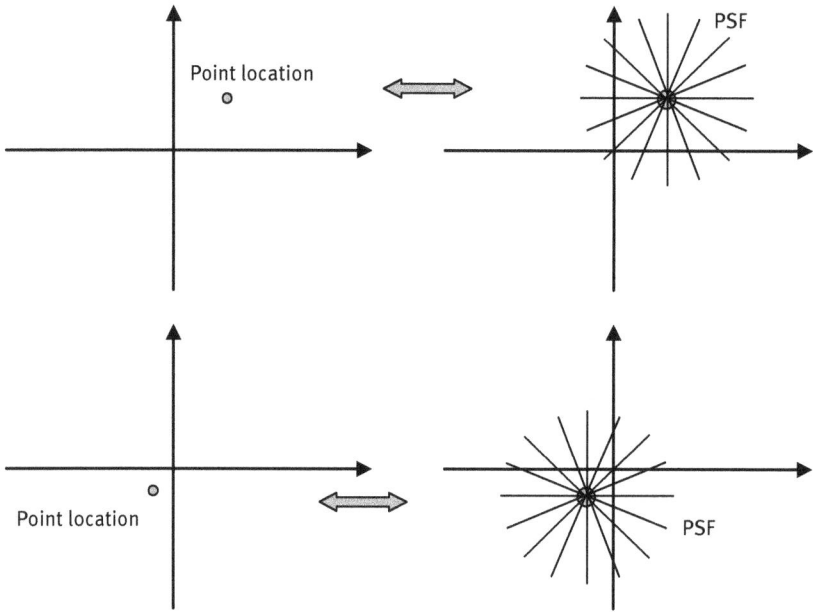

Fig. 3.2: The projection/backprojection PSF is shift invariant.

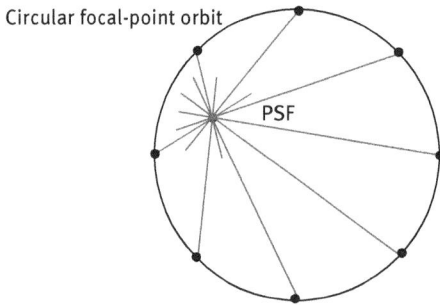

Fig. 3.3: The fan-beam 360° full-scan PSF is the same as that for the parallel-beam scan.

$$B\left(\omega_x, \omega_y\right) = F\left(\omega_x, \omega_y\right) \times \frac{1}{\sqrt{\omega_x^2 + \omega_y^2}}, \tag{3.2}$$

because the 2D Fourier transform of $1/\sqrt{x^2 + y^2}$ is $1/\sqrt{\omega_x^2 + \omega_y^2}$.

We already know that if a 2D ramp filter is applied to the backprojected image $b(x, y)$, the original image $f(x, y)$ can be obtained. The same technique can be used for the fan-beam backprojected image $b(x, y)$. The backprojection-then-filtering algorithm is the same for the parallel-beam and fan-beam imaging geometries. If we apply the 2D ramp filter $\sqrt{\omega_x^2 + \omega_y^2}$ to both sides of the above Fourier domain relationship, the Fourier transform $F(\omega_x, \omega_y)$ of the original image $f(x, y)$ is readily obtained:

$$F(\omega_x, \omega_y) = B(\omega_x, \omega_y) \times \sqrt{\omega_x^2 + \omega_y^2}. \qquad (3.3)$$

Finally, the original image $f(x, y)$ is found by taking the 2D inverse Fourier transform.

3.2 Parallel-beam to fan-beam algorithm conversion

If you want to reconstruct the image by a filter-then-backproject (i.e., filtered backprojection [FBP]) algorithm, a different strategy must be used.

A straightforward approach would be to rebin every fan-beam ray into a parallel-beam ray. For each fan-beam ray sum $g(\gamma, \beta)$, we can find a parallel-beam ray sum $p(s, \theta)$ that has the same orientation as the fan-beam ray with the relations (see Figure 3.4)

$$\theta = \gamma + \beta \qquad (3.4)$$

and

$$s = D \sin \gamma, \qquad (3.5)$$

where D is the focal length. We then assign

$$p(s, \theta) = g(\gamma, \beta). \qquad (3.6)$$

After rebinning the fan-beam data into the parallel-beam format, we then use a parallel-beam image reconstruction algorithm to reconstruct the image. However, this rebinning approach is not preferred because rebinning requires data interpolation when changing coordinates. Data interpolation introduces errors. The idea of rebinning is feasible, but the results may not be accurate enough.

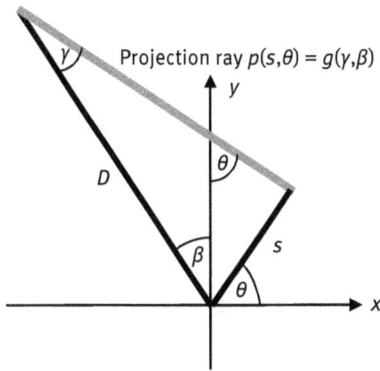

Fig. 3.4: A fan-beam ray can be represented using the parallel-beam geometry parameters.

However, the idea above is not totally useless. Let us do it in a slightly different way. We start out with a parallel-beam image reconstruction algorithm, which is a mathematical expression. On the left-hand side of the expression is the reconstructed image $f(x, y)$. On the right-hand side is an integral expression that contains the projection $p(s, \theta)$ and some other factors associated with s and θ (see Figure 3.5).

Next, we replace the parallel-beam projection $p(s, \theta)$ by its equivalent fan-beam counterpart $g(\gamma, \beta)$ on the right-hand side. Of course, this substitution is possible only if the conditions $\theta = \gamma + \beta$ and $s = D \sin \gamma$ are satisfied. These two relations are easier to see in Figure 3.4, which is derived from Figure 3.1.

In order to satisfy these two conditions, we stick them into the right-hand side of the expression. This procedure is nothing but changing the variables in the integral, where the variables s and θ are changed into γ and β.

As a reminder, in calculus, when you change the variables in an integral, you need a Jacobian factor, which is a determinant calculated with some partial derivatives. This Jacobian is a function of γ and β.

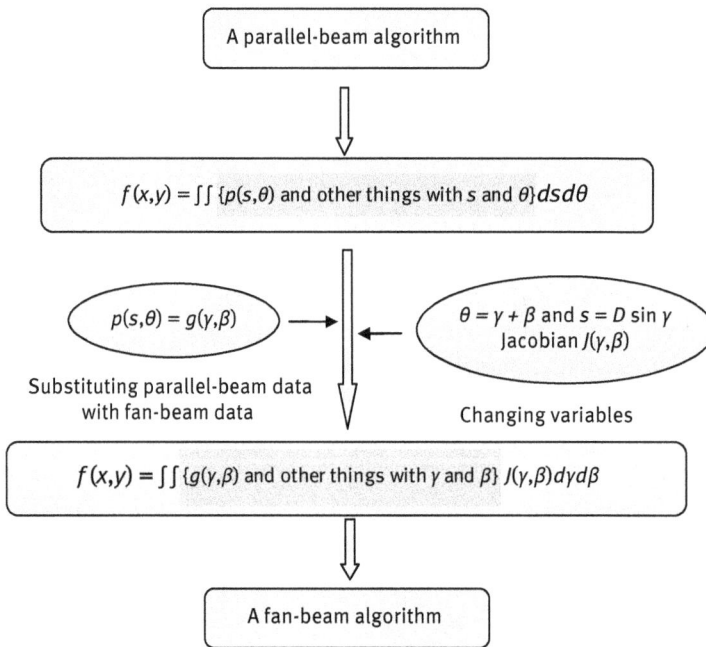

A parallel-beam algorithm

$$f(x,y) = \iint \{p(s,\theta) \text{ and other things with } s \text{ and } \theta\} ds d\theta$$

$p(s,\theta) = g(\gamma,\beta) \quad\longrightarrow\quad$ $\theta = \gamma + \beta$ and $s = D \sin \gamma$ / Jacobian $J(\gamma,\beta)$

Substituting parallel-beam data with fan-beam data

Changing variables

$$f(x,y) = \iint \{g(\gamma,\beta) \text{ and other things with } \gamma \text{ and } \beta\} J(\gamma,\beta) d\gamma d\beta$$

A fan-beam algorithm

Fig. 3.5: The procedure to change a parallel-beam algorithm into a fan-beam algorithm.

After substituting the parallel-beam data $p(s, \theta)$ with fan-beam data $g(\gamma, \beta)$, changing variables s and θ to γ and β, and inserting a Jacobian $J(\gamma, \beta)$, a new fan-beam image reconstruction algorithm is born (see Figure 3.5)!

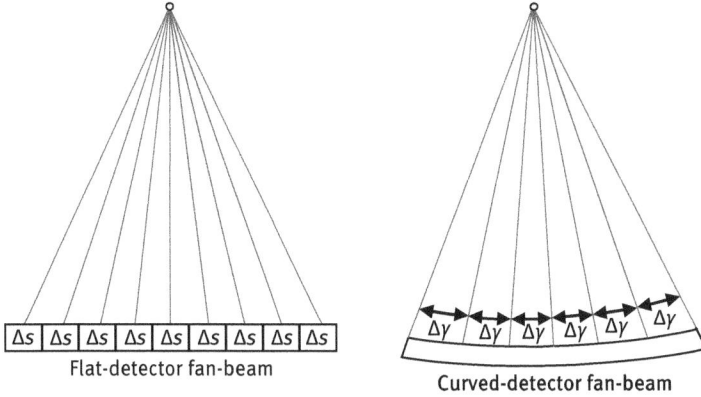

Fig. 3.6: Flat and curved detector fan-beam geometries.

The method outlined in Figure 3.5 is generic. We have a long list of parallel-beam image reconstruction algorithms. They can all be converted into fan-beam algorithms in this way, and accordingly, we also have a long list of fan-beam algorithms. We must point out that after changing the variables, a convolution operation (with respect to variable s) may not turn into a convolution (with respect to variable y) automatically, and some mathematical manipulation is needed to turn it into a convolution form. Like parallel-beam algorithms, the fan-beam algorithms include combinations of ramp filtering and backprojection or combinations of the derivative, Hilbert transform, and backprojection (DHB).

Researchers treat the flat detector fan-beam and curved detector fan-beam differently in reconstruction algorithm development. In a flat detector, the data points are sampled with equal distance Δs intervals, while in a curved detector, the data points are sampled with equal angle Δy intervals (see Figure 3.6). A fan-beam algorithm can be converted from one geometry to the other with proper weighting adjustments.

3.3 Short scan

In parallel-beam imaging, when the detector rotates 2π (i.e., 360°), each projection ray is measured twice, and the redundant data are related by

$$p(s, \theta) = p(-s, \theta + \pi); \tag{3.7}$$

the redundant data are acquired by the two face-to-face detectors (see Figure 3.7). Therefore, it is sufficient to acquire data over an angular range of π.

Likewise, when the fan-beam detector rotates 2π, each projection ray is also measured twice, and the redundant data are related by (see Figure 3.8)

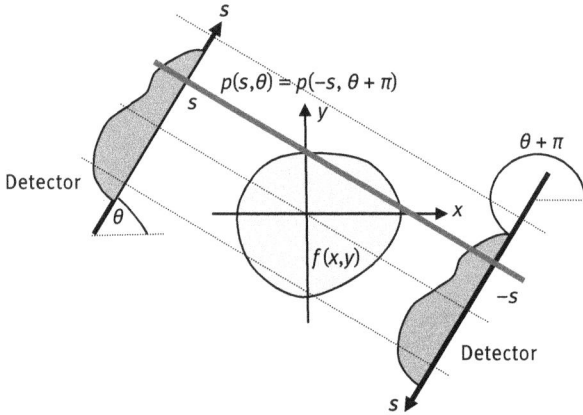

Fig. 3.7: Face-to-face parallel-beam detectors measure the same line integrals.

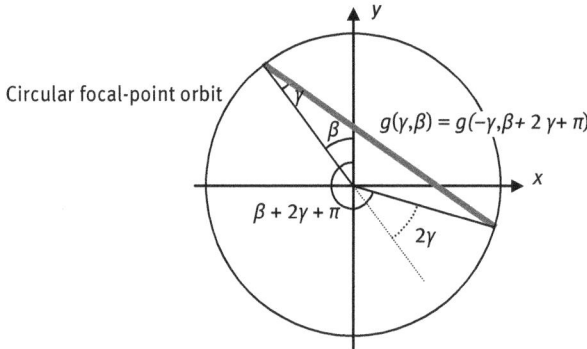

Fig. 3.8: Each ray is measured twice in a fan-beam 360° full scan.

$$g(\gamma, \beta) = g(-\gamma, \beta + 2\gamma + \pi). \tag{3.8}$$

Due to data redundancy, we can use a smaller angular (β) range than 2π for fan-beam data acquisition, hence, the term *short scan*. The minimal range of β is determined by how the data are acquired. This required range can be less than π (see Figure 3.9, left), equal to π (see Figure 3.9, middle), or larger than π (see Figure 3.9, right). The criterion is that we need at least 180° angular coverage for each point in the object in which we are interested.

We need to be cautious that in a fan-beam short scan, some rays are measured once, and other rays are measured twice. Even for the case that the angular range of β is less than π, there are still rays that are measured twice. In fact, any ray that intersects the measured focal-point trajectory twice is measured twice (see Figure 3.10). We require that any ray which passes through the object should be measured at least once. Proper weighting should be used in image reconstruction if data are redundant.

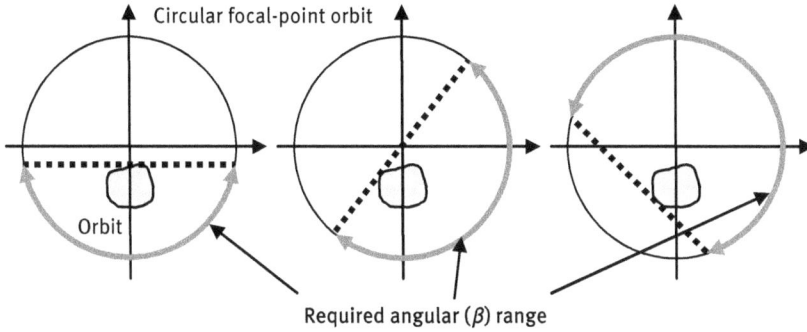

Fig. 3.9: Fan-beam minimum scan angle depends on the location of the object.

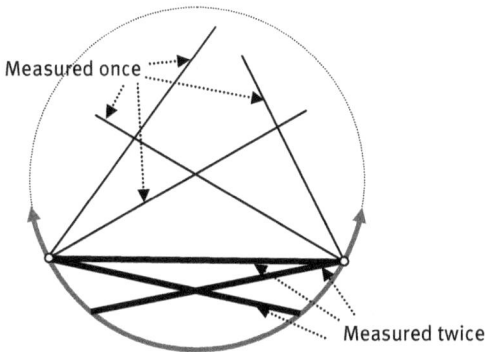

Fig. 3.10: For a fan-beam short scan, some rays are measured once and some rays are measured twice.

For example, if a ray is measured twice, the sum of the weighting factors for these two measurements should be unity. For a fan-beam short scan, the projection/backprojection PSF is no longer shift invariant.

3.4 Mathematical expressions

This section presents the derivation steps for the FBP algorithm for the curved fan-beam detector. The ramp filter used in the filtering step is formulated as a convolution. In this algorithm, the fan-beam backprojector contains a distance-dependent weighting factor, which causes nonuniform resolution throughout the reconstructed image when a window function is applied to the ramp filter in practice. In order to overcome this problem, the ramp filter can be replaced by a derivative operation and the Hilbert transform. The derivation of the fan-beam algorithm using the derivative and the Hilbert transform is given in this section.

3.4.1 Derivation of a filtered backprojection fan-beam algorithm

We begin with the parallel-beam FBP algorithm (see Section 2.6.5), using polar coordinates (r, φ) instead of Cartesian coordinates (x, y). Then $x = r \cos \varphi$, $y = r \sin \varphi$, and $x \cos \theta + y \sin \theta = r \cos(\theta - \varphi)$. We have

$$f(r, \varphi) = \frac{1}{2} \int_0^{2\pi} \int_{-\infty}^{\infty} p(s, \theta) h(r \cos(\theta - \varphi) - s) ds d\theta. \tag{3.9}$$

Changing variables $\theta = \gamma + \beta$ and $s = D \sin \gamma$ with the Jacobian $D \cos \gamma$ yields

$$f(r, \varphi) = \frac{1}{2} \int_0^{2\pi} \int_{-\pi/2}^{\pi/2} g(\gamma, \beta) h(r \cos(\beta + \gamma - \varphi) - D \sin \gamma) D \cos \gamma d\gamma d\beta. \tag{3.10}$$

This is a fan-beam reconstruction algorithm but the inner integral over γ is not yet in the convolution form. Convolution is much more efficient than a general integral in implementation. In the following, we are going to convert the integral over γ to a convolution with respect to γ.

For a given reconstruction point (r, φ), we define D' and γ' as shown in Figure 3.11, then $r \cos(\beta + \gamma - \varphi) - D \sin \gamma = D' \sin(\gamma' - \gamma)$, and

$$f(r, \varphi) = \frac{1}{2} \int_0^{2\pi} \int_{-\pi/2}^{\pi/2} g(\gamma, \beta) h(D' \sin(\gamma' - \gamma)) D \cos \gamma d\gamma d\beta. \tag{3.11}$$

Now, we prove a special property of the ramp filter

$$h(D' \sin \gamma) = \left(\frac{\gamma}{D' \sin \gamma}\right)^2 h(\gamma), \tag{3.12}$$

which will be used in the very last step, the derivation, of the fan-beam formula.

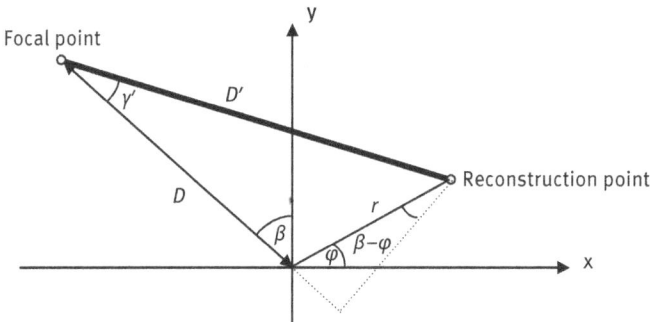

Fig. 3.11: The reconstruction point (r, φ) defines the angle γ' and distance D'.

Using the definition of the ramp filter kernel $h(t) = \int_{-\infty}^{\infty} |\omega| e^{i2\pi\omega t} d\omega$, we have

$$h(D' \sin \gamma) = \int_{-\infty}^{\infty} |\omega| e^{i2\pi D' \sin \gamma} d\omega$$

$$= \left(\frac{\gamma}{D' \sin \gamma}\right)^2 \int_{-\infty}^{\infty} \left|\omega D' \frac{\sin \gamma}{\gamma}\right| e^{i2\pi\omega \frac{D' \sin \gamma}{\gamma} \gamma} d\left(\omega D' \frac{\sin \gamma}{\gamma}\right)$$

$$= \left(\frac{\gamma}{D' \sin \gamma}\right)^2 \int_{-\infty}^{\infty} |\hat{\omega}| e^{i2\pi\omega\gamma} d\hat{\omega}$$

$$= \left(\frac{\gamma}{D' \sin \gamma}\right)^2 h(\gamma). \tag{3.13}$$

If we denote $h_{\text{fan}}(\gamma) = (D/2) \, (\gamma/\sin \gamma)^2 \, h(\gamma)$, then the fan-beam convolution backprojection algorithm is obtained as

$$f(r, \varphi) = \int_0^{2\pi} \frac{1}{(D')^2} \int_{-\pi/2}^{\pi/2} (\cos \gamma) g(\gamma, \beta) h_{\text{fan}} (\gamma' - \gamma) d\gamma d\beta. \tag{3.14}$$

3.4.2 A fan-beam algorithm using the derivative and the Hilbert transform

The general idea of decomposition of the ramp filter into the derivative and the Hilbert transform can be applied to fan-beam image reconstruction. A derivative–Hilbert transform–backprojection algorithm can be obtained by doing a coordinate transformation on the Radon inversion formula (see Section 2.6.6) as follows. Noo, Defrise, Kudo, Clackdoyle, Pan, Chen, Wang, You, and many others have contributed significantly in developing algorithms using the derivative and the Hilbert transform. We first rewrite the Radon inversion formula (see Section 2.6.6) in the polar coordinate system, that is, $f(r, \varphi)$ can be reconstructed as

$$f(r, \varphi) = \frac{1}{2} \int_0^{2\pi} \int_{-\infty}^{\infty} \frac{\partial p(s, \theta)}{\partial s} \frac{1}{2\pi^2(r \cos(\theta - \varphi) - s)} ds d\theta. \tag{3.15}$$

Changing variables from (s, θ) to (γ, β) and using $\dfrac{\partial p}{\partial s} = \dfrac{1}{D \cos \gamma} \dfrac{\partial g}{\partial \gamma}$, we have

$$f(r,\varphi) = \frac{1}{2} \int_0^{2\pi} \int_{-\pi/2}^{\pi/2} \frac{1}{D\cos\gamma} \frac{\partial g(\gamma,\beta)}{\partial\gamma} \frac{1}{2\pi^2 D' \sin(\gamma'-\gamma)} D\cos\gamma \, d\gamma \, d\beta$$

$$= \int_0^{2\pi} \frac{1}{4\pi^2 D'} \int_{-\pi/2}^{\pi/2} \frac{\partial g(\gamma,\beta)}{\partial\gamma} \frac{1}{\sin(\gamma'-\gamma)} \, d\gamma \, d\beta \qquad (3.16)$$

where D' is the distance from the reconstruction point to the focal point at angle β. This D' factor is not desirable in a reconstruction algorithm. A small D' can make the algorithm unstable; this spatially variant factor also costs some computation time. In a 2π scan, each ray is measured twice. If proper weighting is chosen for the redundant measurements, this D' factor can be eliminated.

Let us introduce a weighting function w in the above DHB algorithm:

$$f(r,\varphi) = \frac{1}{4\pi^2} \int_0^{2\pi} \frac{w(\gamma',\beta,r,\varphi)}{D'} \int_{-\pi/2}^{\pi/2} \left(\frac{\partial}{\partial\gamma} + \frac{\partial}{\partial\beta}\right) g(\gamma,\beta) \frac{1}{\sin(\gamma'-\gamma)} \, d\gamma \, d\beta. \qquad (3.17)$$

If we use $\hat{g}$ to denote the result of the derivative and Hilbert transform of the fan-beam data:

$$\hat{g}(\gamma',\beta) = \int_{-\pi/2}^{\pi/2} \left(\frac{\partial}{\partial\gamma} + \frac{\partial}{\partial\beta}\right) g(\gamma,\beta) \frac{1}{\sin(\gamma'-\gamma)} \, d\gamma, \qquad (3.18)$$

then

$$f(r,\varphi) = \frac{1}{4\pi^2} \int_0^{2\pi} \frac{w(\gamma',\beta,r,\varphi)}{D'} \hat{g}(\gamma',\beta) \, d\beta. \qquad (3.19)$$

It can be shown that $\hat{g}/D'$ has the same redundancy property as the original fan-beam data g (see Figure 3.8). Therefore, it is required that the weighting function satisfies the condition

$$w(\gamma',\beta,r,\varphi) + w(\gamma_c',\beta_c,r,\varphi) = 2, \qquad (3.20)$$

with $\gamma_c' = -\gamma'$ and $\beta_c = \beta + 2\gamma' + \pi$. If we define $w(\gamma', \beta, r, \varphi) = D'/(D\cos\gamma')$, then the above condition is satisfied because $D' + D_c' = 2D\cos\gamma'$ (see Figure 3.12). Finally,

$$f(r,\varphi) = \frac{1}{2} \left\{ \frac{1}{4\pi^2} \int_0^{2\pi} \left[\frac{w(\gamma',\beta,r,\varphi)}{D'} + \frac{w(\gamma_c'',\beta_c,r,\varphi)}{D_c''} \right] \hat{g}(\gamma',\beta) \, d\beta \right\}$$

$$= \frac{1}{4\pi^2 D} \int_0^{2\pi} \frac{1}{\cos\gamma'} \hat{g}(\gamma',\beta) \, d\beta. \qquad (3.21)$$

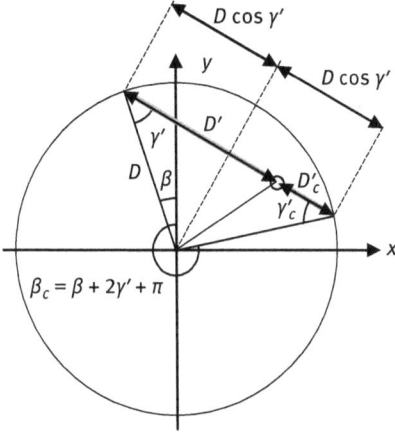

Fig. 3.12: Proper weighting can make the distance-dependent factor disappear from the backprojector in a 360° full scan.

This result can be derived without introducing a weighting function w (see a paper listed at the end of this chapter by You and Zeng [14]).

3.4.3 Expression for the Parker weights

Short scan in fan-beam tomography has wide applications in industry and health care; it can acquire sufficient line-integral measurements by scanning the object 180° plus the full fan angle.

In short-scan fan-beam imaging, the projections $g(\gamma, \beta)$ are acquired when the fan-beam rotation angle β is in the range of $[0°, 180° + 2\delta]$ or in a larger range, where δ is half of the fan angle. The redundant measurements in the range of $[0°, 180° + 2\delta]$ are weighted by the Parker weights:

$$w(\gamma,\beta) = \begin{cases} \sin^2\left(\frac{\pi}{4}\frac{\pi+2\delta-\beta}{\delta+\gamma}\right), & \text{if} \quad \pi - 2\delta \le \beta \le \pi + 2\delta \ \text{and} \ 0 \le \frac{\pi+2\delta-\beta}{\delta+\gamma} \le 2, \\ \sin^2\left(\frac{\pi}{4}\frac{\beta}{\delta+\gamma}\right), & \text{if} \quad 0 \le \beta \le 4\delta \quad \text{and} \ 0 \le \frac{\beta}{\delta-\gamma} \le 2. \\ 1, & \text{otherwise} \end{cases}$$

(3.22)

One then uses the scaled fan-beam projections $w(\gamma, \beta)g(\gamma, \beta)$ in place of $g(\gamma, \beta)$ in the fan-beam FBP image reconstruction algorithm.

Short-scan weighting is not unique. A function $w(\gamma, \beta)$ is a valid short-scan weighting function as long as it satisfies

$$w(\gamma,\beta) + w(-\gamma,\beta+2\gamma+\pi) = 1. \tag{3.23}$$

For discrete data, the weighting function $w(\gamma, \beta)$ must be very smooth, otherwise artifacts will appear. If the weighting function is not smooth enough, a sudden change in the projections can cause a spike in the ramp-filtered projections. In an ideal situation, an opposite spike will also be created and this opposite spike should be able to cancel the other spike. When data are discretely sampled, the positive spike and the negative spike do not match well. The uncanceled residue causes image artifacts. The main strategy of Parker's method is to make the weighting function as smooth as possible so that the weighting function does not introduce any undesired components.

3.4.4 Errors caused by finite bandwidth implementation

In the derivation of the fan-beam FBP algorithm in Section 3.4.1, the bandwidth of the ramp filter is assumed to be infinity:

$$h(\gamma) = \int_{-\infty}^{\infty} |\omega| e^{i2\pi\omega\gamma} d\omega. \tag{3.24}$$

It has a property that

$$h(a\gamma) = \int_{-\infty}^{\infty} |\omega| e^{i2\pi\omega\gamma} d\omega = \frac{1}{a^2} h(\gamma). \tag{3.25}$$

In practice, the data are discretely sampled and the bandwidth is bounded. The band-limited ramp filter is expressed as

$$h_\Omega(\gamma) = \int_{-\Omega}^{\Omega} |\omega| e^{i2\pi\omega\gamma} d\omega. \tag{3.26}$$

Let $x = a\omega$ and we have

$$h_\Omega(a\gamma) = \frac{1}{a^2} \int_{-a\Omega}^{a\Omega} |x| e^{i2\pi x\gamma} dx = \frac{1}{a^2} h_{a\Omega}(\gamma) \neq \frac{1}{a^2} h_\Omega(\gamma). \tag{3.27}$$

Therefore, for band-limited fan-beam data, the derivation of the fan-beam FBP algorithm in Section 3.4.1 is not valid. Fortunately, for the full-scan fan-beam projections, the errors in the FBP reconstruction are not significant and are usually not noticeable. However, the errors can be significant for short-scan fan-beam data when the fan-beam focal length is short and the Parker weights are used. Figure 3.13 illustrates a short-scan fan-beam FBP reconstruction when a short focal length of 270 units is used. The object is a uniform disk with a radius of 230 units. The true image value within the disk is 1. The image display window is [0.76, 1.57].

Fig. 3.13: A short-scan fan-beam FBP reconstruction of a uniform disk, when the fan-beam focal length is very short.

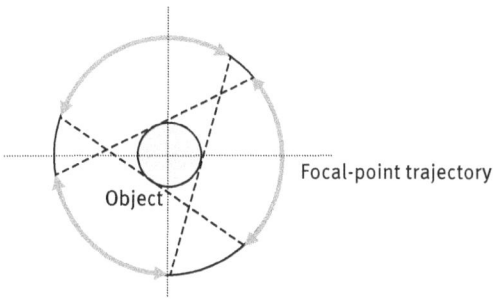

Fig. 3.14: A three-piece fan-beam focal-point trajectory.

3.5 Worked examples

Example 1: Does the following fan-beam geometry acquire sufficient projection data for image reconstruction? The fan-beam focal-point trajectory consists of three disjoint arcs as shown in Figure 3.14.

Solution
Yes. If you draw any line through the circular object, this line will intersect the focal-point trajectory at least once.

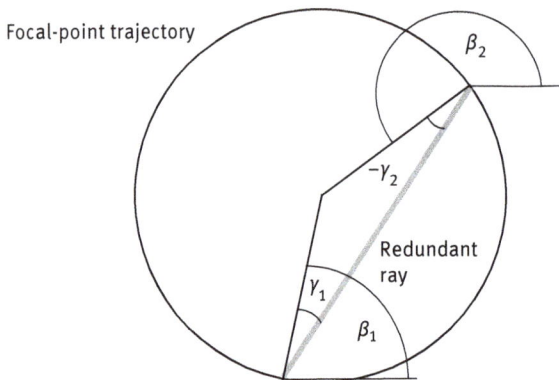

Fig. 3.15: A redundant measurement in a fan-beam scan.

Example 2: On the γ–β plane (similar to the sinogram for the parallel-beam projections), identify the double measurements for a fan-beam 2π scan.

Solution
From Figure 3.15, we can readily find the fan-beam data redundancy conditions for

$$g(\gamma_1, \beta_1) = g(\gamma_2, \beta_2) \qquad (3.28)$$

as

$$\gamma_2 = -\gamma_1 \qquad (3.29)$$

and

$$\beta_2 = \beta_1 + 2\gamma_1 + \pi. \qquad (3.30)$$

Using these conditions, the fan-beam data redundancy is depicted on the γ–β plane in Figure 3.16, where every vertical line corresponds to a redundant slant line.

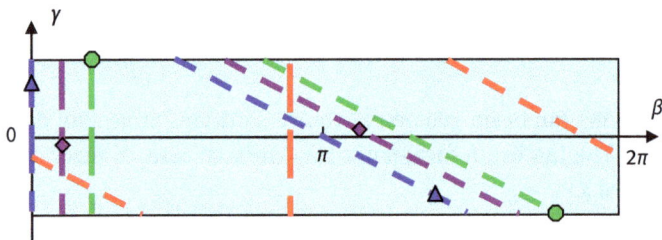

Fig. 3.16: In the γ–β plane representation of the fan-beam data, each vertical line of measurements is the same as the slant line of measurements with the same color.

Example 3: Derive an FBP algorithm for the flat-detector fan-beam geometry.

Solution
We begin with the parallel-beam FBP algorithm, using polar coordinates (r, φ) instead of Cartesian coordinates (x, y). Then $x = r \cos \varphi$, $y = r \sin \varphi$, and $x \cos \theta + y \sin \theta = r \cos(\theta - \varphi)$. We have

$$f(r, \varphi) = \frac{1}{2} \int_0^{2\pi} \int_{-\infty}^{\infty} p(s, \theta) h(r \cos(\theta - \varphi) - s) \, ds \, d\theta. \tag{3.31}$$

For the flat-detector fan-beam projection $g(t, \beta) = p(s, \theta)$, the fan-beam and parallel-beam variables are related by $\theta = \beta + \tan^{-1}(t/D)$ and $s = D(t/\sqrt{D^2 + t^2})$. Changing the parallel-beam variables to the fan-beam variables with the Jacobian $D^3/(D^2 + t^2)^{3/2}$ yields

$$f(r, \varphi) = \frac{1}{2} \int_0^{2\pi} \int_{-\infty}^{\infty} g(t, \beta) h\left((\hat{t} - t) \frac{UD}{\sqrt{D^2 + t^2}}\right) \frac{D^3}{(D^2 + t^2)^{3/2}} \, dt \, d\beta, \tag{3.32}$$

where we have used $r \cos(\theta - \varphi) - s = (\hat{t} - t)(UD/\sqrt{D^2 + t^2})$ with

$$U = \frac{D + r \sin(\beta - \varphi)}{D} \tag{3.33}$$

and

$$\hat{t} = \frac{Dr \cos(\beta - \varphi)}{D + r \sin(\beta - \varphi)} \tag{3.34}$$

(see Figure 3.17).

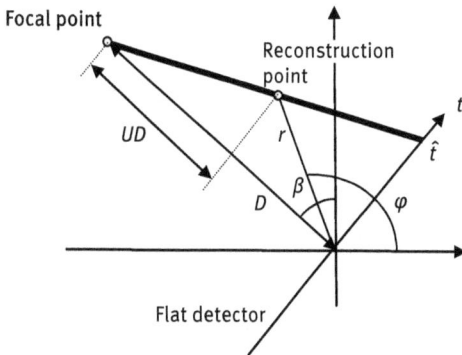

Fig. 3.17: Notation for flat detector fan-beam imaging geometry.

This is a fan-beam reconstruction algorithm, but the inner integral over t is not yet in the convolution form. Using a special property of the ramp filter $h(at) = (1/a^2)\,h(t)$ yields

$$f(r,\varphi) = \frac{1}{2}\int_0^{2\pi}\frac{1}{U^2}\int_{-\infty}^{\infty}\frac{D}{\sqrt{D^2+t^2}}g(t,\beta)h(\hat{t}-t)\,dt\,d\beta. \qquad (3.35)$$

This is a fan-beam convolution backprojection algorithm, where $D/\sqrt{D^2+t^2}$ is the cosine pre-weighting factor, the integral over t is the ramp-filter convolution, and $1/U^2$ is the distance-dependent weighting factor in the backprojection.

Note: The relationship $h(at) = (1/a^2)\,h(t)$ does not hold for a *windowed* ramp filter. Therefore, the fan-beam algorithm derived above does not have uniform resolution in the reconstructed image in practice when a window function is applied to the ramp filter.

3.6 Summary

– The fan-beam geometry is popular in X-ray CT imaging.
– The fan-beam image reconstruction algorithms can be derived from their parallel-beam counterparts via changing of variables.
– There are two types of fan-beam detectors: the flat detectors and curved detectors. Each detector type has its own image reconstruction algorithms.
– If the fan-beam focal-point trajectory is a full circle, it is called full scan. If the trajectory is a partial circle, it is called short scan. Even for a short scan, some of the fan-beam rays are measured twice. The redundant measurements need proper weighting during image reconstruction.
– For some fan-beam image reconstruction algorithms, the backprojector contains a distance-dependent weighting factor. When a window function is applied to the ramp filter, this factor is not properly treated by the window function and the resultant fan-beam FBP is no longer exact in the sense that the reconstructed image has nonuniform resolution and intensity. The short-scan fan-beam FBP algorithm can result in noticeable errors if the focal length is very short.
– The modern derivative and Hilbert transform-based algorithms are able to weigh the short-scan redundant data in a correct way.
– In this chapter, the readers are expected to understand how a fan-beam algorithm can be obtained from a parallel-beam algorithm.

Problems

? *Problem 3.1* The data redundancy condition for the curved detector fan-beam imaging geometry is

$$g(\gamma, \beta) = g(-\gamma, \beta + 2\gamma + \pi).$$

What is the data redundancy condition for the flat detector fan-beam geometry?

Problem 3.2 In this chapter, we assumed that the X-ray source (i.e., the fan-beam focal point) rotates around the object in a circular orbit. If the focal-point orbit is not circular, then the focal length D is a function of the rotation angle β. Extend a fan-beam image reconstruction algorithm developed in this chapter to the situation that the focal-point orbit is noncircular.

Problem 3.3 The method of developing an image reconstruction algorithm discussed in this chapter is not restricted to the fan-beam imaging geometry. For example, we can consider a variable focal-length fan-beam imaging geometry, where the focal length D can be a function of t, which is the coordinate of the detection bin as shown in the figure below. Extend a fan-beam image reconstruction algorithm in this chapter to this variable focal-length fan-beam imaging geometry.

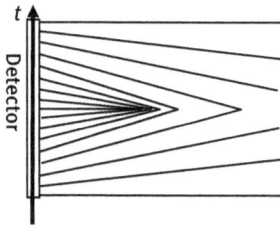

Bibliography

[1]   Besson G (1996) CT fan-beam parameterizations leading to shift-invariant filtering. Inverse Probl 12:815–833.

[2]   Chen GH (2003) A new framework of image reconstruction from fan beam projections. Med Phys 30:1151–1161.

[3]   Chen GH, Tokalkanajalli R, Hsieh J (2006) Development and evaluation of an exact fan-beam reconstruction algorithm using an equal weighting scheme via locally compensated filtered backprojection (LCFBP). Med Phys 33:475–481.

[4]   Dennerlein F, Noo F, Hornegger J, Lauritsch G (2007) Fan-beam filtered-backprojection reconstruction without backprojection weight. Phys Med Biol 52:3227–3240.

[5]   Gullberg GT (1979) The reconstruction of fan-beam data by filtering the back-projection. Comput Graph Image Process 10:30–47.

[6]   Hanajer C, Smith KT, Solmon DC, Wagner SL (1980) The divergent beam X-ray transform. Rocky Mt J Math 10:253–283.

[7]   Natterer N (1993) Sampling in fan-beam tomography. SIAM J Appl Math 53:358–380.

[8]   Horn BKP (1979) Fan-beam reconstruction methods. Proc IEEE 67:1616–1623.

[9]   Noo F, Defrise M, Clackdoyle R, Kudo H (2002) Image reconstruction from fan-beam projections on less than a short scan. Phys Med Biol 47:2525–2546.

[10]   Pan X (1999) Optimal noise control in and fast reconstruction of fan-beam computed tomography image. Med Phys 26:689–697.

[11]   Pan X, Yu L (2003) Image reconstruction with shift-variant filtration and its implication for noise and resolution properties in fan-beam tomography. Med Phys 30:590–600.

[12]   Parker DL (1982) Optimal short scan convolution reconstruction for fan beam CT. Med Phys 9:254–257.

[13]   Silver MD (2000) A method for including redundant data in computed tomography. Med Phys 27:773–774.

[14]   You J, Liang Z, Zeng GL (1999) A unified reconstruction framework for both parallel-beam and variable focal-length fan-beam collimators by a Cormack-type inversion of exponential Radon transform. IEEE Trans Med Imaging 18:59–65.

[15]   You J, Zeng GL (2007) Hilbert transform based FBP algorithm for fan-beam CT full and partial scans. IEEE Trans Med Imaging 26:190–199.

[16]   Yu L, Pan X (2003) Half-scan fan-beam computed tomography with improved noise and resolution properties. Med Phys 30:2629–2637.

[17]   Wang J, Lu H, Li T, Liang Z (2005) An alternative solution to the nonuniform noise propagation problem in fan-beam FBP image reconstruction. Med Phys 32:3389–3394.

[18]   Wei Y, Hsieh J, Wang G (2005) General formula for fan-beam computed tomography. Phys Rev Lett 95:258102.

[19]   Wei Y, Wang G, Hsieh (2005) Relation between the filtered backprojection algorithm and the backprojection algorithm in CT. IEEE Signal Process Lett 12:633–636.

[20]   Wesarg S, Ebert M, Bortfeld T (2002) Parker weights revisited. Med Phys 29:372–378.

[21]   Zeng GL (2015) Fan-beam short-scan FBP algorithm is not exact. Phys Med Biol 60:N131–N139.

[22]   Zeng GL, Gullberg (1991) Short-scan fan beam algorithm for non-circular detector orbits. SPIE Med Imaging V Conf, San Jose, CA, 332–340.

[23]   Zou Y, Pan X, Sidky EY (2005) Image reconstruction in regions-of-interest from truncated projections in a reduced fan-beam scan. Phys Med Biol 50:13–28.

[24]   Zeng GL (2004) Nonuniform noise propagation by using the ramp filter in fan-beam computed tomography. IEEE Trans Med Imaging 23:690–695.

4 Transmission and emission tomography

This book considers real imaging systems in this chapter. If the radiation source is outside the patient, the imaging system acquires transmission data. If the radiation sources are inside the patient, the imaging system acquires the emission data. For transmission scans, the image to be obtained is a map (or distribution) of the attenuation coefficients inside the patient. For the emission scans, the image to be obtained is the distribution of the injected isotopes within the patient. Even for emission scans, an additional transmission scan is sometimes required in order to compensate for the attenuation effect of the emission photons. Some attenuation compensation methods for emission imaging are discussed in this chapter.

4.1 X-ray computed tomography

In this chapter, we relate transmission and emission tomography measurements to line-integral data so that the reconstruction algorithms mentioned in the previous chapters can be used to reconstruct practical data in medical imaging.

X-ray computed tomography (CT) uses transmission measurements to estimate a cross-sectional image within the patient body. X-rays have very high energy, and they are able to penetrate the patient body. However, not every X-ray can make it through the patient's body. Some X-rays get scattered within the body, and their energy gets weakened. During X-ray scattering, an X-ray photon interacts with an electron within the patient, transfers part of its energy to that electron, and dislodges the electron (see Figure 4.1). The X-ray is then bounced to a new direction with decreased energy.

Some other X-rays completely disappear within the body, converting their energy to the tissues in the body, for example, via the photoelectric conversion. The photoelectric effect is a process in which the X-ray photon energy is completely absorbed by an atom within the patient. The absorbed energy ejects an electron from the atom (see Figure 4.2).

Energy deposition within the body can damage DNA if the X-ray dose is too large. Let the X-ray intensity before entering the patient be I_0, and the intensity departing the patient be I_d; I_0 and I_d follow the Beer's law (see Figure 4.3):

$$\frac{I_d}{I_0} = \exp(-p), \tag{4.1}$$

where p is the line integral of the linear attenuation coefficients along the path of the X-rays. A line integral of the attenuation coefficients is obtained by

https://doi.org/10.1515/9783111055404-004

$$p = \ln\left(\frac{I_0}{I_d}\right), \tag{4.2}$$

which is supplied to the image reconstruction algorithm for image reconstruction.

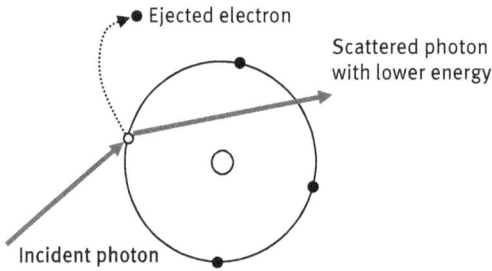

Fig. 4.1: Schematic representation of Compton scattering. The incident photon transfers part of its energy to an electron and is scattered in a new direction.

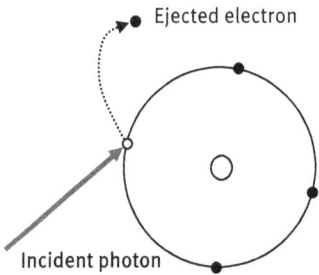

Fig. 4.2: Schematic representation of the photoelectric effect. The incident photon transfers all its energy to an electron and disappears.

Fig. 4.3: The X-ray intensity is reduced after going through the object.

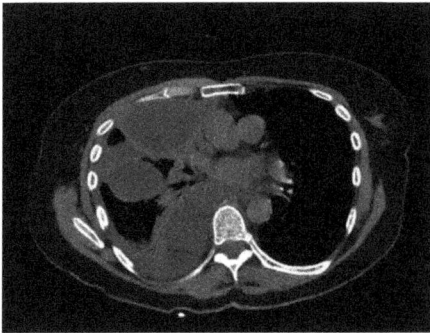

Fig. 4.4: An X-ray CT image.

The goal of X-ray CT is to obtain a cross-sectional image of various attenuation coefficients within the body. A typical X-ray CT image is shown in Figure 4.4. The attenuation coefficient (commonly denoted by notation μ) is a property of a material; it is the logarithm of the input/output intensity ratio per unit length. Bones have higher μ values, and soft tissue has lower μ values. The attenuation coefficient of a material varies with the incoming X-ray energy; it becomes smaller when the X-ray energy gets higher.

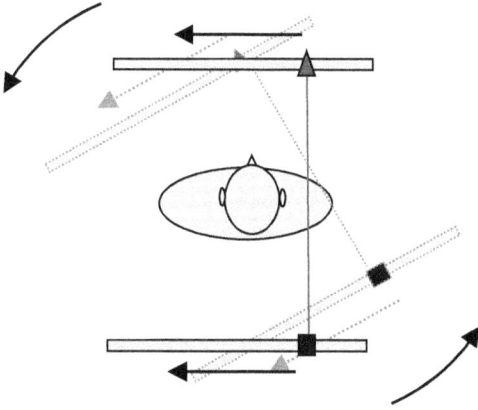

Fig. 4.5: In the first-generation CT, the X-ray tube and the detector translate and rotate.

The first-generation CT, which is no longer in use, had one small detector (see Figure 4.5). The X-ray source and the detector have two motions: linear translation and rotation. The X-ray source sends out a narrow pencil beam to obtain parallel-beam projections. The scanning time was rather long (about 25 min).

The second-generation CT used narrow fan-beam geometry, consisting of 12 detectors (see Figure 4.6). Like the first-generation CT, it has two motions: linear translation and rotation. Due to the fan-beam geometry, the scan time was shortened to about 1 min.

The third-generation CT uses wide fan-beam geometry, consisting of approximately 1,000 detectors (see Figure 4.7). No linear translation motion is necessary, and the scanning time was further reduced to about 0.5 s. The third-generation CT is currently very popular in medical imaging.

The fourth-generation CT has a stationary ring detector. The X-ray source rotates around the subject (see Figure 4.8). This scanning method forms a very fast fan-beam imaging geometry; however, it is impossible to collimate the X-rays on the detector, which causes this geometry to suffer from high rates of scattering.

Modern CT can perform helical scans, which is implemented as translating the patient bed in the axial direction as the X-ray source and the detectors rotate. The modern CT has a 2D multirow detector, and it acquires cone-beam data (see Figure 4.9). Image reconstruction methods for the cone-beam geometry will be covered in the next chapter.

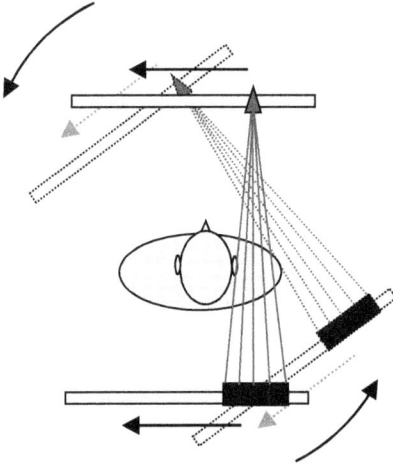

Fig. 4.6: The second-generation CT uses narrow fan-beam X-rays. The X-ray tube and the detector translate across the field of view and rotate around the object.

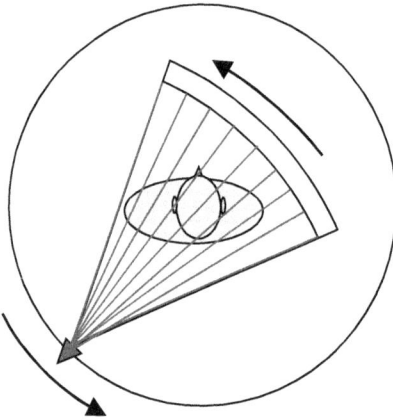

Fig. 4.7: The third-generation CT uses wide fan-beam X-rays. The X-ray tube and the detector rotate around the object; they do not translate anymore.

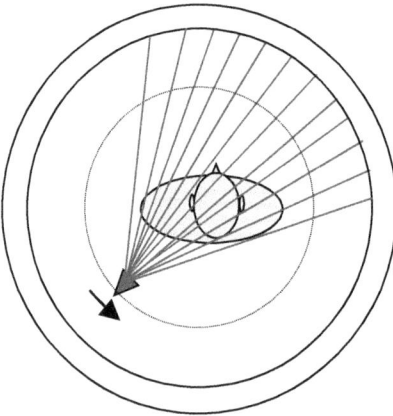

Fig. 4.8: In the fourth-generation CT, the ring detector does not rotate. The X-ray source rotates around the object.

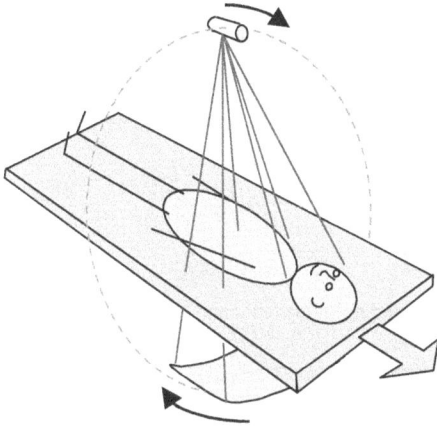

Fig. 4.9: The modern CT can perform cone-beam helix scans with a 2D detector. Some systems have multiple X-ray sources and 2D detectors. The helix orbit is implemented by translating the patient bed while the source and detector rotate.

4.2 Positron emission tomography and single-photon emission computed tomography

In the last section, transmission imaging was discussed. In transmission imaging, the radiation source is placed outside the patient body. The radiation rays (either X-rays or gamma rays) enter the patient from outside, pass through the patient body, exit the patient, and finally get detected by a detector outside.

This section will change the subject to emission imaging, where the radiation sources are inside the patient body. Radiation is generated inside the patient body, emitted from within, and detected by a detector after it escapes from the patient body.

Radioactive atoms with a short half-life are generated in a cyclotron or a nuclear reactor. Radiopharmaceuticals are then made and injected into a patient (in a peripheral arm vein) to trace disease processes. The patient can also inhale or ingest the radiotracer. Radiopharmaceuticals are carrier molecules with a preference for a certain tissue or disease process. The radioactive substance redistributes itself within the body after the injection. The goal of emission tomography is to obtain a distribution map of the radioactive substance.

Unstable atoms emit gamma rays as they decay. Gamma cameras are used to detect the emitted gamma photons (see Figure 4.10). The cameras detect one photon at a time. These measurements approximate the ray sums or line integrals. Unlike the transmission data, we do not need to take the logarithm. SPECT (single-photon emission computed tomography) is based on this imaging principle.

Some isotopes, for example, O-15, C-11, N-13, and F-18, emit positrons (positive electrons) during radioactive decay. A positron exists in nature only for a very short time before it collides with an electron. When the positron interacts with an electron, their masses are annihilated, creating two gamma photons of 511 keV each. These photons are emitted 180° apart. A special gamma camera is used to detect this pair of photons, using coincidence detection technology. Like SPECT, the measurements approximate ray sums or line integrals; no logarithm is necessary to convert the data. This is the principle of PET (positron emission tomography) imaging (see Figure 4.11).

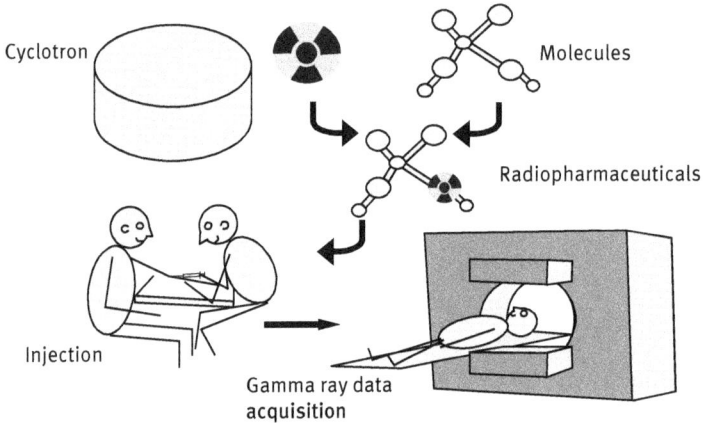

Fig. 4.10: Preparation for a nuclear medicine emission scan.

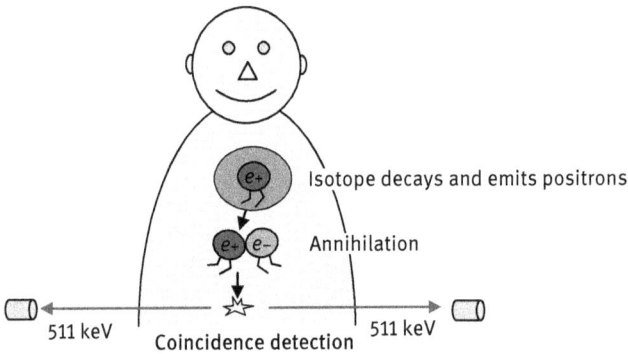

Fig. 4.11: Principle of PET imaging.

The imaging geometry for SPECT is determined by its collimator, which is made of lead septa to permit gamma rays oriented in certain directions to pass through and stop gamma rays with other directions. If a parallel-beam or a fan-beam collimator is used, then the data are acquired in the same corresponding form (see Figure 4.12). Similarly, if a cone-beam or a pinhole collimator is used, the imaging geometry is cone beam (see Figure 4.13). Convergent beam geometries magnify the object so that an image larger than the object can be obtained on the detector.

In PET, each measured event determines a line. The imaging geometry is made by sorting or grouping the events according to some desired rules. For example, we can group them into parallel sets (see three different sets in Figure 4.14). We can also store each event by itself as the list mode.

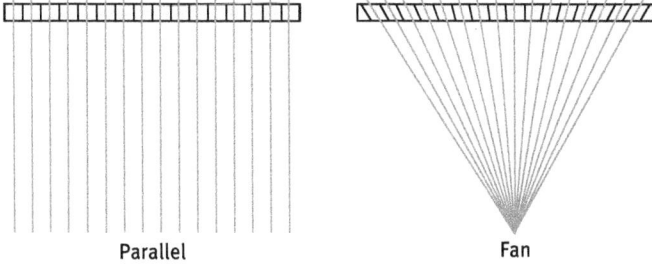

Fig. 4.12: SPECT uses collimators to the selected incoming projection ray geometry.

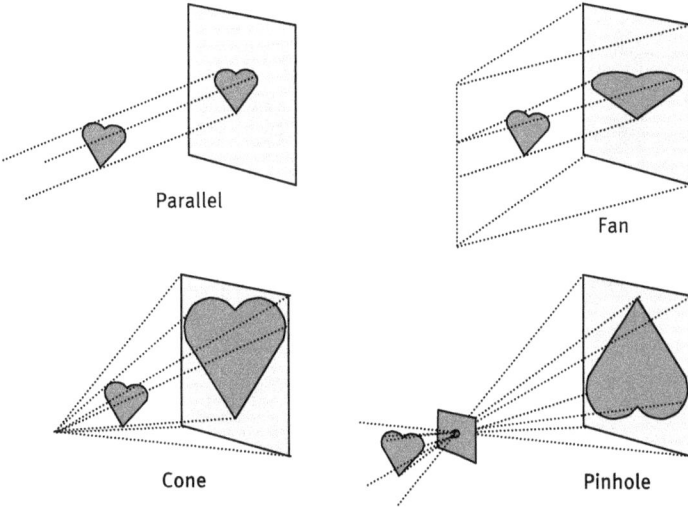

Fig. 4.13: SPECT collimators can be parallel, convergent, or divergent. They produce different sizes of images.

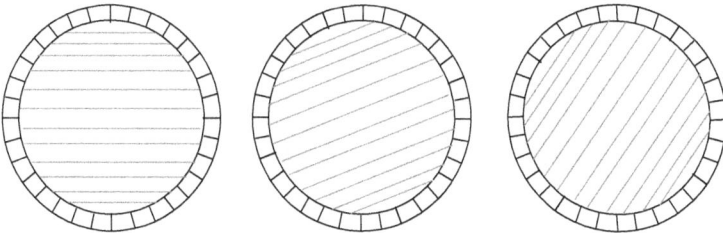

Fig. 4.14: PET data can be grouped into parallel sets.

A typical SPECT image and a PET image are shown in Figure 4.15. When comparing X-ray CT, PET, and SPECT, we observe that X-ray CT has the best resolution and is least noisy, and SPECT has the worst resolution and most noise. Image quality is directly proportional to the photon counts.

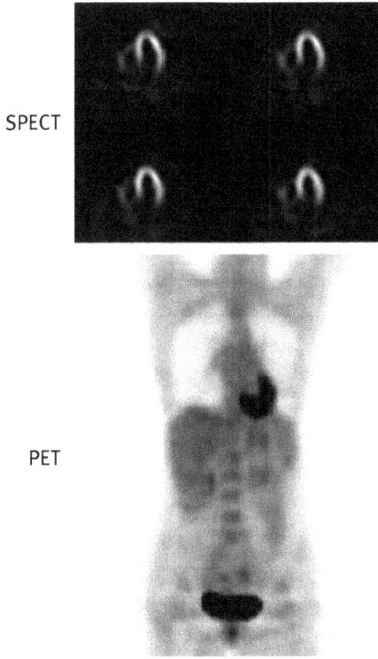

Fig. 4.15: SPECT cardiac images and a PET torso image. The PET image is displayed in the inverse gray scale.

4.3 Noise propagation in reconstruction

In this section, we will see how the noisy data make the image noisy. A filtered backprojection (FBP) algorithm can be considered as a linear system. The input is the projection data and the output is the reconstructed image. The noise in the output is related to the noise in the input.

4.3.1 Noise variance of emission data

Measured projections are noisy. Due to the nature of photon counting, the measurement noise roughly follows the Poisson statistics. A unique feature of Poisson random variable p is that the mean value $\lambda = E(p)$ and the variance $\sigma^2 = \text{Var}(p)$ are the same. In practice, the mean value $E(p)$ is unknown for a measurement p, and we often use the current measurement p to approximate the mean value $E(p)$ and the noise variance $Var(p)$:

$$\text{Emission: } Var(p) \approx p. \tag{4.3}$$

4.3.2 Noise variance of transmission data

For transmission imaging, the line integral of the attenuation coefficient is calculated as $p = \ln(I_0/I_d)$, where I_d is assumed to be Poisson and I_0 can be approximated as a constant. The noise distribution for the post-log data p is complicated. The common practice is to assume the noise in p obeys the Gaussian distribution. We are interested in finding the noise variance of p. Using the linear approximation of the Taylor expansion,

$$p = f(I_d) \approx f(E(I_d)) + \frac{f'(E(I_d))}{1!} I_d, \tag{4.4}$$

with $f(x) = \ln I_0 - \ln x$.

Thus, the variance of p can be approximated as

$$Var(p) \approx \left(\frac{f'(E(I_d))}{1!} \right)^2 Var(I_d) = \frac{1}{(E(I_d))^2} E(I_d) = \frac{1}{E(I_d)} \approx \frac{1}{I_d} = \frac{1}{I_0 e^{-p}}, \tag{4.5}$$

$$\text{Transmission: } Var(p) \approx \frac{1}{I_0 e^{-p}}. \tag{4.6}$$

4.3.3 Noise propagation in an FBP algorithm

We assume that the projection measurements are independent random variables. A general FBP algorithm can be symbolically expressed by the following convolution backprojection algorithm:

$$f(x,y) = \sum_s \sum_\theta b(x,y,s,\theta) h(s - \hat{s}) p(\hat{s},\theta), \tag{4.7}$$

where p is the projection, h is the convolution kennel, and b is the combined factor for all sorts of interpolation coefficients and weighting factors. Then the noise variance of the reconstructed image $f(x, y)$ can be approximately estimated as

$$Var(f(x,y)) \approx \sum_s \sum_\theta b^2(x,y,s,\theta) h^2(s - \hat{s}) Var(p(\hat{s},\theta)), \tag{4.8}$$

where Var($p(\hat{s}, \theta)$) can be substituted by the results in Section 4.3.1 or 4.3.2 depending on whether the projections are emission data or transmission data.

4.4 Attenuation correction for emission tomography

In emission tomography, the gamma ray photons are emitted from within the patient's body. Not all the photons are able to escape from the patient body; thus, they are attenuated when they propagate. The attenuation follows Beer's law, which we have seen in Section 4.1.

4.4.1 PET

In PET, two detectors are required to measure one event with coincidence detection. An event is valid if two detectors simultaneously detect a 511 keV photon. Let us consider the situation depicted in Figure 4.16, where the photons are emitted at an arbitrary location in a nonuniform medium. The photons that reach detector 1 are attenuated by path L_1 with an attenuation factor determined by Beer's law. We symbolically represent this attenuation factor as $\exp\left(-\int_{L_1}\mu\right)$. Similarly, the attenuation factor for path L_2 is $\exp\left(-\int_{L_2}\mu\right)$. The attenuation factor is a number between 0 and 1 and can be treated as a *probability*. The probability that a pair of photons will be detected by both detectors is the product of the probabilities that each photon is detected; that is,

$$\exp\left(-\int_{L_1}\mu\right) \times \exp\left(-\int_{L_2}\mu\right) = \exp\left(-\int_{L_1+L_2}\mu\right) = \exp\left(-\int_{L}\mu\right). \qquad (4.9)$$

Therefore, the overall attenuation factor is determined by the entire path L, regardless of where the location of the gamma source is along this path.

To do an attenuation correction for PET data, a transmission measurement is required with an external transmission (either X-ray or gamma ray) source. This transmission measurement gives the attenuation factor $\exp\left(-\int_L\mu\right)$.

The attenuation-corrected line integral or ray sum of PET data is obtained as

$$p(s, \theta) = \exp\left(\int_{L(s,\theta)}\mu\right) \times [\text{emission measurement of the path } L(s, \theta)], \qquad (4.10)$$

where the reciprocal of the attenuation factor is used to compensate for the attenuation effect. Note that there is no need to reconstruct the attenuation map for PET data attenuation correction.

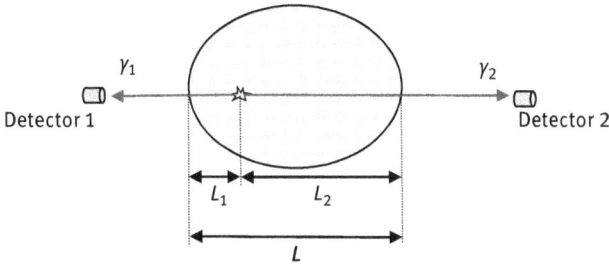

Fig. 4.16: The attenuation of the PET projection is the effect of the total length $L = L_1 + L_2$.

4.4.2 SPECT: Tretiak–Metz FBP algorithm for uniform attenuation

Attenuation correction for SPECT data is much more complicated than the PET data, because an event in SPECT is single photon detection (see Figure 4.17). On the same imaging path, the emission source at a different location has a different attenuation factor. This makes SPECT attenuation correction very difficult. We still do not know how to compensate for the attenuation by processing the projections as easily done as in PET. Some people have tried to perform attenuation compensation in the 2D Fourier domain of the sinogram using the "frequency–distance principle," which is an approximate relationship: The angle in the Fourier domain is related to the distance from the detector in the spatial domain. However, the attenuation effect can be corrected during image reconstruction.

In SPECT, if the attenuator is uniform (i.e., μ = constant within the body boundary), the FBP image reconstruction algorithm is similar to that for the regular unattenuated data. The attenuation-corrected FBP algorithm, developed by Tretiak and Metz, consists of three steps:

(i) Pre-scale the measured projection $p(s, \theta)$ by $e^{\mu d(s,\theta)}$, where the definition of distance $d(s, \theta)$ is given in Figure 4.18. We denote this scaled projection as $\hat{p}(s, \theta)$.

(ii) Filter the pre-scaled data with a notched ramp filter (see Figure 4.19).

(iii) Backproject the data with an exponential weighting factor $e^{-\mu t}$, where t is defined in Figure 4.20 and is dependent on the location of the reconstruction point as well as the backprojection angle θ.

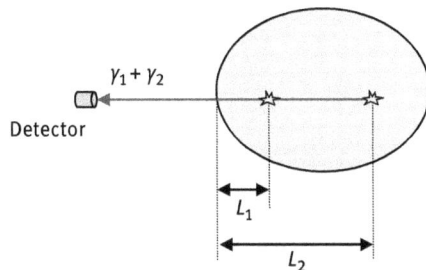

Fig. 4.17: In SPECT, the attenuation is a mixed effect of all lengths.

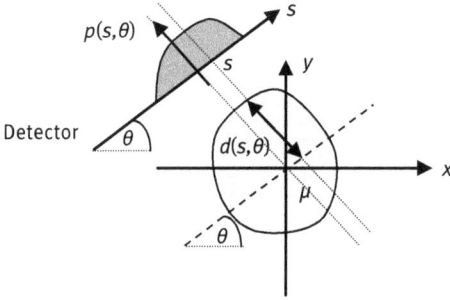

Fig. 4.18: The distance $d(s, \theta)$ is from the boundary of the uniform attenuator to the central line parallel to the detector. A central line is a line passing through the center of rotation.

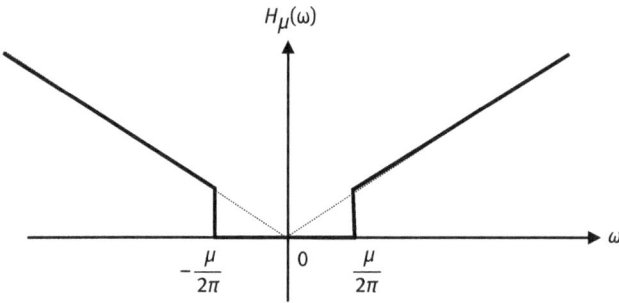

Fig. 4.19: The notched ramp-filter transfer function for image reconstruction in the case of a uniformly attenuated Radon transform. A transfer function is a filter expression given in the Fourier domain, and its inverse Fourier transform is the convolution kernel.

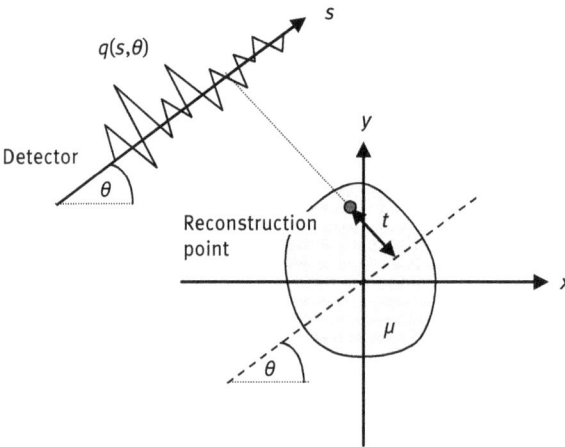

Fig. 4.20: The distance t is from the reconstruction point to the central line parallel to the detector.

Similar to the $\mu = 0$ case, the notched ramp filtering can be decomposed into a derivative and a notched Hilbert transform. In the spatial domain, the Hilbert transform is a convolution with a convolution kernel $1/s$. The notched Hilbert transform, on the other hand, is a convolution with a convolution kernel $(\cos \mu s)/s$.

In the Fourier domain, the notched Hilbert filter function is shown on the left-hand side of the graphic equation in Figure 4.21. The cosine function can be decomposed into $\cos(\mu s) = (e^{i\mu s} + e^{-i\mu s})/2$. The Fourier transform has a property that multiplication by $e^{i\mu s}$ in the s domain corresponds to shifting by $\mu/(2\pi)$ in the ω domain (i.e., the Fourier domain). Therefore, the Fourier transform of $(\cos \mu s)/s$ is the combination of two shifted versions (one shifted to the left by $\mu/(2\pi)$ and the other one shifted to the right by $\mu/(2\pi)$) of the Fourier transform of $1/s$ (see Figure 4.21).

If μ is nonuniform, an FBP-type algorithm exists but is rather sophisticated and will be briefly discussed in Section 4.5.4.

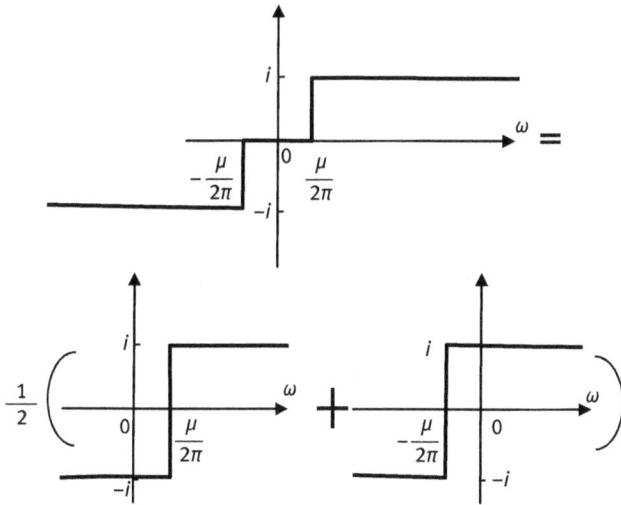

Fig. 4.21: A notched Hilbert transform transfer function can be decomposed into two shifted Hilbert transform transfer functions.

4.4.3 SPECT: Inouye's algorithm for uniform attenuation

The Tretiak–Metz algorithm is in a beautiful FBP form. However, this algorithm generates very noisy images. The significant noise propagation from the projections to the reconstructed image is mainly contributed by the backprojector. In the Tretiak–Metz FBP algorithm, the backprojector has an exponential factor $e^{-\mu t}$, where t can be positive or negative. This exponential factor gives a noise variance amplification factor of $e^{-2\mu t}$. In other words, if the noise variance in the filtered sinogram is σ^2, the noise will propagate into the reconstructed image with a variance approximately $e^{-2\mu t}\sigma^2$. For example, let $\mu = 0.15/\text{cm}$ and $t = -10$ cm; these result in $e^{-2\mu t} = 20$. A 20-fold noise amplification is huge! In the regions farther away from the center (i.e., $|t|$ is larger), the image noise dramatically becomes larger.

The main idea in Inouye's algorithm is to convert the attenuated projections into unattenuated projections, and then use a regular FBP algorithm to reconstruct the image. The regular FBP algorithm does not have an exponential factor in the backprojector. Inouye's algorithm is summarized as follows:

(1) Pre-scale the measured projection $p(s, \theta)$ by $e^{\mu d(s,\theta)}$, where the definition of distance $d(s, \theta)$ is given in Figure 4.18. We denote this scaled projection as $\hat{p}(s, \theta)$. [This step is the same as in the Tretiak–Metz algorithm.]

(2) Take the 2D Fourier transform of $\hat{p}(s, \theta)$, obtaining $\hat{P}(\omega, k)$. [Note that $\hat{p}(s, \theta)$ is periodic with respect to θ, and the Fourier transform in the θ dimension is actually the Fourier series expansion.]

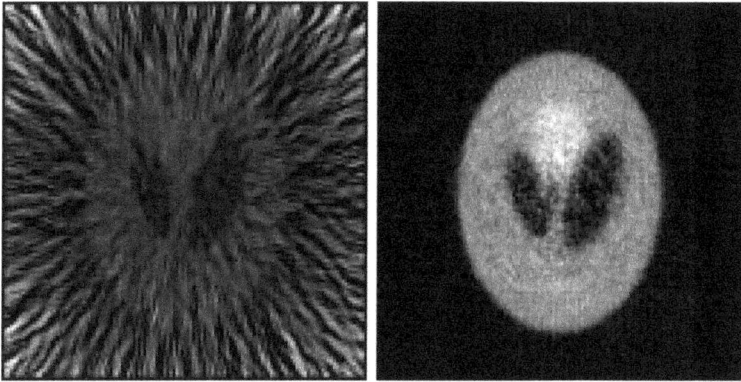

Fig. 4.22: FBP image reconstruction with uniformly attenuated SPECT projections. Left: Reconstruction by the Tretiak–Metz algorithm. Right: Reconstruction by the Inouye algorithm.

(3) Convert $\hat{P}(\omega, k)$ to $P(\omega, k)$ according to

$$P(\omega, k) = \gamma^k \hat{P}\left(\sqrt{\omega^2 + \left(\frac{\mu}{2\pi}\right)^2}, k\right) \tag{4.11}$$

with

$$\gamma = \frac{|\omega|}{\sqrt{\omega^2 + \left(\frac{\mu}{2\pi}\right)^2} + \frac{\mu}{2\pi}}. \tag{4.12}$$

(4) Find the inverse 2D Fourier transform and obtain $p_{\text{new}}(s, \theta)$.

(5) Use a regular FBP to reconstruct the image using the newly estimated projections $p_{\text{new}}(s, \theta)$.

Figure 4.22 shows the comparison of the images reconstructed by the Tretiak–Metz algorithm and the Inouye algorithm, using noisy computer-simulated uniformly attenuated emission projections. The Tretiak–Metz algorithm gives a much noisier image, and the noise gets more severe as the location is farther away from the center.

4.5 Mathematical expressions

4.5.1 Expression for Tretiak–Metz FBP algorithm

Now we give the mathematical expression of the FBP algorithm for SPECT with uniform attenuation correction. This algorithm has been outlined in Section 4.4.2 After the pre-scaling step, the scaled projection $p(s, \theta)$ can be related to the original image $f(x, y)$ as

$$\hat{p}(s, \theta) = \int_{-\infty}^{\infty} e^{\mu t} f(s\vec{\theta} + t\vec{\theta}^{\perp}) dt. \tag{4.13}$$

You can refer to Section 1.5 for notation definitions. The following gives an FBP algorithm using the derivative and the notched Hilbert transform:

$$f(x, y) = \frac{1}{4\pi^2} \int_{0}^{2\pi} e^{-\mu(-x\sin\theta + y\cos\theta)} \int_{-\infty}^{\infty} \frac{\cos(\mu(s - x\cos\theta - y\sin\theta))}{s - x\cos\theta - y\sin\theta} \frac{\partial \hat{p}(s, \theta)}{\partial s} ds d\theta. \tag{4.14}$$

4.5.2 Derivation for Inouye's algorithm

Let $f(x, y)$ be a density function in the x–y plane and f_θ be the function f rotated by θ clockwise. After the pre-scaling step, the scaled projection $\hat{p}(s, \theta)$ can be related to the original image $f(x, y)$ as

$$\hat{p}(s, \theta) = \int_{-\infty}^{\infty} e^{\mu t} f_\theta(s, t) dt. \tag{4.15}$$

Taking the 1D Fourier transform with respect to s yields

$$\begin{aligned}
\hat{p}_{s \to \omega}(\omega, \theta) &= \int_{-\infty}^{\infty} \int_{-\infty}^{\infty} e^{\mu t} f_\theta(s, t) e^{-2\pi i s\omega} dt ds \\
&= \int_{-\infty}^{\infty} \int_{-\infty}^{\infty} f_\theta(s, t) e^{-2\pi i \left(s\omega + t\frac{\mu}{-2\pi i}\right)} ds dt.
\end{aligned} \tag{4.16}$$

Let

$$\omega = \hat{\omega} \cos \alpha,$$
$$\frac{\mu}{-2\pi i} = \hat{\omega} \sin \alpha,$$

(4.17)

which give

$$\hat{\omega} = \sqrt{\omega^2 - \left(\frac{\mu}{2\pi}\right)^2},$$
$$\alpha = \frac{i}{2} \ln \frac{\omega + \frac{\mu}{2\pi}}{\omega - \frac{\mu}{2\pi}}.$$

(4.18)

Changing the s–t coordinate system to the x–y coordinate system with $s = x \cos \theta + y \sin \theta$ and $t = -x \sin \theta + y \cos \theta$, we have

$$\hat{p}_{s \to \omega}(\omega, \theta) = \int_{-\infty}^{\infty} \int_{-\infty}^{\infty} f(x,y) e^{-2\pi i \hat{\omega}[x\cos(\theta + \alpha) + y\sin(\theta + \alpha)]} dx dy$$
$$= F(\hat{\omega}\cos(\theta + \alpha), \hat{\omega} \cos(\theta + \alpha)).$$

(4.19)

Here F is the 2D Fourier transform of the image f. Since $F(\hat{\omega} \cos(\theta + \alpha), \hat{\omega} \cos(\theta + \alpha))$ is a periodic function in θ with a period of 2π, F has a Fourier series expansion. Thus, we can use the Fourier series expansion on the right-hand side of eq. (4.19):

$$\hat{p}_{s \to \omega}(\omega, \theta) = \sum_{n} F_n(\hat{\omega}) e^{in(\theta + \alpha)},$$

(4.20)

where the Fourier coefficients are defined as

$$F_n(\hat{\omega}) = \frac{1}{2\pi} \int_{0}^{2\pi} F(\hat{\omega} \cos \theta, \hat{\omega} \sin \theta) e^{-in\theta} d\theta.$$

(4.21)

Then,

$$\hat{p}_{s \to \omega}(\omega, \theta) = \sum_{n} F_n(\hat{\omega}) e^{in(\theta + \alpha)}$$
$$= \sum_{n} F_n\left(\sqrt{\omega^2 - \left(\frac{\mu}{2\pi}\right)^2}\right) e^{in\theta} e^{in\frac{i}{2}\ln\frac{\omega + \frac{\mu}{2\pi}}{\omega - \frac{\mu}{2\pi}}}$$
$$= \sum_{n} \left(\frac{\omega - \frac{\mu}{2\pi}}{\omega + \frac{\mu}{2\pi}}\right)^{\frac{n}{2}} F_n\left(\sqrt{\omega^2 - \left(\frac{\mu}{2\pi}\right)^2}\right) e^{in\theta}.$$

(4.22)

On the other hand, $\hat{p}_{s \to \omega}(\omega, \theta)$ is also a periodic function with a period of 2π. It has a Fourier series expansion:

$$\hat{p}_{s \rightarrow \omega}(\omega, \theta) = \sum_n \hat{P}(\omega, n) e^{in\theta}. \tag{4.23}$$

By comparing the expansion coefficients on both sides, we have

$$\hat{P}(\omega, n) = \left(\frac{\omega - \frac{\mu}{2\pi}}{\omega + \frac{\mu}{2\pi}} \right)^{\frac{n}{2}} F_n \left(\sqrt{\omega^2 - \left(\frac{\mu}{2\pi} \right)^2} \right). \tag{4.24}$$

If we use $P(\omega, n)$ to represent the 2D Fourier transform of the unattenuated projections $p(s, \theta)$, the above relationship is

$$\hat{P}(\omega, n) = \left(\frac{\omega - \frac{\mu}{2\pi}}{\omega + \frac{\mu}{2\pi}} \right)^{\frac{n}{2}} P \left(\sqrt{\omega^2 - \left(\frac{\mu}{2\pi} \right)^2}, n \right) \tag{4.25}$$

or

$$P(\omega, n) = \left(\frac{\sqrt{\omega^2 + \left(\frac{\mu}{2\pi} \right)^2} + \frac{\mu}{2\pi}}{\sqrt{\omega^2 + \left(\frac{\mu}{2\pi} \right)^2} - \frac{\mu}{2\pi}} \right)^{\frac{n}{2}} \hat{P} \left(\sqrt{\omega^2 + \left(\frac{\mu}{2\pi} \right)^2}, n \right)$$

$$= \left(\frac{\omega}{\sqrt{\omega^2 + \left(\frac{\mu}{2\pi} \right)^2} - \frac{\mu}{2\pi}} \right)^n \hat{P} \left(\sqrt{\omega^2 + \left(\frac{\mu}{2\pi} \right)^2}, n \right). \tag{4.26}$$

The measured attenuated data are used to compute $\hat{P}(\omega, n)$. The above Inouye's relationship is used to estimate $P(\omega, n)$, which gives the estimated unattenuated projections $p(s, \theta)$. The final reconstruction is obtained from $p(s, \theta)$ by a regular FBP algorithm. We would like to clarify that when we say 2D Fourier transform of $p(s, \theta)$, it is really a combination of the 1D Fourier transform with respect to s and the Fourier series expansion with respect to θ.

4.5.3 Rullgård's derivative-then-backprojection algorithm for uniform attenuation

There are many other ways to reconstruct SPECT data with uniform attenuation corrections. We can still play the game of switching the order of filtering and backprojection. For example, we can first take the derivative and then backproject. This results in an intermediate image $\hat{f}(x, y)$ that is closely related to the original image $f(x, y)$:

$$\hat{f}(x, y) = f(x, y) \times \frac{\cosh(\mu x)}{x}, \tag{4.27}$$

which is a 1D convolution (i.e., line-by-line convolution in the x-direction). The deconvolution of this expression to solve for $f(x, y)$ is not an easy task because the function

$(\cosh(\mu x))/x$ tends to infinity as x goes to infinity. It is impossible to find a function u (x) such that

$$\delta(x) = u(x) \times \frac{\cosh(\mu x)}{x} \tag{4.28}$$

for $-\infty < x < \infty$. However, it is possible to find such a function $u(x)$ to make the above expression hold in a small interval, say $(-1, 1)$. Outside this small interval, $u(x) \times ((\cosh(\mu x))/x)$ is undefined. This "second best" solution is useful in image reconstruction because our objects are always supported in a small finite region. Unfortunately, we do not yet have a closed-form expression for such a function $u(x)$, and $\hat{f}(x, y) = f(x, y) \times ((\cosh(\mu x))/x)$ can only be deconvolved numerically.

The advantage of this derivative-then-backprojection algorithm is its ability to exactly reconstruct a region of interest with truncated data.

4.5.4 Novikov–Natterer FBP algorithm for nonuniform attenuation SPECT

In SPECT imaging with a nonuniform attenuator $\mu(x, y)$, if attenuation correction is required, a transmission scan should be performed in addition to the emission scan. The transmission projections are used to reconstruct the attenuation map $\mu(x, y)$. The following is a reconstruction algorithm that can correct for the nonuniform attenuation. Despite its frightening appearance, it is merely an FBP algorithm:

$$f(x,y) = \frac{1}{4\pi^2} Re\left\{\int_0^{2\pi} \frac{\partial}{\partial q}\left[e^{a_\theta(q,t)-g(q,\theta)}\int_{-\infty}^{\infty}\frac{(e^g p)(l,\theta)}{q-l}dl\right]\Big|_{q=s}d\theta\right\}, \tag{4.29}$$

where Re means taking the real part, $p(s, \theta)$ is the measured attenuated projection, $s = x \cos\theta + y \sin\theta$, $t = -x\sin\theta + y\cos\theta$, $a_\theta(s,t) = \int_t^\infty \mu\left(s\vec\theta+\tau\vec\theta^\perp\right)d\tau$, $g(s,\theta)=\frac12[(\mathbf{R}+i\mathbf{HR})\mu](l,\theta)$, $i=\sqrt{-1}$, $\mathbf{R}$ is the Radon transform operator, and $\mathbf{H}$ is the Hilbert transform operator with respect to variable l. This algorithm was independently developed by Novikov and Natterer.

Unfortunately, this algorithm contains an exponential factor in the backprojector, which amplifies the noise and makes the reconstruction very noisy. The Tretiak–Metz algorithm is a special case of this algorithm.

4.6 Worked examples

Example 1: In radiology, did the X-ray CT scanners provide images of the distribution of the linear attenuation coefficients within the patient body?

Answer
Not quite. The reconstructed linear attenuation coefficients μ are converted to the so-called CT numbers, defined as

$$\text{CT number } h = 1{,}000 \times \frac{\mu - \mu_{water}}{\mu_{water}}. \tag{4.30}$$

The CT numbers are in Hounsfield units (HU). For water, $h = 0$ HU; for air, $h = -1{,}000$ HU; and for bone, $h = 1{,}000$ HU.

Example 2: Is there a central slice theorem for the exponential Radon transform?

Answer
In 1988, Inouye derived a *complex* central slice theorem to relate the uniformly attenuated projections to the object in the Fourier domain. He used a concept of "imaginary" frequency, which was attenuation coefficient. Let the exponential Radon transform be

$$\hat{p}(s, \theta) = \int_{-\infty}^{\infty} e^{\mu t} f\left(\vec{s\theta} + \vec{t\theta^{\perp}}\right) dt. \tag{4.31}$$

Let the 1D Fourier transform of $\hat{p}(s, \theta)$ with respect to s be $P_{\mu,}(\omega, \theta)$ and the 2D Fourier transform of the original object be $F(\omega_x, \omega_y)$. Inouye's complex central slice theorem is expressed as follows:

$$P_\mu(\omega, \theta) = F(\omega \cos(\theta + v), \omega \cos(\theta + v)), \tag{4.32}$$

where $v = \dfrac{i}{2} \ln \dfrac{\omega - \mu/(2\pi)}{\omega + \mu/(2\pi)}$ is an imaginary frequency.

Example 3: In Figure 4.19, the frequency components below $\mu/(2\pi)$ are discarded during image reconstruction. How do the low-frequency components of the image get reconstructed?

Answer
We will use Inouye's result to answer this question. Let the attenuation-free Radon transform be

$$p(s, \theta) = \int_{-\infty}^{\infty} f\left(s\vec{\theta} + t\vec{\theta}^{\perp}\right) dt, \tag{4.33}$$

and the exponential Radon transform be

$$\hat{p}(s, \theta) = \int_{-\infty}^{\infty} e^{\mu t} f\left(s\vec{\theta} + t\vec{\theta}^{\perp}\right) dt. \tag{4.34}$$

The exponential Radon transform is the attenuated projection scaled by the factor $e^{\mu d(s, \theta)}$.

Then we take the 2D Fourier transform of $p(s, \theta)$ and $\hat{p}(s, \theta)$, and obtain $P(\omega, k)$ and $\hat{P}(\omega, k)$, respectively. Inouye's relationship is given as

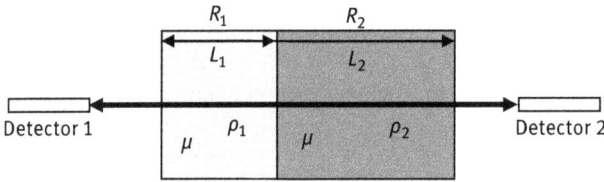

Fig. 4.23: Concentrations are different in the two compartments.

$$P(\omega, k) = \left(\frac{\sqrt{\omega^2 + \left(\frac{\mu}{2\pi}\right)^2} + \frac{\mu}{2\pi}}{\sqrt{\omega^2 + \left(\frac{\mu}{2\pi}\right)^2} - \frac{\mu}{2\pi}}\right)^{k} \hat{P}\left(\sqrt{\omega^2 + \left(\frac{\mu}{2\pi}\right)^2}, k\right) \tag{4.35}$$

$$\text{for } \omega \geq 0 \text{ and } k \geq 0.$$

This relationship implies that the frequency ω has been shifted to $\sqrt{\omega^2 + (\mu/(2\pi))^2}$. The attenuation procedure during data generation actually shifts the "frequency." The lowest frequency in the image corresponds to the frequency at $\mu/(2\pi)$ in the projections. Therefore, all frequency components are preserved.

Example 4: Consider a 2D PET imaging problem shown in Figure 4.23, where the object consists of two compartments R_1 and R_2. The radionuclide concentration is ρ_1 in R_1 and ρ_2 in R_2. The attenuation coefficient μ is the same for both compartments. Calculate the PET coincident measurement p and the attenuation-corrected measurement p_c.

Solution
The coincidence measurement is

$$p = (L_1\rho_1 + L_2\rho_2)e^{-\mu(L_1+L_2)}. \tag{4.36}$$

The attenuation-corrected measurement is

$$p_c = L_1\rho_1 + L_2\rho_2. \tag{4.37}$$

4.7 Summary

- The working principle for X-ray CT measurement is Beer's law. One needs to take the logarithm to convert the CT measurements into line integrals.
- PET and SPECT measure line integrals directly. However, these measurements suffer from photon attenuation within the patient.
- Attenuation correction for PET can be achieved by the pre-scaling method. The scaling factor is obtained by transmission measurements.
- Attenuation correction for SPECT is difficult and cannot be done by pre-scaling. Attenuation correction is a built-in feature in SPECT reconstruction algorithms. FBP algorithms exist for uniform attenuator and for nonuniform attenuator as well. However, the FBP algorithm for the nonuniform attenuator is very complicated to implement.
- The readers are expected to understand the differences between transmission and emission tomography, and also the different attenuation effects in PET and SPECT.

Problems

Problem 4.1 The line-integral data p from transmission tomography can be obtained by $p = \ln(I_0/I_d)$, where I_0 and I_d are the numbers of the photons entering and leaving the patient body, respectively. The noise characteristics of the photon numbers are Poisson distributed. In practice, the entering number I_0 is very large and can be treated as a constant. Prove that the mean value (i.e., the expected value) and the variance of the line integral p can be approximated as p and $1/I_d$, respectively.

Problem 4.2 The object to be imaged is a 2D uniform disk with a radius R. The linear attenuation coefficient of the disk is a constant μ and its radioactivity line density is a constant ρ. The center of the disk is at the center of the detector rotation. Find the expressions of the attenuated projection data $p(s, \theta)$ for SPECT and PET cases, respectively.

Problem 4.3 Prove that the inverse Fourier transform of the transfer function $H_\mu(\omega)$ of the filter shown in Figure 4.19 is the convolution kernel $\dfrac{\cos(\mu s)}{s} * \delta'(s)$

$$= -\frac{(\mu s)\sin(\mu s) + \cos(\mu s)}{s^2}.$$

Bibliography

[1] Bellini S, Piacenti M, Caffario C, Rocca F (1979) Compensation of tissue absorption in emission tomography. IEEE Trans Accoust Speech, Signal Process ASSP 27:213–218.

[2] Bukhgeim AA, Kazantsev SG (2002) Inversion formula for the fan-beam attenuated Radon transform in a unit disk, The Sobolev Institute of Mathematics, Russian Academy of Science Siberian Branch.

[3] Guillement JP, Novikov R (2004) A noise property analysis of single-photon emission computed tomography data. Inverse Probl 20:175–198.

[4] Gullberg GT (1979) The attenuated Radon transform: Theory and application in medicine and biology, Ph.D. Dissertation, Lawrence Berkeley Laboratory, University of California, Berkeley, CA.

[5] Gullberg GT, Budinger TF (1981) The use of filtering methods to compensate for constant attenuation in single photon emission computed tomography. IEEE Trans Biomed Eng 28:142–157.

[6] Hawkins WG, Leichner PK, Yang NC (1988) The circular harmonic transform for SPECT reconstruction and boundary conditions on the Fourier transform of the sinogram. IEEE Trans Med Imaging 7:135–148.

[7] Hsieh J (2003) Computed Tomography: Principles, Design, Artifacts, and Recent Advances, SPIE, Bellingham, WA.

[8] Huang Q, Zeng GL, Wu J (2006) An alternative proof of Bukjgeim and Kazantsev's inversion formula for attenuated fan-beam projections. Med Phys 33:3983–3987.

[9] Huang Q, You J, Zeng GL, Gullberg GT (2009) Reconstruction from uniformly attenuated SPECT projection data using the DBH method. IEEE Trans Med Imaging 28:17–29.

[10] Inouye T, Kose K, Hasegawa A (1988) Image reconstruction algorithm for single-photon-emission computed tomography with uniform attenuation. Phys Med Biol 34:299–304.

[11] Kunyansky LA (2001) A new SPECT reconstruction algorithm based upon the Novikov's explicit inversion formula. Inverse Probl 17:293–306.

[12] Markoe A (1984) Fourier inversion of the attenuated X-ray transform. SIAM J Math Anal 15:718–722.

[13] Metz CE, Pan X (1995) A unified analysis of exact methods of inverting the 2-D exponential Radon transform, with implications for noise control in SPECT. IEEE Trans Med Imaging 17:643–658.

[14] Natterer F (2001) Inversion of the attenuated Radon transform. Inverse Probl 17:113–119.

[15] Noo F, Wagner JM (2001) Image reconstruction in 2D SPECT with 180° acquisition. Inverse Probl 17:1357–1371.

[16] Novikov RG (2002) An inversion formula for the attenuated X-ray transformation. Ark Math 40:145–167.

[17] Pan X, Kao C, Metz CE (2002) A family of π-scheme exponential Radon transforms and the uniqueness of their inverses. Inverse Probl 18:825–836.

[18] Puro A (2001) Cormack-type inversion of exponential Radon transform. Inverse Probl 17:179–188.

[19] Rullgård H (2002) An explicit inversion formula for the exponential Radon transform using data from 180°. Ark Math 42:145–167.

[20] Rullgård H (2004) Stability of the inverse problem for the attenuated Radon transform with 180° data. Inverse Probl 20:781–797.

[21] Tang Q, Zeng GL, Gullberg GT (2007) A Fourier reconstruction algorithm with constant attenuation compensation using 180° acquisition data for SPECT. Phys Med Biol 52:6165–6179.

[22] Tretiak O, Metz CE (1980) The exponential Radon transform. SIAM J Appl Math 39:341–354.

[23] You J, Zeng GL, Liang Z (2005) FBP algorithms for attenuated fan-beam projections. Inverse Probl 21:1179–1192.

[24] Wang G, Lin T, Cheng P, Shinozake DM, Kim HG (1991) Scanning cone-beam reconstruction algorithms for X-ray microtomography. Proc SPIE 1556:99–112.

[25] Wang G, Yu H, DeMan B (2008) An outlook on X-ray CT research and development. Med Phys 35:1051–1064.

[26] Webster JG (1998) Medical Instrumentations: Application and Design, 3rd ed., Wiley, New York.

[27] Wolbarst AB (1993) Physics of Radiology, Appleton and Lange, Norwalk, CT.

[28] Zeng GL, Gullberg GT (2009) Exact emission SPECT reconstruction with truncated transmission data. Phys Med Biol 54:3329–3340.

5 Three-dimensional image reconstruction

This chapter is focused on 3D tomographic imaging. In 3D, we will consider the parallel line-integral projections, parallel plane-integral projections, and cone-beam line-integral projections, separately. For the 3D parallel line-integral projections and parallel plane-integral projections, there exist the central slice theorems, from which the image reconstruction algorithms can be derived. However, for the cone-beam projections the situation is different; there is no central slice theorem for cone beam. We have to somehow establish a relationship between the cone-beam projections and the 3D image itself. Since the cone-beam image reconstruction is an active research area, this chapter spends a significant effort on discussing cone-beam reconstruction algorithms, among which the Katsevich algorithm is the latest and the best one.

5.1 Parallel line-integral data

In many cases, 3D image reconstruction can be decomposed into a series of slice-by-slice 2D image reconstructions if the projection rays can be divided into groups, where each group contains only those rays that are confined within a transaxial slice (see Figure 5.1, left).

In other cases, the projection rays run through transaxial slices, where the slice-by-slice 2D reconstruction approach does not work (see Figure 5.1, middle and right).

The foundation for 2D parallel-beam image reconstruction is the central slice theorem (Section 2.2). The central slice theorem in 3D states that the 2D Fourier transform P $(\omega_u, \omega_v, \vec{\theta})$ of the projection $P(u, v, \vec{\theta})$ of a 3D function $f(x, y, z)$ is equal to a slice through the origin of the 3D Fourier transform $F(\omega_x, \omega_y, \omega_z)$ of that function which is parallel to the detector (see Figure 5.2). Here, $\vec{\theta}$ is the normal direction of the u–v plane and the ω_u–ω_v plane. The direction $\vec{\theta}$ represents a group of rays that are parallel to $\vec{\theta}$.

Based on this central slice theorem, we can determine some specific trajectories of $\vec{\theta}$ so that we are able to fill up the $(\omega_x, \omega_y, \omega_z)$ Fourier space. One such option is shown in Figure 5.3, where the trajectory of $\vec{\theta}$ is a great circle. A great circle is a circle with unit radius that lies on the surface of the unit sphere (see Figure 5.4). Each unit vector $\vec{\theta}$ on the great circle corresponds to a measured $P(\omega_u, \omega_v, \vec{\theta})$ plane in the $(\omega_x, \omega_y, \omega_z)$ Fourier space. After the unit vector $\vec{\theta}$ completes the great circle, the measured $P(\omega_u, \omega_v, \vec{\theta})$ planes fill up the entire $(\omega_x, \omega_y, \omega_z)$ Fourier space. In fact, due to symmetry, if the unit vector $\vec{\theta}$ completes *half* of the great circle, a complete data set is obtained.

The above example can be generalized, as stated in Orlov's condition: A complete data set can be obtained if every great circle intersects the trajectory of the unit vector $\vec{\theta}$, which is the direction of the parallel rays. The trajectory can be curves on the sphere and can also be regions on the sphere. Some examples of the $\vec{\theta}$ trajectories

https://doi.org/10.1515/9783111055404-005

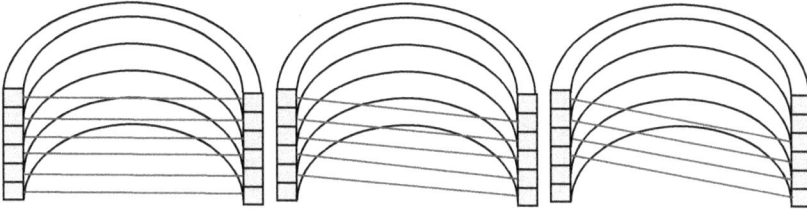

Fig. 5.1: The measurement rays can be in the planes perpendicular to the axial direction and can also be in the slant planes.

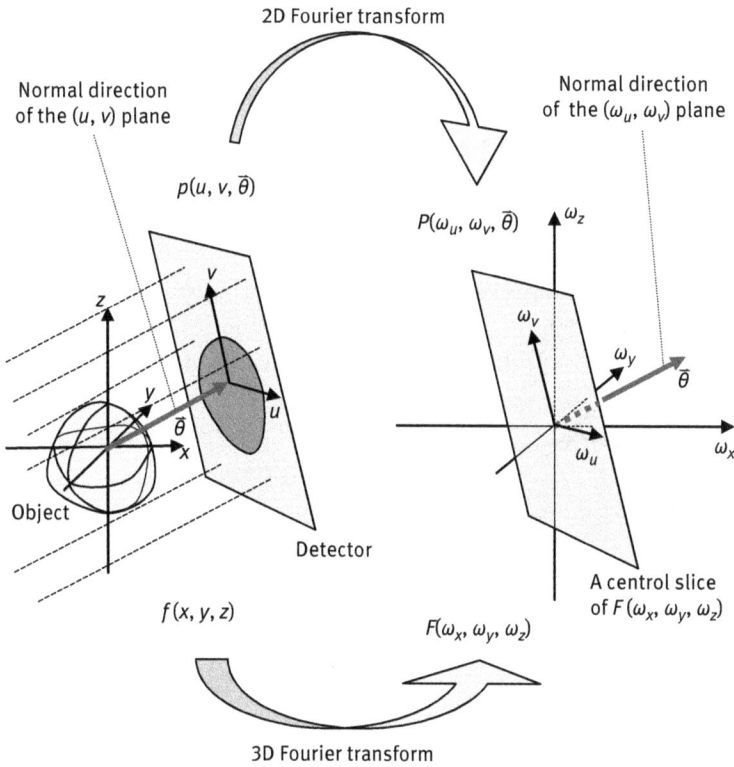

Fig. 5.2: The central slice theorem for the 3D line-integral projections.

(shaded with red color) are shown in Figure 5.5, where the first three satisfy Orlov's condition, and the last two do not.

The image reconstruction algorithm depends on the trajectory of the direction vector $\vec{\theta}$ geometry. The basic algorithm development can follow the guidelines below.

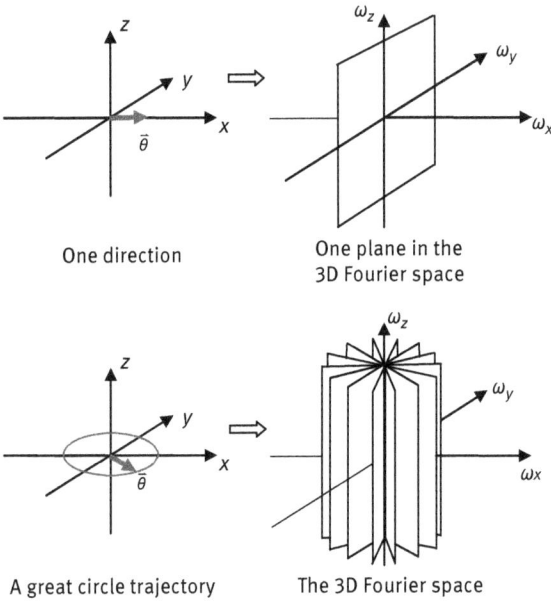

One direction   One plane in the
3D Fourier space

A great circle trajectory   The 3D Fourier space

Fig. 5.3: One measuring direction gives a measured plane in the Fourier space. A great circle trajectory provides full Fourier space measurements.

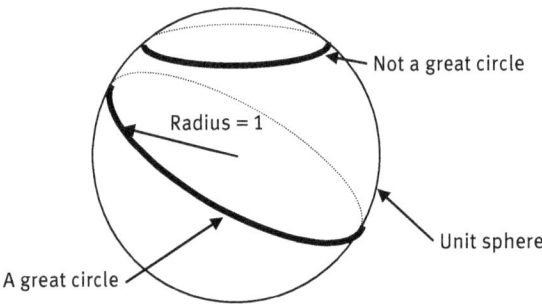

Fig. 5.4: A great circle is a unit circle on the unit sphere.

5.1.1 Backprojection-then-filtering

If the data are sufficiently measured, the 3D image can be exactly reconstructed. Like 2D image reconstruction, one can reconstruct an image either by performing the backprojection first or by performing the backprojection last. If the backprojection is performed first, the algorithm is a backprojection-then-filtering algorithm, and it is described in the following steps:

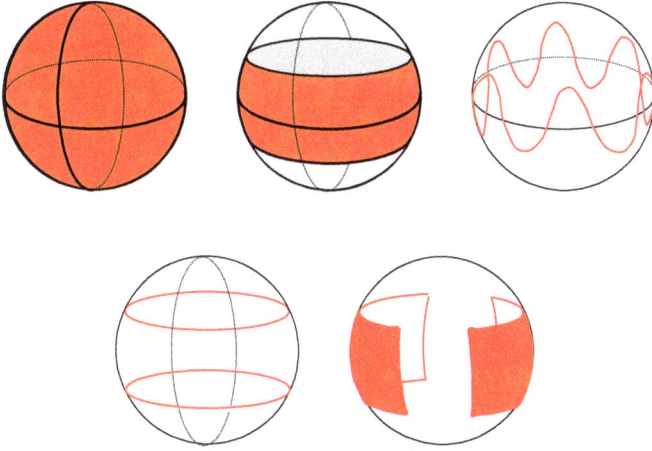

Fig. 5.5: The directional vector trajectories are illustrated as curves or shaded areas on the unit sphere. The trajectories on the top row satisfy Orlov's condition; the trajectories on the bottom row do not satisfy the condition.

(i) Use an arbitrary point source and find the 3D projection/backprojection point spread function (PSF) h (defined in Figure 3.2). If the original image is $f(x, y, z)$ and the backprojected image is $b(x, y, z)$, then

$$b = f *** h, \tag{5.1}$$

where "$***$" denotes the 3D convolution. For example, if the trajectory of the directional vector $\vec{\theta}$ is the full unit sphere as shown in the leftmost case in the first row of Figure 5.5, the PSF is

$$h(x, y, z) = \frac{1}{x^2 + y^2 + z^2} = \frac{1}{r^2}, \tag{5.2}$$

where r is the distance to the point source. In 2D, this projection/backprojection PSF is $1/r$. This implies that in 3D, the PSF is sharper than that in 2D because the PSF falls off at a faster rate as r increases.

(ii) Take the 3D Fourier transform of the relationship $b = f *** h$ and obtain

$$B(\omega_x, \omega_y, \omega_z) = F(\omega_x, \omega_y, \omega_z) H(\omega_x, \omega_y, \omega_z). \tag{5.3}$$

After this Fourier transform, b, f, and h become B, F, and H, respectively, and convolution becomes multiplication. If we again use the example in the leftmost case in the upper row of Figure 5.5, the transfer function H is

$$H(\omega_x, \omega_y, \omega_z) = \frac{\pi}{\sqrt{\omega_x^2 + \omega_y^2 + \omega_z^2}}. \tag{5.4}$$

Thus, a 3D ramp filter

$$\frac{1}{H(\omega_x, \omega_y, \omega_z)} = \frac{\sqrt{\omega_x^2 + \omega_y^2 + \omega_z^2}}{\pi} \tag{5.5}$$

can be used for image reconstruction in this case.

Solve for F as

$$F(\omega_x, \omega_y, \omega_z) = B(\omega_x, \omega_y, \omega_z) \frac{\sqrt{\omega_x^2 + \omega_y^2 + \omega_z^2}}{\pi}. \tag{5.6}$$

Finally, the image $f(x, y, z)$ is obtained by taking the 3D inverse Fourier transform of F.

In general, 3D line-integral data are measured with heavy redundancy. Therefore, the image reconstruction algorithm is not unique because you can always weigh redundant data differently.

5.1.2 Filtered backprojection

In the filtered backprojection (FBP) algorithm, the projection $p(u, v, \vec{\theta})$ is first filtered by a 2D filter (or a 2D convolution), obtaining $q(u, v, \vec{\theta})$. A backprojection of the filtered data $q(u, v, \vec{\theta})$ gives the reconstruction of the image $f(x, y, z)$.

Due to data redundancy, the 2D filter is not unique. The filter is usually different depending on different data orientation $\vec{\theta}$. One way to obtain a Fourier domain filter is through the central slice theorem.

In Section 5.1.1, we had a projection/backprojection PSF $h(x, y, z)$ and its Fourier transform $H(\omega_x, \omega_y, \omega_z)$. If we let $G(\omega_x, \omega_y, \omega_z) = 1/H(\omega_x, \omega_y, \omega_z)$, then G is the Fourier domain filter in the backprojection-then-filtering algorithm. The Fourier domain 2D filter for projection $p(u, v, \vec{\theta})$ can be selected as the central slice of $G(\omega_x, \omega_y, \omega_z)$ with the normal direction $\vec{\theta}$ (see Figure 5.2). Note that the filter in general depends on the direction $\vec{\theta}$.

5.2 Parallel plane-integral data

In 3D, the parallel plane-integral $p(s, \vec{\theta})$ of an object $f(x, y, z)$ is referred to as the Radon transform (see Figure 5.6). In 2D, the Radon transform is the parallel line-integral $p(s, \theta)$ of $f(x, y)$. In a general n-D space, the $(n-1)$-D hyperplane integral of an n-D function f is called the Radon transform. On the other hand, 1D integral of the object is called the line integral, ray sum, X-ray transform, or ray transform. In 2D, the Radon transform and the X-ray transform are the same thing.

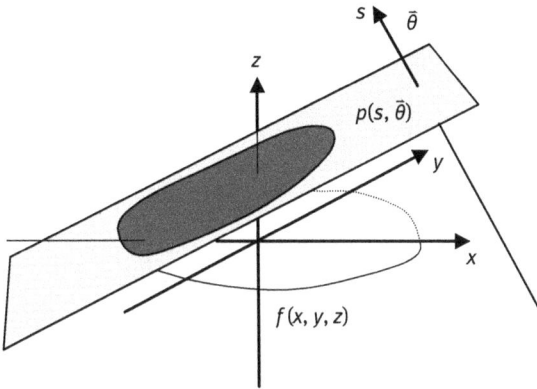

Fig. 5.6: In 3D, the plane integral of an object is the Radon transform.

Unlike the line-integral data, the plane-integral data are not popular in medical imaging. Nevertheless, the Radon transform in 3D is still worthwhile to investigate because it has a simple and nice inversion and can be used to solve other related imaging problems.

To study the Radon transform in 3D, we imagine a 1D detector that is able to measure plane integrals $p(s, \vec{\theta})$ with the planes orthogonal to the detector. The detector is along the direction $\vec{\theta}$. The central slice theorem for the Radon transform in 3D states that the 1D Fourier transform $P(\omega, \vec{\theta})$ of the projection $p(s, \vec{\theta})$ of a 3D function $f(x, y, z)$ is equal to a 1D profile through the origin of the 3D Fourier transform $F(\omega_x, \omega_y, \omega_z)$ of that function which is parallel to the detector (see Figure 5.7). Here, $\vec{\theta}$ is the direction of the 1D detector and the 1D profile in the $(\omega_x, \omega_y, \omega_z)$ space.

We observe from Figure 5.7 that each detector position only measures the frequency components along one line in the $(\omega_x, \omega_y, \omega_z)$ space. The direction $\vec{\theta}$ must go through a half unit sphere to get enough measurements for image reconstruction. After the data are acquired, the image reconstruction algorithm is simple and is in the form of FBP. This form is also referred to as the Radon inversion formula.

In order to reconstruct the image, first take the second-order derivative of $p(s, \vec{\theta})$ with respect to variable s. This step is called filtering. Then backproject the filtered data to the 3D image array. You will not find an image reconstruction algorithm simpler than this.

The 3D backprojector in the 3D Radon inversion formula backprojects a point into a 2D plane. There is a trick to perform the 3D backprojection with the Radon data. This trick is to perform the 3D backprojection in two steps, and each step is a 2D backprojection. In the first step, a point is backprojected into a line (see Figure 5.8(i)). All data points along a line are backprojected into a set of parallel lines, and these lines are in a 2D plane. In the second step, each line is backprojected into a 2D plane (see Figure 5.8 (ii)). The backprojection directions in these two steps are orthogonal to each other.

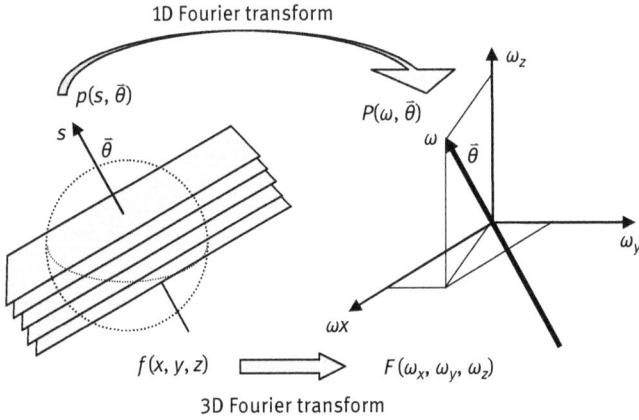

Fig. 5.7: The central slice theorem for the 3D Radon transform.

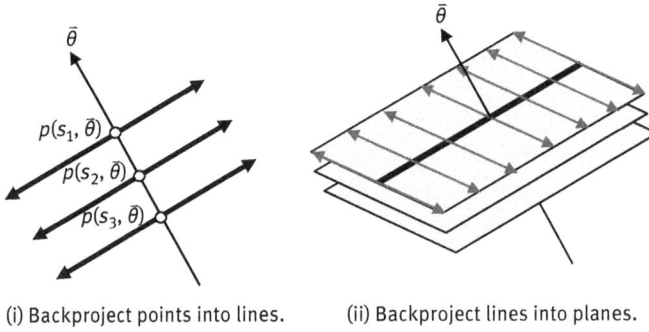

(i) Backproject points into lines.   (ii) Backproject lines into planes.

Fig. 5.8: Three-dimensional Radon backprojection is implemented as two steps: a point to a line and then a line to a plane.

5.3 Cone-beam data

Cone-beam image reconstruction is considerably more complex than that of parallel line-integral and parallel plane-integral data. There is no equivalent central slice theorem known to us. Cone-beam imaging geometry (see the two lower figures in Figure 4.13) is extremely popular, for example, in X-ray computed tomography (CT) and in pinhole SPECT (single-photon emission computed tomography); we will spend some effort to talk about its reconstruction methods.

First, we have a cone-beam data sufficiency condition (known as Tuy's condition): every plane that intersects the object of interest must contain a cone-beam focal point.

This condition is very similar to the fan-beam data sufficiency condition: Every line that intersects the object of interest must contain a fan-beam focal point.

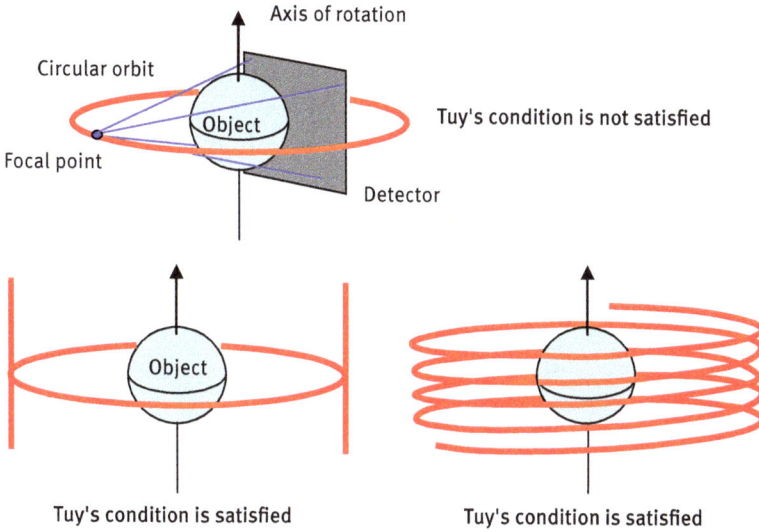

Fig. 5.9: The circular orbit does not satisfy Tuy's conditions. The circle-and-line and the helix orbits satisfy the conditions.

In Figure 5.9, the circular cone-beam focal-point orbit does not satisfy Tuy's condition. If we draw a plane cutting through the object above (or below) the orbit plane and parallel to the orbit plane, this plane will never intersect the circular orbit. The helical and circle-and-line orbits shown in Figure 5.9 satisfy Tuy's condition, and they can be used to acquire cone-beam projections for exact image reconstruction. Modern CT uses helical orbit to acquire projection data (see Figure 4.8).

5.3.1 Feldkamp's algorithm

Feldkamp's cone-beam algorithm is dedicated to the circular focal-point trajectory. It is an FBP algorithm and is easy to use. Because the circular trajectory does not satisfy Tuy's condition, Feldkamp's algorithm can only provide approximate reconstructions. Artifacts can appear especially at locations away from the orbit plane. The artifacts include reduction in activity in the regions away from the orbit plane, cross-talk between adjacent slices, and undershoots.

Feldkamp's algorithm is practical and robust. Cone angle, as defined in Figure 5.10, is an important parameter in cone-beam imaging. If the cone angle is small, say less than 10°, this algorithm gives fairly good images. At the orbit plane, this algorithm is exact. This algorithm also gives an exact reconstruction if the object is constant in the axial direction (e.g., a tall cylinder).

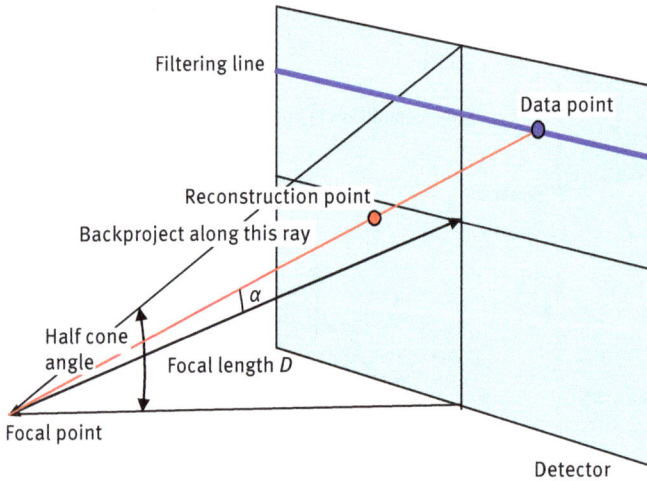

Fig. 5.10: The coordinate system for Feldkamp's cone-beam algorithm.

Feldkamp's cone-beam algorithm (Section 5.4) is nothing but a modified fan-beam FBP algorithm (Section 3.4.1). It consists of the following steps:

(i)   Pre-scale the projections by a cosine function cos α (see Figure 5.10 for angle α).

(ii) Row-by-row ramp filter the pre-scaled data.

(iii) Cone-beam backproject the filtered data with a weighting function of the distance from the reconstruction point to the focal point.

5.3.2 Grangeat's algorithm

Feldkamp's algorithm converts the cone-beam image reconstruction problem to the fan-beam image reconstruction problem; Grangeat's algorithm, on the other hand, converts it to the 3D Radon inversion problem (Section 5.2). Feldkamp's algorithm is derived for the circular orbit; Grangeat's algorithm can be applied to any orbit. If the orbit satisfies Tuy's condition, Grangeat's algorithm can provide an exact reconstruction.

Grangeat's method first tries to convert cone-beam ray sums to plane integrals, by calculating the line integrals on the cone-beam detector (see Figure 5.11).

We observe that the line integral on the detector plane gives a weighted plane integral of the object with a special nonuniform weighting function $1/r$. Here r is the distance to the cone-beam focal point. We must remove this $1/r$ weighting before we can obtain a regular unweighted plane integral.

From Figure 5.12 we observe that the angular differential da multiplied by the distance r equals the tangential differential dt, that is, $r\,da = dt$. If we perform the angular derivative on the $1/r$ weighted plane integral, we will cancel out this $1/r$ weighting

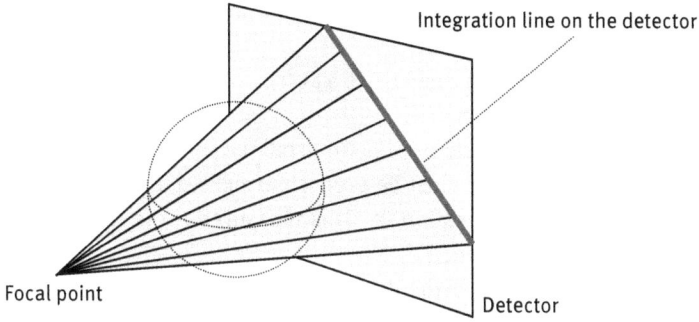

Fig. 5.11: Integration along a line on the cone-beam detector gives a weighted plane integral of the object.

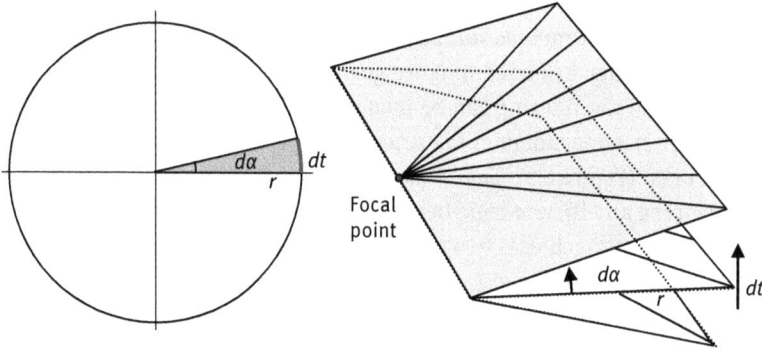

Fig. 5.12: The differential dt in the tangent direction is equal to the angular differential $d\alpha$ times the distance r.

factor by the factor r, obtaining the derivative of the unweighted plane integral with respect to variable t, which is in the normal direction of the plane, that is,

$$\frac{\partial\left(\frac{1}{r}\text{ weighted plane integral}\right)}{\partial\alpha} = \frac{\partial(\text{Radon transform})}{\partial t}. \tag{5.7}$$

Recall that the Radon inversion formula is the second-order derivative of the plane integral with respect to t, followed by the 3D Radon backprojection. Therefore, a cone-beam image reconstruction algorithm can be implemented as follows:

(i)   Form all possible (all orientations and all locations) line integrals on each detector plane (see Figure 5.11), obtaining $(1/r)$-weighted plane integrals.
(ii) Perform the angular derivative on results from (i).
(iii) Rebin the results from (ii) to the $(s, \vec{\theta})$ Radon space (see Figure 5.6).

(iv) Take the derivative of the results of (iii) with respect to t, in the normal direction of the plane.
(v) Perform the 3D Radon backprojection (see Figure 5.8).

We now expand on Step (iii). For a practical focal-point orbit, the $(s, \vec{\theta})$ Radon space is not uniformly sampled. Data redundancy must be properly weighted. For example, if the value at a particular Radon space location $(s, \vec{\theta})$ is measured three times, then after rebinning this value needs to be divided by 3.

Grangeat's algorithm is not an FBP algorithm, and it requires data rebinning, which can introduce large interpolation errors.

5.3.3 Katsevich's algorithm

Katsevich's cone-beam algorithm was initially developed for the helical orbit cone-beam geometry and was later extended to more general orbits. Katsevich's algorithm is in the form of FBP, and the filtering can be made shift invariant. By shift invariant we mean that the filter is independent of the reconstruction location.

Using a helical orbit (Figure 5.9), Tuy's data sufficiency condition is satisfied. The main issue in developing an efficient cone-beam FBP algorithm is to properly normalize the redundant data. Katsevich uses two restrictions to handle this issue.

It can be shown that for any point (x, y, z) within the volume surrounded by the helix, there is a unique line segment that passes through the point (x, y, z) where both endpoints touch two different points on the helix and are separated by less than one pitch, say, at s_b and s_t shown in Figure 5.13. This particular line segment is referred to as the π-segment (or π-line). The first restriction is the use of the cone-beam measurements that are measured only from the helix orbit between s_b and s_t.

The second restriction is the filtering direction, which handles the normalization of redundant data. In order to visualize the data redundancy problem, let us look at three cone-beam image reconstruction problems: (a) data are sufficient and not redundant; (b) data are insufficient; and (c) data are sufficient but redundant.
(a) The scanning cone-beam focal-point orbit is an arc (i.e., an incomplete circle). The point to be reconstructed is in the orbit plane and on the line that connects the two endpoints of the arc. The line connecting the arc's endpoints is the π-segment of the reconstruction point. For our special case that the object is only a point, the cone-beam measurement of this point at one focal-point position can provide a set of plane integrals of the object. Those planes all contain the line that connects the focal point and the reconstruction point. After the focal point goes through the entire arc, all plane integrals are obtained. Recall the central slice theorem for the 3D Radon transform; an exact reconstruction requires that the planar integrals of the object are available for all directions $\vec{\theta}$, which is indicated in Figure 5.14 (top).

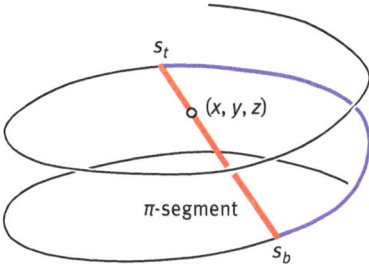

Fig. 5.13: For any point (x, y, z) inside the helix, there is one and only one π-segment.

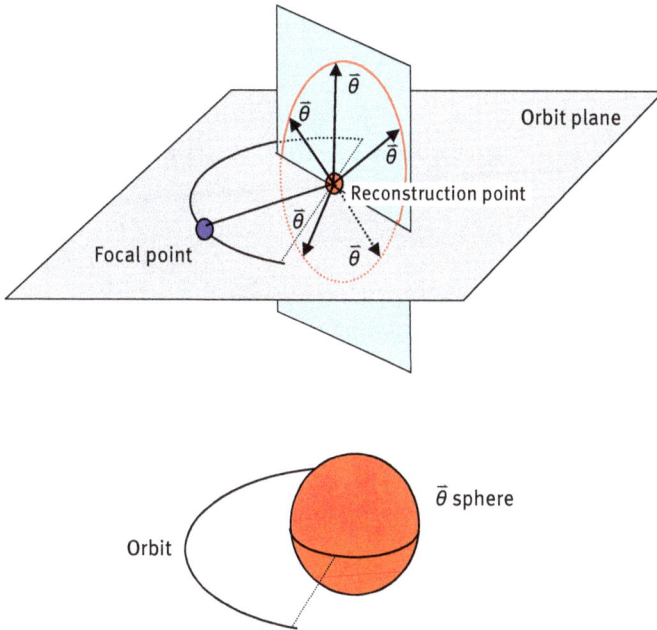

Fig. 5.14: Top: At one focal-point position, the directional vectors trace a unit circle. A directional vector represents a measured plane integral. The plane containing this unit circle is perpendicular to the line that connects the reconstruction point and the focal point. Bottom: When the focal point travels through the entire arc, the directional vectors trace a full unit sphere.

The unit vector $\vec{\theta}$ traces a circle (let us call it a $\vec{\theta}$-circle) in a plane that is perpendicular to the orbit plane, and the line connecting the focal point and the reconstruction point is normal to this plane. Thus, every focal point on the arc orbit corresponds to a $\vec{\theta}$-circle. Let the focal point travel through the entire arch, then the corresponding $\vec{\theta}$-circles form a complete unit sphere, which we can call a $\vec{\theta}$ sphere (see Figure 5.14, bottom).

(b) If the same arc orbit as in (a) is used and the object to be reconstructed is a point above the orbit plane, then the data are insufficient. When we draw the $\vec{\theta}$-circle for each focal-point position, the $\vec{\theta}$-circle is not in a vertical plane, but in a plane that has a slant angle. If we let the focal point travel through the entire arc, the corresponding $\bar{\theta}$-circles do not form a complete unit sphere anymore – both the Arctic Circle and the Antarctic Circle are missing (see Figure 5.15).

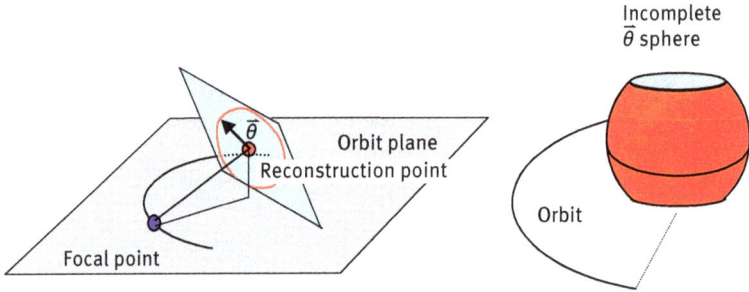

Fig. 5.15: If the reconstruction point is above the orbit plane, the unit sphere is not completely covered by all the directional vectors.

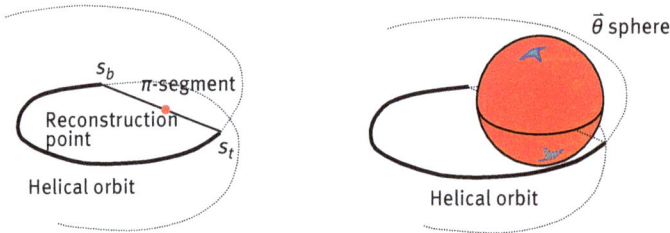

Fig. 5.16: For the helical orbit, the unit sphere is completely measured. At the North Pole and South Pole area, a small region is covered (i.e., measured) three times.

(c) Now we consider a helical orbit with the reconstruction point inside. We determine the π-segment for the reconstruction point and find the π-segment endpoints. The segment of the helical orbit between the π-segment endpoints are shown in Figure 5.16.

In Figure 5.16, the $\vec{\theta}$ sphere is fully measured. In fact, it is over-measured. The small triangle-like regions near the North and South Poles are measured three times. Let us look at this situation in another way. Draw a plane passing through the reconstruction point. In most cases, the plane will intersect the piece of helix orbit at one point. However, there is a small chance that the plane can intersect the piece of helix at three points (see a side view of the helix in Figure 5.17).

If a plane intersects the orbit three times, we must normalize the data by assigning a proper weight to each measurement. The sum of the weights must be 1. Common knowledge teaches us that we should use all available data and weigh the redundant measurement by the inverse of its noise variance. However, in order to derive a shift-invariant FBP algorithm, we need to do something against common sense. In Katsevich's algorithm, a measurement is weighted by either +1 or –1. If a plane is measured once, we must make sure that it is weighted by +1. If a plane is measured three times, we need to make sure that two of them are weighted by +1 and the third one of them is weighted by –1. In other words, we keep one and throw the other two away. Ouch! Further discussion about the weighting and filtering will be given in the next section.

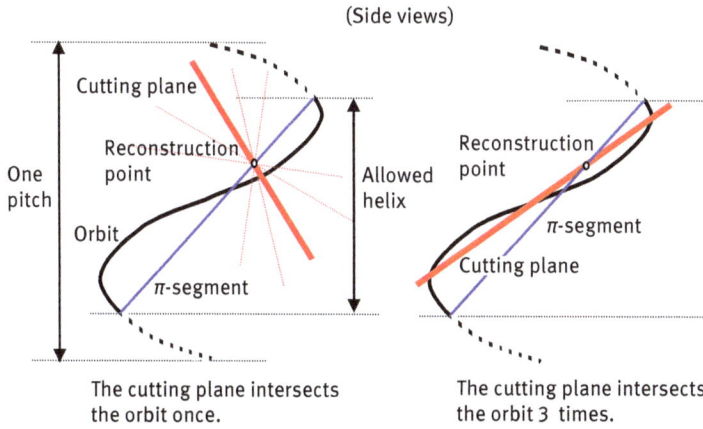

Fig. 5.17: The π-segment defines a section of helix. A cutting plane is a plane passing through the reconstruction point. The cutting plane either intersects the section of helix one or three times.

In Katsevich's algorithm, the normalization issue is taken care of by selecting a proper filtering direction. Here, filtering means performing the Hilbert transform. After the filtering direction and the normalization issues have been taken care of, Katsevich's algorithm is implemented with the following steps:

(i) Take the derivative of the cone-beam data with respect to the orbit parameter along the helix orbit.

(ii) Perform the Hilbert transform along the directions that have been carefully selected.

(iii) Perform a cone-beam backprojection with a weighting function, similar to the backprojection in Feldkamp's algorithm.

Some later versions of the algorithm have replaced the derivative with respect to the orbit parameters by partial derivative with respect to the detector coordinates.

5.4 Mathematical expressions

Some 3D image reconstruction algorithms are presented here without proofs. For the 3D parallel line-integral projections, we have a backprojection-then-filtering algorithm and FBP algorithms, which are not unique. For the parallel plane-integral projections (i.e., the Radon transform), we also have a backprojection-then-filtering algorithm and an FBP algorithm which is the Radon inversion formula.

For cone-beam projections, Felkamp's circular orbit algorithm and the Katsevich's helical orbit algorithm are highlighted because they are in the form of convolution and cone-beam backprojection. Tuy's relation and Grangeat's relation are also discussed in this section.

5.4.1 Backprojection-then-filtering for parallel line-integral data

For this type of algorithm, the projections are backprojected to the image domain first, obtaining $b(x, y, z)$. Then the 3D Fourier transform is applied on $b(x, y, z)$, obtaining $B(\omega_x, \omega_y, \omega_z)$. Next, a Fourier domain filter $G(\omega_x, \omega_y, \omega_z)$ is used to multiply $B(\omega_x, \omega_y, \omega_z)$, obtaining $F(\omega_x, \omega_y, \omega_z) = B(\omega_x, \omega_y, \omega_z) \, G(\omega_x, \omega_y, \omega_z)$. Finally, the 3D inverse Fourier transform is applied to $F(\omega_x, \omega_y, \omega_z)$ to obtain the reconstructed image $f(x, y, z)$. Here, the filter transfer function $G(\omega_x, \omega_y, \omega_z)$ is imaging geometry dependent. Some of the imaging geometries are shown in Figure 5.5, where the $\vec{\theta}$ trajectories are displayed as shaded regions. Let Ω denote the occupied region by the $\vec{\theta}$ trajectories on the unit sphere.

When $\Omega = \Omega_{4\pi}$ is 4π, that is, $\Omega_{4\pi}$ is the entire unit sphere, G is a ramp filter:

$$G\left(\omega_x, \omega_y, \omega_z\right) = \sqrt{\omega_x^2 + \omega_y^2 + \omega_y^2}/\pi. \tag{5.8}$$

If Ω is not the full sphere, this ramp filter will be modified by the geometry of Ω. Then the general form of G is

$$G\left(\omega_x, \omega_y, \omega_z\right) = \sqrt{\omega_x^2 + \omega_y^2 + \omega_z^2}/D\left(\vec{\theta}\right), \tag{5.9}$$

where $D(\vec{\theta})$ is half of the arc length of the intersection of a great circle with Ω. The normal direction of the great circle in the Fourier domain is $\vec{\theta}$, where $\vec{\theta}$ is the direction from the origin to the point $(\omega_x, \omega_y, \omega_z)$.

When $\Omega = \Omega_\psi$ is the region shown in Figure 5.18, $D(\vec{\theta})$ is the arc length γ, which is orientation $\vec{\theta}$ dependent. Using the geometry, we have $\gamma = \pi$ if $\theta \le \psi$, and we have $\sin \frac{\gamma}{2} = \frac{\sin \psi}{\sin \theta}$ if $\theta > \psi$.

5.4.2 FBP algorithm for parallel line-integral data

In the FBP algorithm, we need to find a 2D filter transfer function for each orientation $\vec{\theta} \in \Omega$. If $\Omega = \Omega_{4\pi}$, this filter is a ramp filter

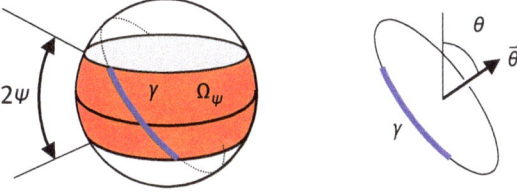

Fig. 5.18: The definition of the Ω_ψ and the arc length γ, which is a part of a great circle.

$$Q(\omega_u, \omega_v) = \sqrt{\omega_u^2 + \omega_v^2}/\pi,$$
(5.10)

which is the same for all orientations $\vec{\theta} \in \Omega$.

If Ω is not the full sphere $\Omega_{4\pi}$, this ramp filter $Q(\omega_u, \omega_v)$ becomes orientation $\vec{\theta}$ dependent as $Q_{\vec{\theta}}(\omega_u, \omega_v)$ (note: a subscript is added) and can be obtained by selecting the "central slice" of $G(\omega_x, \omega_y, \omega_z)$ with the normal direction $\vec{\theta}$. Here, the u-axis and v-axis are defined by unit vectors $\vec{\theta}_v$ and $\vec{\theta}_v$, respectively. The three vectors, $\vec{\theta}$, $\vec{\theta}_u$, and $\vec{\theta}_v$, form an orthogonal system in 3D, where $\vec{\theta}$ represents the direction of a group of parallel lines perpendicular to a detector, $\vec{\theta}_u$ and $\vec{\theta}_v$ are on the detector plane, and $\vec{\theta}_u$ is also in the x–y plane of the global (x, y, z) system in 3D.

If we consider the case of $\Omega = \Omega_\psi$ shown in Figure 5.18, $Q_{\vec{\theta}}(\omega_u, \omega_v)$ has two expressions in two separate regions (see Figure 5.19):

$$Q_{\vec{\theta}}(\omega_u,\ \omega_v) = \frac{\sqrt{\omega_u^2 + \omega_v^2}}{\pi}$$
(5.11)

if

$$0 \le \sqrt{\omega_u^2 + \omega_v^2 \cos^2 \theta} \le \sqrt{\omega_u^2 + \omega_v^2} \sin \psi;$$
(5.12)

$$Q_{\vec{\theta}}(\omega_u, \omega_v) = \frac{\sqrt{\omega_u^2 + \omega_v^2}}{2 \sin^{-1}\left(\frac{\sqrt{\omega_u^2 + \omega_v^2} \sin \psi}{\sqrt{\omega_u^2 + \omega_v^2 \cos^2\theta}}\right)}$$
(5.13)

if

$$\sqrt{\omega_u^2 + \omega_v^2} \sin \psi < \sqrt{\omega_u^2 + \omega_v^2 \cos^2 \theta} \le \sqrt{\omega_u^2 + \omega_v^2}.$$
(5.14)

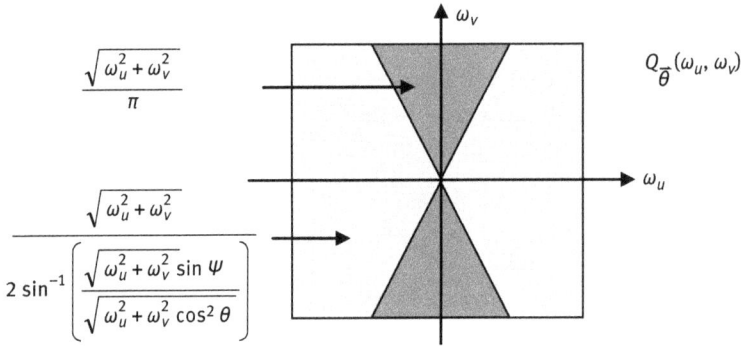

Fig. 5.19: The 2D filter transfer function Q for the imaging geometry Ω_ψ.

Now the expression (5.13) is explained below. We need to set up some 3D coordinate systems: the regular (x, y, z) system and the rotated (u, v, t) system. The unit vectors for the (u, v, t) system are defined as

$$\vec{\theta}_u = (-\sin\phi, \ \cos\phi, 0), \tag{5.15}$$

$$\vec{\theta}_v = (-\cos\theta \cos\phi, \ -\cos\theta \sin\phi, \ \sin\theta), \tag{5.16}$$

$$\vec{\theta}_t = (\sin\theta \cos\phi, \ \sin\theta \sin\phi, \ \cos\theta), \tag{5.17}$$

where the u-axis is in the x–y plane. The t-axis is parallel to the projection rays. The u–v plane is the virtual 2D detector. The spatial domain systems are shown in Figure 5.20. The corresponding Fourier domain coordinate systems are shown in Figure 5.21.

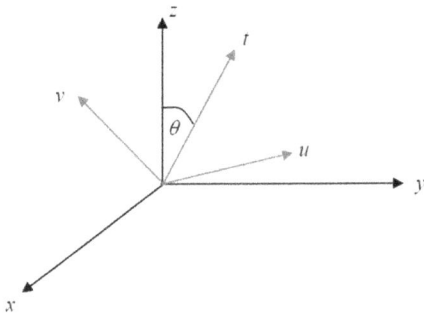

Fig. 5.20: Two 3D coordinate systems: the regular (x, y, z) system and the rotated (u, v, t) system.

Let us consider a great circle with the normal vector pointing at the direction $\vec{\theta}$. The arc length y as shown in Figure 5.18 is given by

$$\sin\frac{y}{2} = \frac{\sin\psi}{\sin\theta}, \tag{5.18}$$

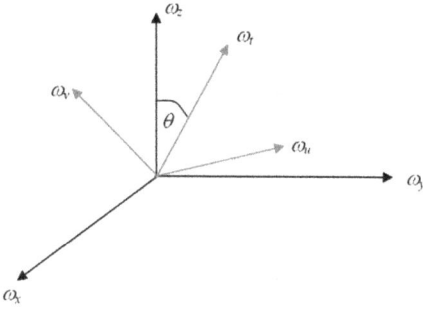

Fig. 5.21: The Fourier domain coordinate systems corresponding to the two 3D coordinate systems shown in Figure 5.20.

that is,

$$y = 2\sin^{-1}\left(\frac{\sin\psi}{\sin\theta}\right). \tag{5.19}$$

Let us consider a different great circle whose normal vector pointing at a different direction $\vec{\theta_\omega}$, and the corresponding arc length is $y = 2\sin^{-1}(\sin\psi/\sin\theta)$.

In the following, we consider a special point $\vec{\omega}$ in the $\omega_u - \omega_v$ plane in the 3D Fourier domain as shown in Figure 5.22. The ω_t component of $\vec{\omega}$ is zero. We have

$$\vec{\omega} = \omega_u \vec{\theta}_u + \omega_v \vec{\theta}_v, \tag{5.20}$$

$$\|\vec{\omega}\| = \sqrt{\omega_u^2 + \omega_v^2}. \tag{5.21}$$

For this special point $\vec{\omega}$, the arc length y given by eq. (5.19) becomes

$$y = 2\sin^{-1}\left(\frac{\sin\psi}{\sin\theta_{\vec{\omega}}}\right) = 2\sin^{-1}\left(\frac{\sin\psi}{\sqrt{1 - \cos^2\theta_{\vec{\omega}}}}\right) = 2\sin^{-1}\left(\frac{\sin\psi}{\sqrt{1 - \left(\frac{\omega_z}{\|\vec{\omega}\|}\right)^2}}\right), \tag{5.22}$$

where ω_z is the z-component of $\vec{\omega} = \omega_u\vec{\theta}_u + \omega_v\vec{\theta}_v$, and thus

$$\omega_z = \omega_u 0 + \omega_v \sin\theta = \omega_v \sin\theta. \tag{5.23}$$

Therefore,

$$
y = 2\sin^{-1}\left(\frac{\sin\psi}{\sqrt{1-\left(\dfrac{\omega_v\sin\theta}{\sqrt{\omega_u^2+\omega_v^2}}\right)^2}}\right) = 2\sin^{-1}\left(\frac{\sqrt{\omega_u^2+\omega_v^2}\,\sin\psi}{\sqrt{\omega_u^2+\omega_v^2-\omega_v^2\sin^2\theta}}\right)
$$

$$
= 2\sin^{-1}\left(\frac{\sqrt{\omega_u^2+\omega_v^2}\,\sin\psi}{\sqrt{\omega_u^2+\omega_v^2\cos^2\theta}}\right), \tag{5.24}
$$

which readily leads to (5.13).

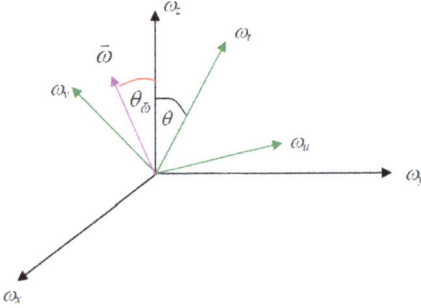

Fig. 5.22: The Fourier domain coordinate systems correspond to the two 3D coordinate systems shown in Figure 5.20. A special point is marked as $\vec{\omega}$.

5.4.3 Three-dimensional Radon inversion formula (FBP algorithm)

The 3D Radon inversion formula can only be applied to 3D plane-integral data:

$$
f(x,y,z) = \frac{-1}{8\pi^2}\iint\limits_{4\pi}\frac{\partial^2 p\left(s,\vec{\theta}\right)}{\partial s^2}\bigg|_{s=\vec{x}\cdot\vec{\theta}}\sin\theta\, d\theta\, d\phi, \tag{5.25}
$$

where $\vec{\theta} = (\sin\theta\cos\phi, \sin\theta\sin\phi, \cos\theta)$ and $\vec{x} = (x, y, z)$. The coordinate systems are shown in Figure 5.23. If we backproject over a 2π solid angle, then $-1/(8\pi^2)$ should be replaced by $-1/(4\pi^2)$.

5.4.4 Three-dimensional backprojection-then-filtering algorithm for Radon data

Let the backprojected image be

$$b(x,y,z) = \iint_{2\pi} p\left(s,\vec{\theta}\right)\Big|_{s=\vec{x}\cdot\vec{\theta}} \sin\theta d\theta d\phi. \tag{5.26}$$

For the 3D Radon case, the Fourier transform of the projection/backprojection PSF is $1/(\omega_x^2 + \omega_y^2 + \omega_z^2)$. Therefore, the Fourier transform of the image $f(x, y, z)$ can be obtained as

$$F\left(\omega_x,\omega_y,\omega_z\right) = B\left(\omega_x,\omega_y,\omega_z\right) \times \left(\omega_x^2 + \omega_y^2 + \omega_z^2\right). \tag{5.27}$$

In the spatial domain, the backprojection-then-filtering algorithm for Radon data can be expressed as

$$f(x,y,z) = \Delta\, b(x,y,z) = \frac{\partial^2 b(x,y,z)}{\partial x^2} + \frac{\partial^2 b(x,y,z)}{\partial y^2} + \frac{\partial^2 b(x,y,z)}{\partial z^2}, \tag{5.28}$$

where Δ is the Laplacian operator.

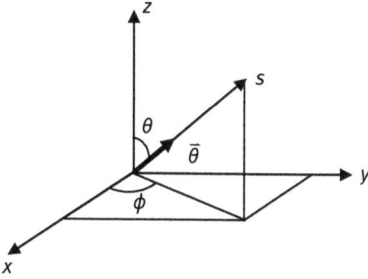

Fig. 5.23: The coordinate system for the 3D Radon inversion formula.

5.4.5 Feldkamp's algorithm

First, let us write down the fan-beam FBP reconstruction algorithm for the flat detector and express the image in polar coordinates (see Figure 5.24):

$$f(r,\varphi) = \frac{1}{2}\int_0^{2\pi} \left(\frac{D}{D-s}\right)^2 \int_{-\infty}^{\infty} \frac{D}{\sqrt{D^2+l^2}} g(l,\beta) h(l'-l)\, dld\beta, \tag{5.29}$$

where $h(l)$ is the convolution kernel of the ramp filter, D is the focal length, $g(l, \beta)$ is the fan-beam projection, l is the linear coordinate on the detector, $s = r\sin(\varphi - \beta)$, and $l' = \frac{Dr\cos(\varphi-\beta)}{D-r\sin(\varphi-\beta)}$. In this formula, $D/\sqrt{D^2+l^2}$ is the cosine of the incidence angle. When this algorithm is implemented, we first multiply the projections by this cosine function. Then we apply the ramp filter to the pre-scaled data. Finally, we perform the fan-beam backprojection with a distance-dependent weighting $[D/(D–s)]^2$, where s is the distance from the reconstruction point to the virtual detector, which is placed at the center of rotation for convenience.

Feldkamp's algorithm is almost the same as this fan-beam algorithm, except that the backprojection is a cone-beam backprojection. The ramp-filtering is performed in the row-by-row fashion. There is no filtering performed in the axial direction. Let the axial direction be the z-direction (see Figure 5.25), then

$$f(r,\varphi,z) = \frac{1}{2}\int_0^{2\pi}\left(\frac{D}{D-s}\right)^2 \int_{-\infty}^{\infty}\frac{D}{\sqrt{D^2+l^2+\hat{z}^2}}g(l,\hat{z},\beta)h(l'-l)\,dl\,d\beta. \qquad (5.30)$$

In this formula, $g(l, \hat{z}, \beta)$ is the cone-beam projection, $D/\sqrt{D^2+l^2+\hat{z}^2}$ is the cosine of the incidence angle, and $\hat{z}$ and l' are defined in Figure 5.25.

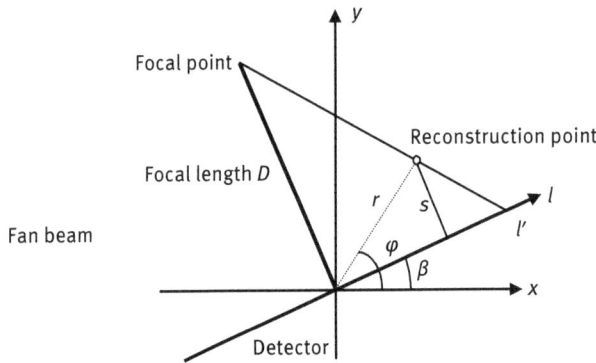

Fig. 5.24: The coordinate system for the flat-detector fan-beam imaging geometry.

5.4.6 Tuy's relationship

Tuy published a paper in 1983. In this paper, he established a relationship between the cone-beam data and the original image, which plays a similar role to the central slice theorem. Let us derive this relationship in the section.

The object to be imaged is f. Let the cone-beam focal-point trajectory be denoted by a vector $\vec{\Phi}$ and let $\vec{a}$ be the unit vector, indicating the direction of a projection ray. Therefore, the cone-beam data can be expressed by the following expression:

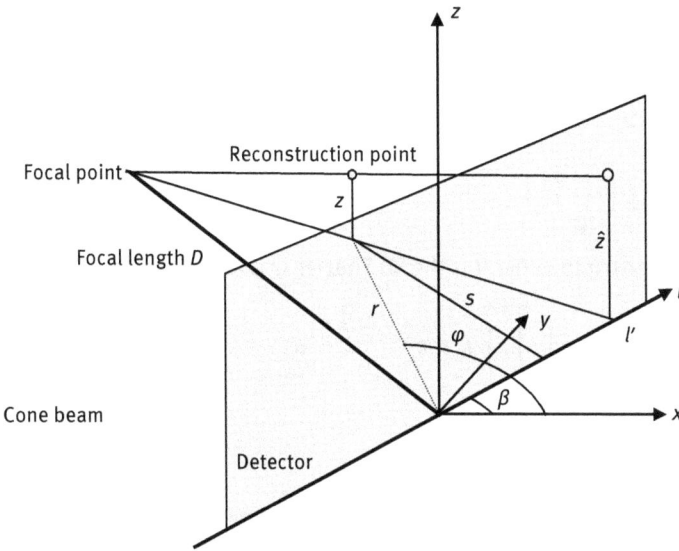

Fig. 5.25: The coordinate system for Feldkamp's cone-beam algorithm.

$$g\left(\vec{\Phi}, \vec{a}\right) = \int_{0}^{\infty} f\left(\vec{\Phi} + t\vec{a}\right) dt, \, ||\vec{a}|| = 1. \tag{5.31}$$

We now replace the unit vector $\vec{a}$ by a general 3D vector $\vec{x}$, and the above 2D projection becomes an extended 3D function:

$$g\left(\vec{\Phi}, \vec{x}\right) = \int_{0}^{\infty} f\left(\vec{\Phi} + t\vec{x}\right) dt. \tag{5.32}$$

Taking the 3D Fourier transform of this function with respect to $\vec{x}$ and using notation $\vec{\beta}$ as the frequency domain variable, we have

$$G\left(\vec{\Phi}, \vec{\beta}\right) = \int_{-\infty}^{\infty} \int_{-\infty}^{\infty} \int_{-\infty}^{\infty} g\left(\vec{\Phi}, \vec{x}\right) e^{-2\pi i \vec{x} \cdot \vec{\beta}} d\vec{x} = \int_{-\infty}^{\infty} \int_{-\infty}^{\infty} \int_{-\infty}^{\infty} \int_{0}^{\infty} f\left(\vec{\Phi}, t\vec{x}\right) e^{-2\pi i \vec{x} \cdot \vec{\beta}} dt d\vec{x}. \tag{5.33}$$

Let $\vec{y} = \vec{\Phi} + t\vec{x}$; $d\vec{y} = |t|^{3} d\vec{x}$. The above expression becomes

$$G\left(\vec{\Phi}, \vec{\beta}\right) = \int_{-\infty}^{\infty} \int_{-\infty}^{\infty} \int_{-\infty}^{\infty} \int_{0}^{\infty} f(\vec{y}) e^{-\frac{2\pi i}{t}(\vec{y} - \vec{\Phi}) \cdot \vec{\beta}} \frac{1}{|t|^{3}} dt d\vec{y}. \tag{5.34}$$

Let $s = (1/t); ds = (1/t^2) dt$. We then have

$$G\left(\vec{\Phi}, \vec{\beta}\right) = \int_{-\infty}^{\infty} \int_{-\infty}^{\infty} \int_{-\infty}^{\infty} f(\vec{y}) e^{-2\pi i s\left(\vec{y} - \vec{\Phi}\right) \cdot \vec{\beta}} |s|^3 \frac{1}{s^2} d\vec{y} ds, \tag{5.35}$$

$$G\left(\vec{\Phi}, \vec{\beta}\right) = \int_{-\infty}^{\infty} \left(\int_{-\infty}^{\infty} \int_{-\infty}^{\infty} \int_{-\infty}^{\infty} f(\vec{y}) e^{-2\pi i \vec{y} \cdot \left(s\vec{\beta}\right)} d\vec{y} \right) |s| e^{2\pi i s\left(\vec{\Phi} \cdot \vec{\beta}\right)} ds. \tag{5.36}$$

Recognizing that the inner triple integral is the 3D Fourier transform of f, that is,

$$F(s) = \int_{-\infty}^{\infty} \int_{-\infty}^{\infty} \int_{-\infty}^{\infty} f(\vec{y}) e^{-2\pi i \vec{y} \cdot \left(s\vec{\beta}\right)} d\vec{y}, \tag{5.37}$$

we have

$$G\left(\vec{\Phi}, \vec{\beta}\right) = \int_{-\infty}^{\infty} F\left(s\vec{\beta}\right) |s| e^{2\pi i s\left(\vec{\Phi} \cdot \vec{\beta}\right)} ds. \tag{5.38}$$

Next, we change the limit from [0, ∞) to (−∞, ∞) and obtain

$$G\left(\vec{\Phi}, \vec{\beta}\right) = \int_{-\infty}^{\infty} F\left(s\vec{\beta}\right) |s| e^{2\pi i s\left(\vec{\Phi} \cdot \vec{\beta}\right)} ds + \int_{-\infty}^{\infty} F\left(s\vec{\beta}\right) \frac{is}{i} e^{2\pi i s\left(\vec{\Phi} \cdot \vec{\beta}\right)} ds. \tag{5.39}$$

Using 3D Radon transform's central slice theorem, $F(s \vec{\beta})$ is the Fourier transform of the plane integral of the original image f in the direction $\vec{\beta}$. The factor $|s|$ is the ramp filter in the Fourier domain, and the factor (is) corresponds to the derivative in the spatial domain. Each of the two terms at the right-hand side of the above expression is in the form of an inverse Fourier transform, which is, in fact, a convolution. Therefore,

$$G\left(\vec{\Phi}, \vec{\beta}\right) = p_{\vec{\beta}}(t) * h(t) - i p_{\vec{\beta}}(t) * \delta'(t), \tag{5.40}$$

where $i = \sqrt{-1}, t = \vec{\Phi} \cdot \vec{\beta}$, $h(t)$ is the ramp-filter convolution kernel

$$h(t) = \int_{-\infty}^{\infty} |s| e^{2\pi i st} ds \tag{5.41}$$

and the Radon transform of the original image f is defined as

$$p_{\vec{\beta}}(t) = \int_{-\infty}^{\infty} \int_{-\infty}^{\infty} \int_{-\infty}^{\infty} f(\vec{x}) \delta\left(\vec{x} \cdot \vec{\beta} - t\right) d\vec{x}. \tag{5.42}$$

The left-hand side of eq. (5.40) is related to the cone-beam projections, and the right-hand side is related to the plane integral of the original image.

Tuy's algorithm is stated as

$$f(\vec{x}) = \frac{1}{2\pi} \iint_{4\pi} \frac{1}{\left| \vec{\Phi}'(\lambda) \cdot \vec{\beta} \right|} \frac{\partial G\left(\vec{\Phi}(\lambda), \vec{\beta}\right)}{\partial \lambda} d\vec{\beta}, \tag{5.43}$$

with $\vec{\Phi}(\lambda) \cdot \vec{\beta} = \vec{x} \cdot \vec{\beta}$. In fact, the factor $(2\pi i)$ in Tuy's algorithm does not make the reconstruction $f(\vec{x})$ imaginary. Because the combination of the first term in eq. (5.40) and the factor $\left| \vec{\Phi}'(\lambda) \cdot \vec{\beta} \right|$ is odd in $\vec{\beta}$, the real part in $\iint_{4\pi} \frac{1}{\left| \vec{\Phi}'(\lambda) \cdot \vec{\beta} \right|} \frac{\partial G\left(\vec{\Phi}(\lambda), \vec{\beta}\right)}{\partial \lambda} d\vec{\beta}$ will vanish. Thus, Tuy's algorithm will reconstruct a real image.

5.4.7 Grangeat's relationship

Grangeat established a relationship between the derivative of Radon data and the derivative of the line integrals of the data on the flat cone-beam detection plane.

In the following, we only consider a fixed focal-point position $\vec{\Phi}$. We arbitrarily select a straight line on the detector and sum up the cone-beam projections along this line. Let us set up a coordinate system on the detector plane (see Figure 5.26). The u-axis is along the integral line on the detector and the v-axis is orthogonal to the u-axis. We denote the cone-beam projection data on the u–v coordinate system as $g(u, v)$. If we use the object f with the spherical system, then the projection $g(u, v)$ can be expressed as (see Figure 5.27)

$$g(u, v) = \int_{-\infty}^{\infty} f\left(\vec{\beta}, \theta, r\right) dr, \tag{5.44}$$

where u and θ are related as

$$u = \sqrt{D^2 + v^2} \tan \theta \tag{5.45}$$

As in Feldkamp's algorithm, the cone-beam data are pre-scaled by a cosine function $D/\sqrt{D^2 + u^2 + v^2}$. Thus, the data summation along the u-axis is actually

$$s(v) = \int_{-\infty}^{\infty} g(u, v) \frac{D}{\sqrt{D^2 + u^2 + V^2}} du = \int_{-\infty}^{\infty} \int_{-\infty}^{\infty} f\left(\vec{\beta}, \theta, r\right) \frac{D}{\sqrt{D^2 + u^2 + v^2}} dr du. \tag{5.46}$$

We now change the variable u to variable θ, using $u = \sqrt{D^2 + v^2} \tan \theta$, $\cos \theta = \left(\sqrt{D^2 + v^2} / \sqrt{D^2 + u^2 + v^2} \right)$, and $du/d\theta = \left((D^2 + v^2)/\cos^2\theta \right)$, and we obtain

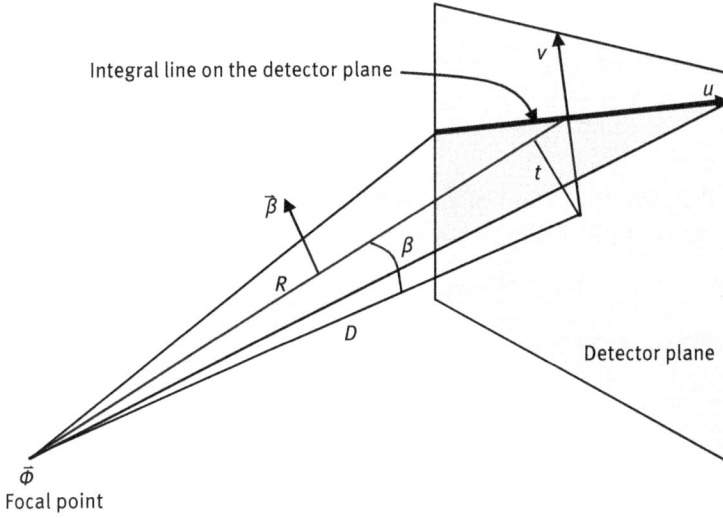

Fig. 5.26: A line is drawn on the detector. The cone-beam data are summed on this line as $s(v)$.

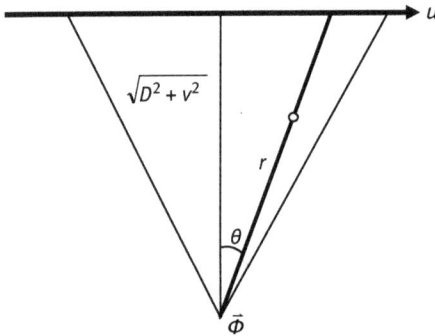

Fig. 5.27: The formation of cone-beam projection.

$$s(v) = \int_{-\infty}^{\infty} \int_{-\pi/2}^{\pi/2} f\left(\vec{\beta}, \theta, r\right) \frac{D}{\sqrt{D^2 + u^2 + v^2}} \frac{\sqrt{D^2 + v^2}}{\cos^2\theta} \, dr d\theta = \int_{-\infty}^{\infty} \int_{-\pi/2}^{\pi/2} f\left(\vec{\beta}, \theta, r\right) \frac{D}{\cos\theta} \, dr d\theta. \quad (5.47)$$

We would like to relate this $s(v)$ to the Radon transform of f:

$$p_{\vec{\beta}}(t) = \int_{-\infty}^{\infty} \int_{-\pi/2}^{\pi/2} f\left(\vec{\beta}, \theta, r\right) r \, dr d\theta, \quad (5.48)$$

where the parameter t is defined in Figure 5.26. We will now make this connection with the idea presented in Section 5.3.2, from which we know that (see Figure 5.28)

$$\frac{\partial f}{\partial \beta} = R\frac{\partial f}{\partial t} = r\cos\theta\,\frac{\partial f}{\partial t}.\tag{5.49}$$

Using this relationship, we have

$$\frac{\partial p_{\vec{\beta}}(t)}{\partial t} = \int_{-\infty}^{\infty}\int_{-\pi/2}^{\pi/2}\frac{\partial}{\partial t}f\left(\vec{\beta},\theta,r\right)r\,dr\,d\theta$$

$$= \int_{-\infty}^{\infty}\int_{-\pi/2}^{\pi/2}\frac{1}{r\cos\theta}\frac{\partial}{\partial\beta}f\left(\vec{\beta},\theta,r\right)r\,dr\,d\theta = \frac{1}{D}\int_{-\infty}^{\infty}\int_{-\pi/2}^{\pi/2}\frac{D}{\cos\theta}\frac{\partial}{\partial\beta}f\left(\vec{\beta},\theta,r\right)dr\,d\theta$$

$$= \frac{1}{D}\frac{\partial s(v)}{\partial\beta}.\tag{5.50}$$

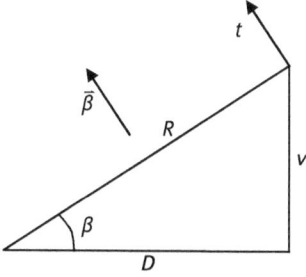

Fig. 5.28: The direction of t is the tangent direction.

Finally, we will replace the partial derivative with respect to β with the partial derivative with respect to v. From Figure 5.28, we see that

$$v = D\tan\beta \quad\text{and}\quad \frac{dv}{d\beta} = \frac{D}{\cos^2\beta}.\tag{5.51}$$

The Grangeat's relationship is obtained as

$$\frac{\partial p_{\vec{\beta}}(t)}{\partial t} = \frac{1}{D}\frac{\partial s(v)}{\partial\beta} = \frac{1}{D}\frac{\partial s(v)}{\partial v}\frac{\partial v}{d\beta}\frac{dv}{d\beta} = \frac{1}{D}\frac{\partial s(v)}{\partial\beta}\frac{D}{\cos^2\beta} = \frac{s'(v)}{\cos^2\beta}.\tag{5.52}$$

5.4.8 Katsevich's algorithm

We denote the helix focal point by a vector

$$\vec{a}(s) = \left(R\cos s, R\sin s, \frac{h}{2\pi}s\right), \quad s\in I_\pi(\vec{x}),\tag{5.53}$$

where R is the radius, h is the helix pitch, $\vec{x} = (x, y, z)$ is the reconstruction point, s is the orbit parameter, and $I_\pi(\vec{x}) = [s_b, s_t]$, which is determined by the π-segment of the point $\vec{x}$.

The cone-beam projection is represented in a local coordinate system as $g(\vec{\theta}, \vec{a})$, where $\vec{\theta}$ is a function of the local coordinate system $(\vec{a}, \vec{\beta})$ (see Figure 5.29) with a parameter γ and is defined as

$$\vec{\theta}(\gamma) = (\cos \gamma)\vec{a} + (\sin \gamma)\vec{\beta}, \quad -\frac{\pi}{2} < \gamma < \frac{\pi}{2}. \tag{5.54}$$

The unit vector $\vec{a}$ is defined as the direction from the focal point to the reconstruction point. The unit vector $\vec{\beta}$ is the filtering direction on the detector plane (see Figure 5.30).

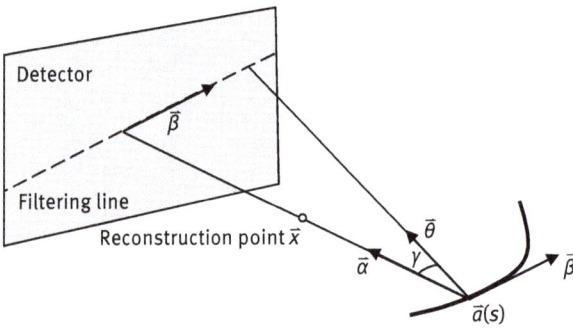

Fig. 5.29: The coordinate system for Katsevich's helical orbit cone-beam algorithm.

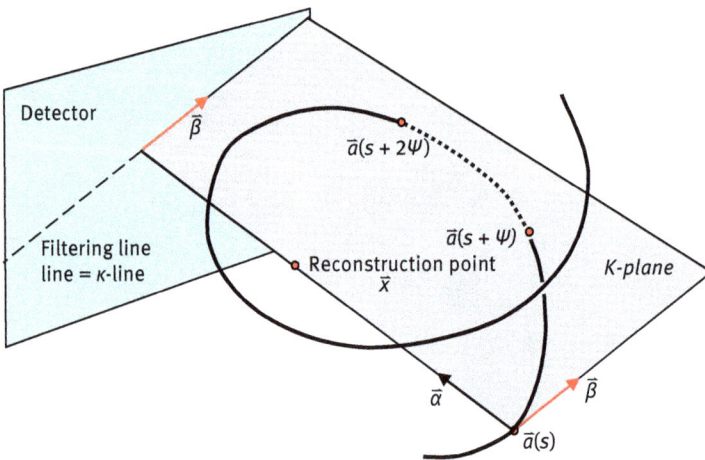

Fig. 5.30: The κ-plane contains the reconstruction point and three uniformly spaced points on the helix within the section governed by the π-segment.

Generally speaking, filtering can be any kind of processing on the data. In Katsevich's formula, when we say "filtering" we specifically mean the Hilbert transform. Vectors $\vec{a}$ and $\vec{\beta}$ are orthogonal to each other. Katsevich's formula is given as follows:

$$f(\vec{x}) = \frac{-1}{2\pi^2} \int\limits_{I_\pi(\vec{x})} \frac{1}{\|\vec{x} - \vec{a}(s)\|} \int\limits_{-\pi/2}^{\pi/2} \left. \frac{\partial g\left(\vec{\theta}(\gamma), \vec{a}(q)\right)}{\partial q} \right|_{q=s} \frac{1}{\sin \gamma} d\gamma ds, \tag{5.55}$$

where the integral over γ is the Hilbert transform and the integral over s is the cone-beam backprojection. Katsevich derived this nice and clean formula because he thought of a trick to assign the weights for redundant plane integrals by selecting a special filtering direction $\vec{\beta}$. It can be shown that his assignment of $\vec{\beta}$ makes the weight +1 for a plane that is measured one time and makes the weight +1 or –1 for a plane that is measured three times (two of them get +1 and one of them gets –1). The solution of $\vec{\beta}$ is not unique. Different selections of the filtering direction give different algorithms.

Here is one way to find a special filtering direction $\vec{\beta}$. Let us define a κ-plane as follows. The focal-point location is $\vec{a}(s)$. A point in the field of view $\vec{x}$ is to be reconstructed. We then find an angle ψ in $(-\pi, \pi)$ such that the four points $\vec{x}$, $\vec{a}(s)$, $\vec{a}(s + \psi)$, and $\vec{a}(s + 2\psi)$ are in one plane (see Figure 5.30). This plane exists but is not unique. The angle ψ with the smallest magnitude $|\psi|$ that can construct this plane will be chosen, and the corresponding plane that contains these four points is referred to as a κ-plane $\kappa(s, \psi)$.

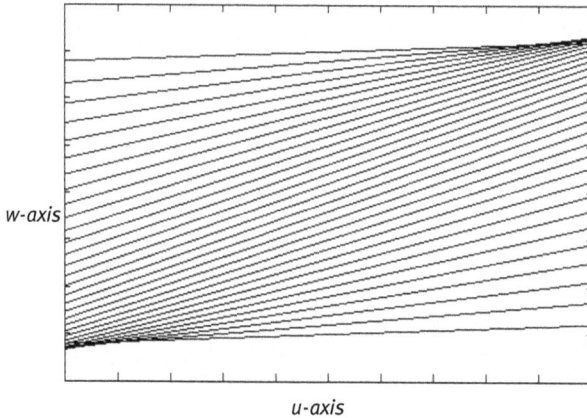

w-axis

u-axis

Fig. 5.31: The κ-lines on the flat cone-beam detector. Each line corresponds to a fixed angle ψ. Each reconstruction point associates with one and only one κ-line.

The intersection of a κ-plane $\kappa(s, \psi)$ with the detector is called a κ-line. The filtering direction $\vec{\beta}$ is along the κ-line. If you let the point $\vec{x}$ vary, you will get a bunch of these κ-planes $\kappa(s, \psi)$, and you will get a bunch of κ-lines on one detector (see Figure 5.31).

On a flat detector, the κ-lines are straight lines. Let us assign a u–w coordinate system to the detector plane as indicated in Figure 5.31 with the w-axis being the helix axis (i.e., the z-axis). The κ-lines can be described by the u–w relation for a fixed ψ:

$$w = \frac{Dh}{2\pi R}\left(\psi + \frac{\psi}{\tan\psi}\frac{u}{D}\right), \tag{5.56}$$

where R is the radius of the helix, h is the pitch of the helix, and D is the distance between the detector plane and the focal point. There are many lines depicted in Figure 5.31, and each line corresponds to a fixed ψ value.

If a curved detector is used, the κ-lines are no longer straight lines on the detector but are curves. For a fixed ψ, the α–w relationship is given as

$$w = \frac{Dh}{2\pi R}\left(\psi\cos\alpha + \frac{\psi}{\tan\psi}\sin\alpha\right), \tag{5.57}$$

where the angle α is defined in Figure 5.32. We can plot the curved κ-lines using this relationship for a set of ψ values to obtain a counterpart of Figure 5.31 for the curved detector (see Figure 5.33).

Finally, we will explain a little more on the curved detector implementation of Katsevich's algorithm, which is again given below:

$$f(\vec{x}) = \frac{-1}{2\pi^2}\int_{I_n(\vec{x})}\frac{1}{\|\vec{x}-\vec{a}(s)\|}\int_{-\pi/2}^{\pi/2}\left.\frac{\partial g\left(\vec{\theta}(\gamma),\vec{a}(q)\right)}{\partial q}\right|_{q=s}\frac{1}{\sin\gamma}d\gamma ds. \tag{5.58}$$

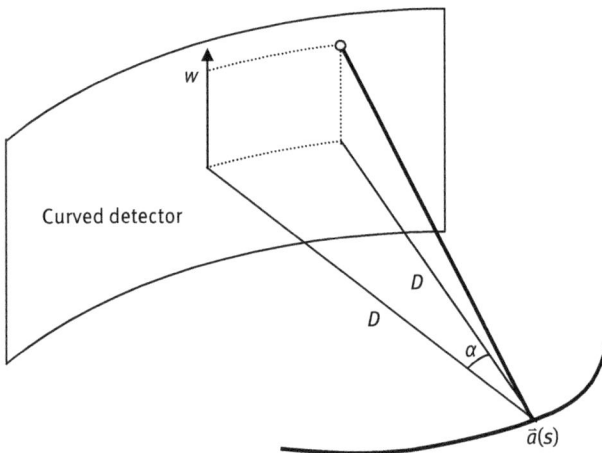

Fig. 5.32: The coordinate system for a curved detector cone-beam geometry.

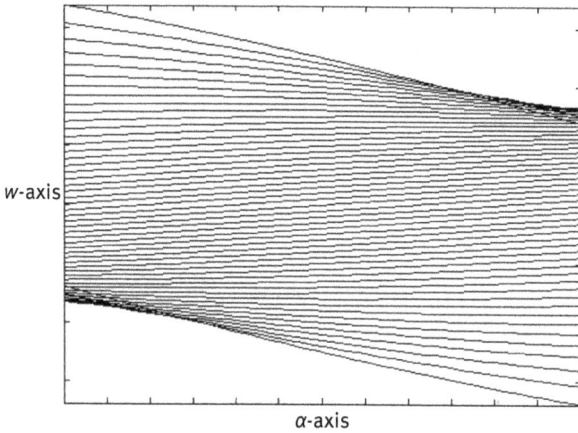

Fig. 5.33: The κ-curves on the curved cone-beam detector. Each curve corresponds to a fixed angle ψ. Each reconstruction point associates with one and only one κ-curve.

Step 1: We take the derivative at a constant direction $\vec{\theta}$ with respect to the orbit parameter s, that is, to evaluate $\frac{\partial g\left(\vec{\theta}, \vec{a}(s)\right)}{\partial s}$.

In practice, the projection data are sampled at discrete focal-point locations with a discrete detector. The derivative will be implemented as a finite difference using pairs of consecutive projections at focal-point locations $\vec{a}(s_k)$ and $\vec{a}(s_{k+1})$. When evaluating the difference, the pair of projection rays $g(\vec{\theta}, \vec{a}(s_k))$ and $g(\vec{\theta}, \vec{a}(s_{k+1}))$ must have the same (global) direction $\vec{\theta}$ and the same w coordinate on the detector, as illustrated in Figure 5.34.

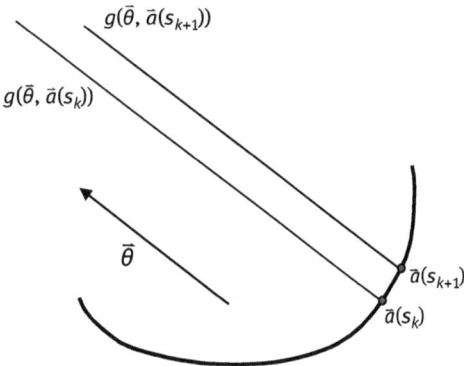

Fig. 5.34: Two neighboring detector views are required to implement the derivative with respect to the orbit parameter s as the difference. When taking the finite difference, the two rays must be parallel and have the same w coordinate.

In the detector α–w coordinates, $g(\vec{\theta}, \vec{a}(s_k))$ and $g(\vec{\theta}, \vec{a}(s_{k+1}))$ will have different α values. If you let us abuse the notation a little and use the detector coordinate system, we let

$$g\left(\alpha - \frac{\Delta s}{2}, w, \vec{a}(s_k)\right) = g\left(\vec{\theta}, \vec{a}(s_k)\right) \tag{5.59}$$

and

$$g\left(\alpha - \frac{\Delta s}{2}, w, \vec{a}(s_{k+1})\right) = g\left(\vec{\theta}, \vec{a}(s_{k+1})\right), \tag{5.60}$$

then we approximate the derivative as

$$\frac{\partial g\left((\alpha, w, \vec{a}\left(s_{k+\frac{1}{2}}\right)\right)}{\partial s} \approx \frac{\partial g\left(\alpha + \frac{\Delta s}{2}, w, \vec{a}(s_{k+1})\right) - g\left(\alpha - \frac{\Delta s}{2}, w, \vec{a}(s_k)\right)}{\Delta s}, \tag{5.61}$$

where $\Delta s = s_{k+1} - s_k$.

Step 2: The result of Step 1 is weighed by the cosine function $d/\sqrt{D^2 + w^2}$, obtaining

$$\frac{D}{\sqrt{D^2 + w^2}} \frac{\partial g\left(\alpha, w, \vec{a}\left(s_{k+\frac{1}{2}}\right)\right)}{\partial s}. \tag{5.62}$$

This scaling is also called the length-correction weighting. This is a point-by-point scaling on the detector.

Step 3: To prepare for Hilbert transform operation, we rebin the κ-curves to horizontal lines, mapping the α–w detector representation to α–ψ representation (see Figure 5.35). In other words, we move each data point on the α–w detector up or down according to $w = \frac{Dh}{2\pi R}\left(\psi \cos \alpha + \frac{\psi}{\tan \psi} \sin \alpha\right)$ so that the κ-curves become horizontal lines.

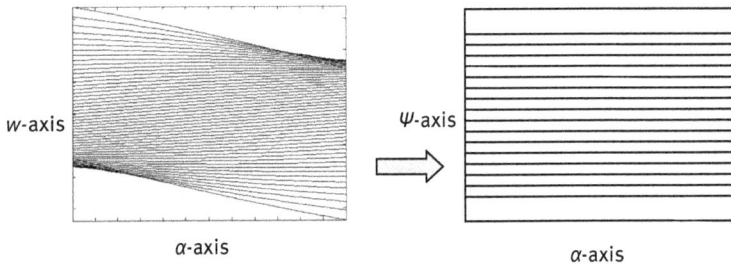

w-axis

α-axis

ψ-axis

α-axis

Fig. 5.35: Rebin the κ-curves into horizontally parallel lines.

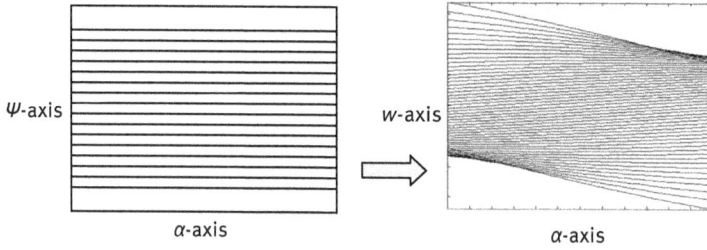

Fig. 5.36: Rebin the parallel lines back to κ-curves.

Step 4: After rebinning, the regular line-by-line 1D Hilbert transform is performed.

Step 5: This step is the inverse of Step 3. It rebins the filtered (i.e., Hilbert transformed) data back to the original detector a–w coordinates with curved κ-lines (see Figure 5.36).

Step 6: We weigh the data on the curved detector plane by a cosine function cos α, for all different a locations. The angle α is defined in Figure 5.32.

Step 7: This is the final step, and it performs the ray-by-ray 3D cone-beam backprojection with the weighting factor $1/\|\vec{x} - \vec{a}(s)\|$ using a helix orbit. This step is similar to the backprojection in the Feldkamp algorithm, where a circular orbit is used.

5.5 Radon transform and ray transform in *n* dimensions

Even though the focus of this chapter is on the three-dimensional real space, the concepts of Radon transform and ray transform can be readily extended into the n-dimensional real space. In 2D, the Radon transform and the ray transform are the same. They are both the parallel line integrals.

In 3D, they are different. The 3D Radon transform is the parallel planar integrals, while the 3D ray transform is the parallel line integrals. In n-D, the Radon transform is the parallel $(n$-1$)$-D hyperplane integrals of an n-D function f. Regardless the value of n, the Radon transform, $p(s, \vec{\theta})$, is always a 1D function in s, which is the along the direction of the 1D detector. Here, $\vec{\theta}$ is a unit directional vector in the n-D space and indicates the direction of the 1D detector. On the other hand, the ray transform, $p(\vec{s}, \vec{\theta})$, is $(n$-1$)$-D function in $\vec{s}$, which is the location on the $(n$-1$)$-D hyperplane detector. Here, $\vec{\theta}$ is a unit vector in the n-D space and indicates the normal direction of the $(n$-1$)$-D hyperplane detector.

The inversion formula for the n-D Radon transform is in the form of FBP. If n is an odd number, the filter is the $(n$-1$)$th-order 1D derivative operator with respect to the variable s. For the 3D Radon transform, the tomographic filter is the second-order derivative with respect to s, as presented in Section 5.4.3.

If n is an even number, the filter is a 1D ramp filter combined with the $(n\text{-}2)$th-order 1D derivative operator with respect to the variable s. For the 2D Radon transform, the tomographic filter is just the ramp filter. The ramp filter is the combination of the first-order derivative and the Hilbert transform.

As for the n-D ray transform, the inversion formula will depend on the imaging geometry similar to the situation in Section 5.4.2.

5.6 Worked examples

Example 1: A SPECT camera is mounted with a parallel-beam collimator which has a 30° tilt angle as shown in Figure 5.37. The camera rotates around the patient to collect projections. Does this imaging geometry satisfy Orlov's condition?

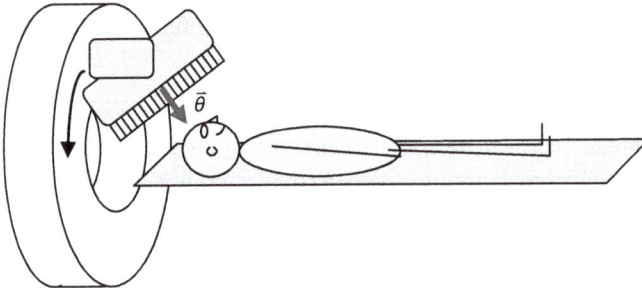

Fig. 5.37: The SPECT detector is tilted.

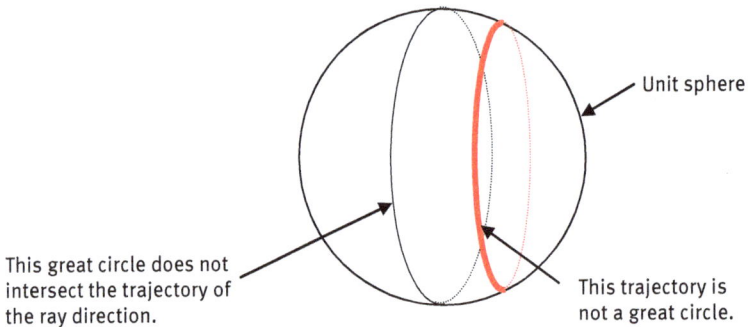

This great circle does not intersect the trajectory of the ray direction.

Unit sphere

This trajectory is not a great circle.

Fig. 5.38: Orlov's condition is not met.

solution

No. If we draw the trajectory of the projection direction $\vec{\theta}$ on a unit sphere, we see that the trajectory is a small circle (see Figure 5.38). This does not satisfy Orlov's condition.

Example 2: State the central slice theorem with a mathematical expression for the 3D line-integral projections and for the 3D plane-integral projections.

Solution

(i) *Three-dimensional plane-integral data*

Let $\vec{\theta} = (\sin\theta\cos\phi, \sin\theta, \cos\theta)$, then the central slice theorem for the 3D Radon transform (i.e., plane-integral projections) is

$$P\left(\omega, \vec{\theta}\right) = F(\omega\sin\theta\cos\phi, \omega\sin\theta\sin\phi, \omega\cos\theta), \tag{5.63}$$

where $P(w, \vec{\theta})$ is the 1D Fourier transform of the plane integrals $p(s, \vec{\theta})$ with respect to s, and $F(\omega_x, \omega_y, \omega_z)$ is the 3D Fourier transform of the object $f(x, y, z)$.

(ii) *Three-dimensional line-integral data*

For the 3D line-integral projections, we need a coordinate system on the detector plane which is perpendicular to $\vec{\theta}$. Let

$$\begin{aligned}\vec{\theta}_u &= (-\sin\phi, \cos\phi, 0) \quad \text{and} \\ \vec{\theta}_v &= (-\cos\theta\cos\phi, -\cos\theta\sin\phi, \sin\theta),\end{aligned} \tag{5.64}$$

then $\vec{\theta}$, $\vec{\theta}_u$, and $\vec{\theta}_v$ form an orthogonal system, $\vec{\theta}_u$ is in the direction of the u-axis, and $\vec{\theta}_v$ is in the direction of the v-axis. Thus, the central slice theorem can be stated as

$$\begin{aligned}P\left(\omega_u, \omega_v, \vec{\theta}\right) &= F(-\omega_u\sin\phi - \omega_v\cos\theta\cos\phi, \omega_u\cos\phi \\ &\quad - \omega_v\cos\theta\sin\phi, \omega_v\sin\theta) \\ &= F\left(\omega_u\vec{\theta}_u + \omega_v\vec{\theta}_v\right),\end{aligned} \tag{5.65}$$

where $P(\omega_u, \omega_v, \vec{\theta})$ is the 2D Fourier transform of the line integrals $p(u, v, \vec{\theta})$ with respect to u and v.

Example 3: Run Feldkamp's algorithm and observe the reconstruction artifacts by varying the cone angle.

Solution

We programmed Feldkamp's algorithm and did a set of computer simulations using a circular focal-point orbit and a Defrise phantom, which consists of five flat uniform ellipsoids (see Figure 5.39). Six different cone angles (2°, 4°, 8°, 16°, 32°, and 64°) were used in the simulations. The central sagittal view for each simulation is displayed in Figure 5.40. It is observed that for large cone angles, Feldkamp's algorithm introduces severe artifacts, especially in the regions away from the central plane (i.e., the orbit plane). The images are quite good for small cone angles.

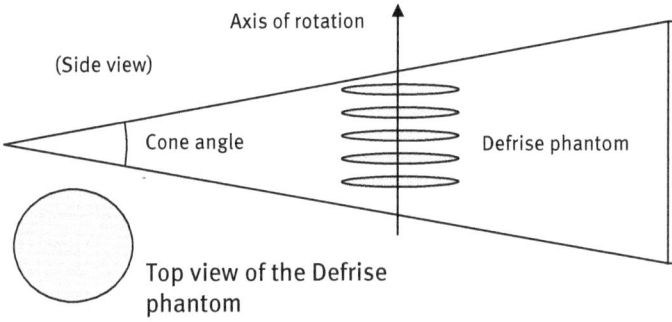

Fig. 5.39: A Defrise phantom is used in computer simulations.

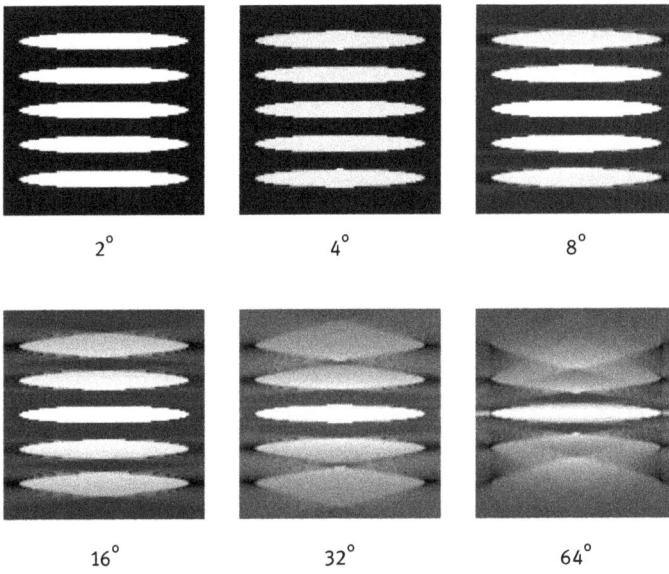

Fig. 5.40: Computer simulation results with Feldkamp's algorithm using different cone angles.

5.7 Summary

- In 3D, the parallel line integrals are referred to as the ray transform, and the parallel plane integrals are referred to as the Radon transform.
- For the Radon projection data, we require that all directions in a 2π solid angle should be measured.
- For the Radon projection data, the image reconstruction algorithm (i.e., the Radon inversion formula) is very simple – a second-order derivative followed by the Radon backprojection. Of course, one can switch the order of derivative and backprojection. If the backprojection is performed first, it should be followed by a Laplacian operator, which is a summation of the second-order partial derivative with respect to x, y, and z, respectively.
- An efficient way to perform Radon backprojection is to do it in two steps, and each step is a series of 2D backprojections.
- For the ray transform data, we require that Orlov's data sufficiency condition be satisfied. The ray directions trace a trajectory on the unit sphere. If every great circle intersects this trajectory, Orlov's condition is met.
- The image reconstruction algorithm for the ray transform depends on the ray direction trajectory. Due to data redundancy, the reconstruction algorithm is not unique. One can either do filtering first or backprojection first.
- Feldkamp et al. developed a simple and robust FBP algorithm for cone-beam circular orbit imaging. Even though this algorithm is a modification of a fan-beam's FBP algorithm and is not exact, it has wide applications in many fields. The reconstruction errors are not significant if the cone angle is small enough.
- One can use Tuy's condition to verify if the cone-beam imaging geometry is able to provide sufficient projections. Nonplanar orbits, such as the helix orbit or the circle-and-line orbit, are required to satisfy Tuy's condition. Tuy developed a relationship between the cone-beam data and the original image; he also developed a cone-beam inversion formula, but it is difficult to use.
- Grangeat's relationship is that the angular derivative of the cone-beam-weighted planar integral is equal to the derivative of the Radon planar integral. In Grangeat's cone-beam image reconstruction algorithm, the image is reconstructed using the Radon inversion formula. A drawback of Grangeat's cone-beam reconstruction method is the rebinning from the cone-beam data to the Radon data. The rebinning step can cause large errors.
- Katsevich's cone-beam image reconstruction algorithm is truly an FBP algorithm with shift-invariant filtering and cone-beam backprojection. One drawback of Katsevich's algorithm is its difficulty in selection of filtering directions. Another drawback is that the cone-beam projection data are not fully used.
- The readers are expected to understand the Radon inversion formula and Feldkamp's cone-beam image reconstruction algorithm in this chapter.

Problems

? *Problem 5.1* Calculate the 3D parallel line integrals $p(u, v, \vec{\theta})$ and parallel plane integrals $p(s, \vec{\theta})$ of a uniform ball, in which the line density and area density are both 1. The center of the ball is at the origin of the coordinate system, and the radius of the uniform ball is R.

Problem 5.2 A cone-beam focal-point orbit is a circle with two lines as shown. The radius of the circular orbit is R. The object to be imaged is a ball of radius r. Determine the length of the linear orbits so that Tuy's condition can be satisfied.

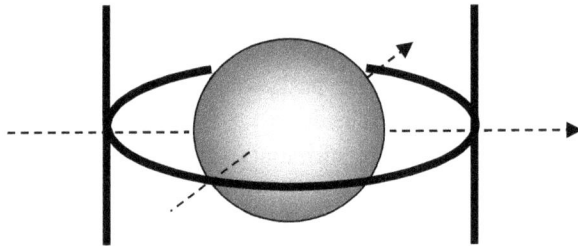

Problem 5.3 Prove that Feldkamp's algorithm can give an exact reconstruction for the object $f(x, y, z)$ that is constant in the axial direction (i.e., z-direction). In other words, for any given point (x_0, y_0), the function $f(x_0, y_0, z)$ does not vary with variable z.

Bibliography

[1] Axelsson C, Danielsson PE (1994) Three-dimensional reconstruction from cone-beam data in $O(N^3\log N)$ time. Phys Med Biol 39:477–491.

[2] Barrett HH, Swindell W (1981) Radiological Imaging, Academic Press, New York.

[3] Chen GH (2003) An alternative derivation of Katsevich's cone-beam reconstruction formula. Med Phys 30:3217–3226.

[4] Clack R (1992) Towards a complete description of three-dimensional filtered backprojection. Phys Med Biol 37:645–660.

[5] Clack R, Defrise M (1994) Overview of reconstruction algorithms for exact cone-beam tomography. Proc SPIE 2299:230–241.

[6] Colsher JG (1980) Fully three-dimensional positron emission tomography. Phys Med Biol 25:103–115.

[7] Crawford CR (1991) CT filtration aliasing artifacts. IEEE Trans Med Imaging 10:99–102.

[8] Deans SR (1983) The Radon Transform and Some of Its Applications, John Wiley, New York.

[9] Defrise M (1995) A factorization method for the 3D X-ray transform. Inverse Probl 11:983–994.

[10] Defrise M, Clack R (1994) A cone-beam reconstruction algorithm using shift-variant filtering and cone-beam backprojection. IEEE Trans Med Imaging 13:186–195.

[11] Defrise M, Clack R, Townsend D (1993) Solution to the three-dimensional image reconstruction problem from two-dimensional projections. J Opt Soc A 10:869–877.

[12] Defrise M, Clack R, Townsend D (1995) Image reconstruction from truncated, two-dimensional parallel projections. Inverse Probl 11:287–313.

[13] Defrise M, Kinahan PE, Townsend DW, Michel C, Sibomana M, Newport DF (1997) Exact and approximate rebinning algorithms for 3-D PET data. IEEE Trans Med Imaging 16:145–158.

[14] Feldkamp LA, Davis LC, Kress JW (1984) Practical cone beam algorithm. J Opt Soc Am A 1:612–619.

[15] Grangeat P (1991) Mathematical framework of cone beam 3D reconstruction via the first derivative of the Radon transform. In: Herman G, Luis AK, Natterer F (eds) Mathematical Methods in Tomography, Lecture Notes Math, Springer, Berlin, Heidelberg, 1497:66–97.

[16] Katsevich A (2002a) Theoretically exact filtered backprojection-type inversion algorithm for spiral CT. SIAM J Appl Math 62:2012–2026.

[17] Katsevich A (2002b) Analysis of an exact inversion algorithm for spiral cone-beam CT. Phys Med Biol 47:2583–2597.

[18] Katsevich A (2003) An improved exact filtered backprojection algorithm for spiral computed tomography. Adv Appl Math 32:681–697.

[19] Kinahan PE, Rogers JG, Harrop R, Johnson RR (1988) Three-dimensional image reconstruction in object space. IEEE Trans Nucl Sci 35:635–640.

[20] Kudo H, Saito T (1994) Derivation and implementation of a cone-beam reconstruction algorithm for non-planar orbit. IEEE Trans Med Imaging 13:196–211.

[21] Natterer F (1986) The Mathematics of Computerized Tomography, John Wiley, New York.

[22] Noo F, Pack J, Heuscher D (2003) Exact helical reconstruction using native cone-beam geometries. Phys Med Biol 48:3787–3818.

[23] Orlov SS (1976a) Theory of three-dimensional image reconstruction I. Conditions for a complete set of projections. Sov Phys Crystallogr 20:429–433.

[24] Orlov SS (1976b) Theory of three-dimensional image reconstruction II. The recovery operator. Sov Phys Crystallogr 20:429–433.

[25] Pack J, Noo F, Clackdoyle (2005) Cone-beam reconstruction using the backprojection of locally filtered projections. IEEE Trans Med Imaging 24:70–85.

[26] Proksa R, Kohler T, Grass M, Timmer J (2000) The n-PI-method for helical cone-beam CT. IEEE Trans Med Imaging 19:848–863.

[27] Ra JB, Jim CB, Cho ZH, Hilal SK, Correll J (1992) A true 3D reconstruction algorithm for the spherical positron tomography. Phys Med Biol 27:37–50.

[28] Schaller S, Noo F, Sauer F, Tam KC, Lauritsch G, Flohr T (2000) Exact Radon rebinning algorithm for the long object problem in helical cone-beam CT. IEEE Trans Med Imaging 19:822–834.

[29] Smith BD (1985) Image reconstruction from cone-beam projection: Necessary and sufficient conditions and reconstruction methods. IEEE Trans Med Imaging MI 4:14–25.

[30] Stazyk M, Rogers J, Harrop R (1992) Analytic image reconstruction in PVI using the 3D Radon transform. Phys Med Biol 37:689–704.

[31] Stearns CW, Chesler DA, Brownell GL (1990) Accelerated image reconstruction for a cylindrical positron tomography using Fourier domain methods. IEEE Trans Nucl Sci 37:773–777.

[32] Stearns CW, Crawford CR, Hu H (1994) Oversampled filters for quantitative volumetric PET reconstruction. Phys Med Biol 39:381–388.

[33] Taguchi K, Aradate H (1998) Algorithm for image reconstruction in multi-slice helical CT. Med Phys 25:550–561.

[34] Tam KC, Samarasekera S, Sauer F (1998) Exact cone-beam CT with a spiral scan. Phys Med Biol 43:1015–1024.

[35] Turbell H, Danielsson PE (2000) Helical cone-beam tomography. Int J Imaging Syst Technol 11:91–100.

[36] Tuy HK (1983) An inverse formula for cone-beam reconstruction. SIAM J Appl Math 43:546–552.

[37] Wang G, Vannier MW (1993) Helical CT image noise-analytical results. Med Phys 20:1653–1640.

[38] Wang G, Lin TH, Cheng P, Shinozaki DM (1993) A general cone-beam reconstruction algorithm. IEEE Trans Med Imaging 12:486–496.

[39] Wang G, Ye Y, Yu H (2007) Approximate and exact cone-beam reconstruction with standard and non-standard spiral scanning. Phys Biol Med 52:R1–R13.

[40] Ye Y, Zhao S, Yu H, Wang G (2005) A general exact reconstruction for cone-beam CT via backprojection-filtration. IEEE Trans Med Imaging 24:1190–1198.

[41] Zeng GL, Gullberg GT (1992) A cone-beam tomography algorithm for orthogonal circle-and-line orbit. Phys Med Biol 37:563–577.

[42] Zeng GL, Clack R, Gullberg GT (1994) Implementation of Tuy's inversion formula. Phys Med Biol 39:493–507.

[43] Zhuang TL, Leng S, Nett BE, Chen GH (2004) Fan-beam and cone-beam image reconstruction via filtering the backprojection image of differentiated data. Phys Med Biol 49:5489–5503.

[44] Zou Y, Pan X (2004) An extended data function and its generalized backprojection for image reconstruction in helical cone-beam CT. Phys Med Biol 49:N383–N387.

[45] Zou Y, Pan X, Xia D, Wang G (2005) PI-line-based image reconstruction in helical cone-beam computed tomography with a variable pitch. Med Phys 32:2639–2648.

6 Iterative reconstruction

Previous chapters dealt with analytic image reconstruction algorithms. This chapter, on the other hand, introduces iterative image reconstruction algorithms. Due to high-speed computers, iterative algorithms get more and more attention in medical image reconstruction. This chapter describes the imaging problem as a system of linear equations and reconstructs an image by minimizing an objective function. Many algorithms are available to solve the system of linear equations or to minimize an objective function. The objective function can be set up by using the likelihood function and can also include the prior knowledge about the image. The likelihood function models the noise distribution in the projection measurements. The maximum-likelihood expectation-maximization (ML-EM) algorithm or ordered-subset expectation-maximization (OS-EM) algorithm is the most popular iterative image reconstruction algorithm in emission tomography, and this chapter has devoted significant efforts to it. Many strategies for noise control are discussed. This chapter also presents a research hot spot – image reconstruction with highly undersampled data, which is often referred to as compressed sensing and is, in fact, nothing but another application of Bayesian image reconstruction.

6.1 Solving a system of linear equations

Instead of using an analytical algorithm to reconstruct an image, image reconstruction can also be obtained by solving a system of linear equations. In doing so, the image is first discretized into pixels or voxels (volumetric pixels) as illustrated in Figure 6.1.

Here, the image pixels x_j ($j = 1, 2, \ldots$) are labeled in a 1D sequential order, as are all projections p_i ($i = 1, 2, \ldots$). For the simple example in Figure 6.1, we can relate the image pixels and the projections using a system of linear equations:

$$
\begin{aligned}
x_1 + x_2 + x_3 &= p_1, \\
x_4 + x_5 + x_6 &= p_2, \\
x_7 + x_8 + x_9 &= p_3, \\
x_3 + x_6 + x_9 &= p_4, \\
x_2 + x_5 + x_8 &= p_5, \\
x_1 + x_4 + x_7 &= p_6, \\
2(\sqrt{2}-1)x_4 + (\sqrt{2}-1)x_7 + 2(\sqrt{2}-1)x_8 &= p_7, \\
\sqrt{2}x_1 + \sqrt{2}x_5 + \sqrt{2}x_9 &= p_8, \\
2(\sqrt{2}-1)x_2 + (\sqrt{2}-1)x_3 + 2(\sqrt{2}-1)x_6 &= p_9.
\end{aligned}
\tag{6.1}
$$

This system can be rewritten in the matrix form as

https://doi.org/10.1515/9783111055404-006

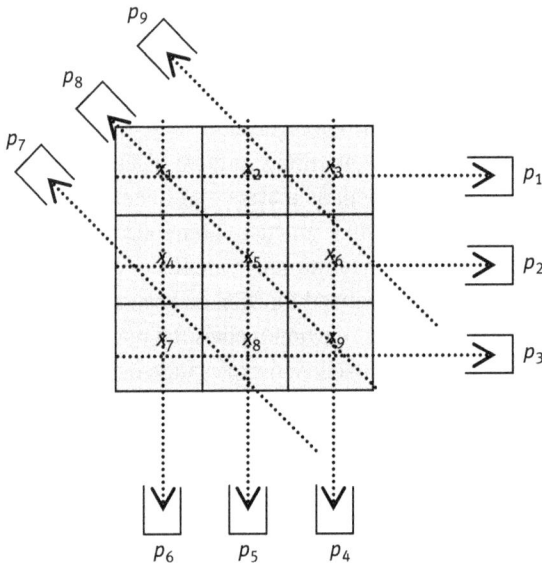

Fig. 6.1: An example with nine unknowns and nine measurements.

$$AX = P, \tag{6.2}$$

where $X = [x_1, x_2, \ldots, x_9]^T$, $P = [p_1, p_2, \ldots, p_9]^T$, and A is the coefficient matrix of the system. The element a_{ij} in A represents the weight of the contribution of the jth pixel x_j to the ith projection p_i. In this example, the contribution is the segment length of the projection ray within the pixel. If the inverse matrix A^{-1} of A exists, the reconstructed image is given as

$$X = A^{-1}P. \tag{6.3}$$

Line length is not the only way to model the "contribution." Some imaging physics (e.g., attenuation and point spread function) can also be included as well.

For a practical imaging problem, the matrix A is not square. In this case, a generalized inverse of the matrix can be used. For example, we can find a least-squares solution:

$$X = \left(A^T A\right)^{-1} A^T P, \quad \text{if the system is overdetermined;} \tag{6.4}$$

$$X = A^T \left(A A^T\right)^{-1} P, \quad \text{if the system is under-determined.} \tag{6.5}$$

A generalized inverse can be obtained via a least-squares minimization. In the case that the system is overdetermined (i.e., the number of projection rays is greater than the number of image pixels), we let

$$x^2 = \|AX - P\|^2 = (AX - P)^T(AX - P)$$
$$= X^A A^T AX - 2X^T A^T P + P^T P \tag{6.6}$$

and set the partial derivatives (i.e., gradient) to zero:

$$\vec{0} = \nabla x^2 = 2A^T AX - 2A^T P. \tag{6.7}$$

Rearranging the terms, we have

$$A^T AX = A^T P, \tag{6.8}$$

which is a set of normal equations because $(AX - P)$ is orthogonal (i.e., normal) to the rows of A, that is, $A^T(AX - P) = \vec{0}$. Solving the normal equations immediately yields a generalized solution

$$X = (A^T A)^{-1} A^T P. \tag{6.9}$$

On the other hand, in the case that the system is underdetermined (i.e., the number of image pixels is greater than the number of projection rays), the system will have infinite number of solutions for $AX = P$, assuming that the system is consistent. In this case, we would like to choose the minimum norm solution. Therefore, we use the method of Lagrange multipliers to minimize $\|X\|^2$ subject to $AX = P$. We thus set up a Lagrange function

$$L = \|X\|^2 + \Lambda^T(AX - P), \tag{6.10}$$

with a row matrix $\Lambda^T = [\lambda_1, \lambda_2, \ldots, \lambda_m]$ containing the Lagrange multipliers $\lambda_1, \lambda_2, \ldots, \lambda_m$ and m being the number of projection rays.

Setting the partial derivatives (i.e., gradient) of the Lagrange function to zero yields

$$\vec{0} = 2X + A^T \Lambda \text{ and } AX = P. \tag{6.11}$$

Pre-multiplying with matrix A, $\vec{0} = 2X + A^T \Lambda$ becomes

$$\vec{0} = 2AX + AA^T \Lambda. \tag{6.12}$$

Solving for Λ and using $AX = P$, we have

$$\Lambda = -2(AA^T)^{-1}AX = -2(AA^T)^{-1}P. \tag{6.13}$$

Finally, solving for X from $\vec{0} = 2X + A^T \Lambda$ gives

$$X = -\frac{1}{2}A^T \Lambda = A^T(AA^T)^{-1}P. \tag{6.14}$$

Even if the matrix A is square, its inverse A^{-1} may not exist. When A is not full rank, A^{-1} does not exist. In fact, the matrix A for the example in Figure 6.1 is not full rank. One can easily check that the sum of the rows 1, 2, and 3 is the same as the sum of the rows 4, 5, and 6. If the matrix A is not full rank, we cannot even calculate $(AA^T)^{-1}$ or $(A^TA)^{-1}$. In all applications, the matrix A is not full rank and not square either, and the projections are not consistent due to noise. If the matrix A is rank deficient, you could use the singular value decomposition (SVD) to find a pseudo-inverse.

The SVD technique is a powerful and stable method to find a generalized inverse and diagnose the system condition. Now we use SVD to find the least-squares solution for $AX = P$ as follows.

Assume that matrix A has m rows and n columns and is denoted as $A_{m \times n}$. Using SVD, the matrix $A_{m \times n}$ can be decomposed into

$$A_{m \times n} = U_{m \times m} \Sigma V_{n \times n}^T, \tag{6.15}$$

where

$$V^T V = I_{n \times n}, \tag{6.16}$$

$$U^T U = I_{m \times m}, \tag{6.17}$$

and

$$\sum_{m \times n} = \begin{bmatrix} \text{diag}\{\sigma_l\} & 0 \\ 0 & 0 \end{bmatrix} \tag{6.18}$$

with the singular values arranged in the descending order:

$$\sigma_1 \geq \sigma_2 \geq \cdots \geq \sigma_i \geq \cdots \geq 0. \tag{6.19}$$

A generalized inverse (or pseudo-inverse) is defined as

$$A^+ = V \Sigma^+ U^T, \tag{6.20}$$

where

$$\Sigma_{n \times m}^+ = \begin{bmatrix} D_r & 0 \\ 0 & 0 \end{bmatrix}, \tag{6.21}$$

and the diagonal matrix D_r with a cutoff index r is defined as

$$D_r = \text{diag}\left\{ \frac{1}{\sigma_1}, \frac{1}{\sigma_2}, \ldots, \frac{1}{\sigma_r}, 0, \ldots, 0 \right\}. \tag{6.22}$$

The reconstructed image is given as

$$X = A^+ P = V \Sigma^+ U^T P. \tag{6.23}$$

In the SVD method, the user selects the cutoff index r. If a very small r is chosen, the resultant reconstructed image only contains low-frequency components. If a very large r is chosen, the resultant reconstructed image will contain high-frequency components and the image is noisy as well.

More often than not, the matrix A is too large to store in the computer; it can only be generated one row at a time when this row is used in solving the system of equations. Not every method that is able to solve a system of linear equations can be used here. For example, the methods based on diagonalizing the matrix A or transforming matrix A into an upper triangle matrix are not applicable. Any method that modifies the matrix A cannot be used. We can only use methods that use matrix A and its transposed matrix A^T. Therefore, iterative methods that only use A and A^T (but do not modify them) make sense in finding an approximate solution to our imaging problem.

An analytic reconstruction can be thought of as an "open-loop" system, while an iterative algorithm can be thought of as a "closed-loop" system. Each loop, referred to as an iteration, usually consists of a projection operation, a comparison of the projected data with the measured data, and a backprojection operation. The backprojection maps the data discrepancies from the projection space to the image space. The backprojected discrepancies will be used to modify the currently estimated image at each iteration (see Figure 6.2).

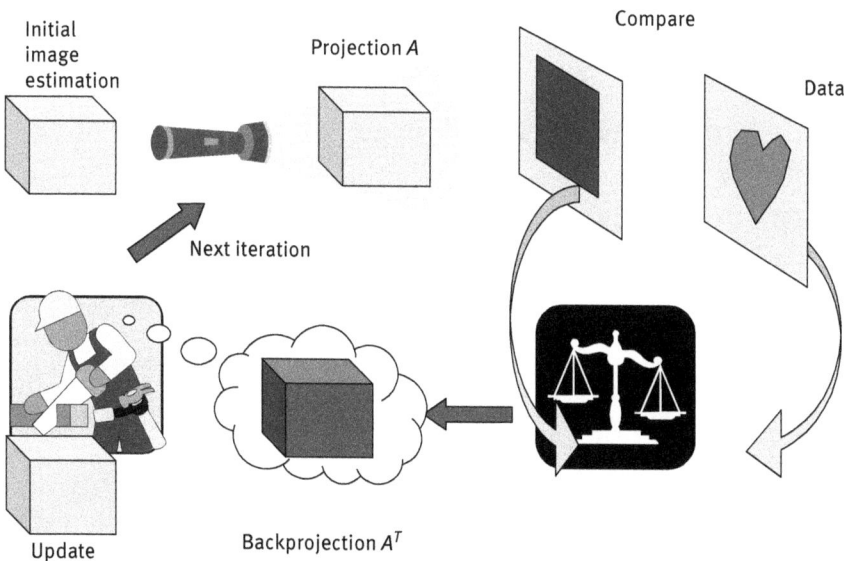

Fig. 6.2: A general procedure of an iterative image reconstruction algorithm.

6.2 Algebraic reconstruction technique

The main idea of the algebraic reconstruction technique (ART) algorithm (which is also known as the Kaczmarz method) is to make the estimated image satisfy one equation at a time as illustrated in Figure 6.3, where three lines (L_1, L_2, and L_3) represent three equations, and their intersection is the solution. In this example, the image only consists of two pixels.

In Figure 6.3, $\vec{x}^0$ is the initial guess of the solution. The first step is to project this point $\vec{x}^0$ perpendicularly onto L_1, obtaining $\vec{x}^1$. Next, project $\vec{x}^1$ perpendicularly onto L_2 to obtain $\vec{x}^2$, and so on, projecting each point onto a line (which is one equation) one at a time. Eventually, the algorithm will converge to the solution of the system of equations (see Figure 6.3, top). If the equations are not consistent, the algorithm will bounce around and never converge (see Figure 6.3, bottom). One iteration is defined as the procedure of going through all the equations once.

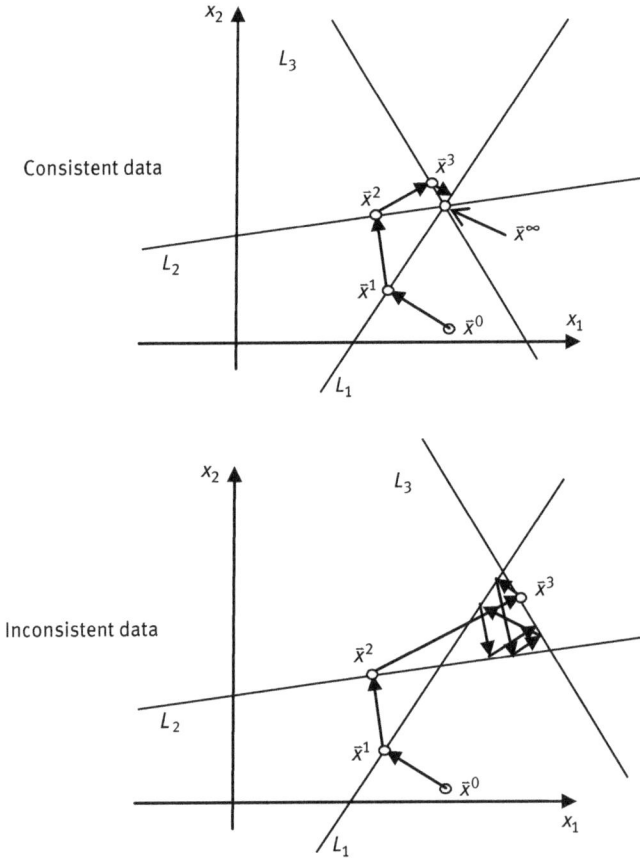

Fig. 6.3: The ART algorithm tries to satisfy each equation at each update.

The ART algorithm executed one projection ray at a time, and the image is updated after each ray is considered. Symbolically, the algorithm can be written as

$$\vec{x}^{\text{next}} = \vec{x}^{\text{current}} - \text{Backproject}_{\text{ray}} \left\{ \frac{\text{Project}_{\text{ray}} \left(\vec{x}^{\text{current}} \right) - \text{Measurement}_{\text{ray}}}{\text{Normalization factor}} \right\}. \qquad (6.24)$$

6.3 Gradient descent algorithms

First, an objective function χ^2 is formed based on the system of imaging equations:

$$\chi^2 = \|AX - P\|^2, \qquad (6.25)$$

which is a quadratic function (see Figure 6.4). Due to the noise, the equations are inconsistent. Thus, the minimum value of the objective function χ^2 usually has a non-zero, positive value.

6.3.1 The gradient descent algorithm

The strategy of gradient descent algorithms is to evaluate the gradient of the objective function χ^2 and use the gradient information to find the minimum of the objective function. The gradient in 1D is the derivative of the function. A positive gradient means an upward direction, and a negative gradient means a downward direction. The gradient descent algorithms take the opposite direction of the direction that is indicated by the gradient and use a small enough step size so that the algorithms can find the minimum (see Figure 6.5). The general form of a gradient descent algorithm looks like

$$\vec{x}^{\text{next}} = \vec{x}^{\text{current}} - a_{\text{current}} \vec{\Delta} \left(\vec{x}^{\text{current}} \right), \qquad (6.26)$$

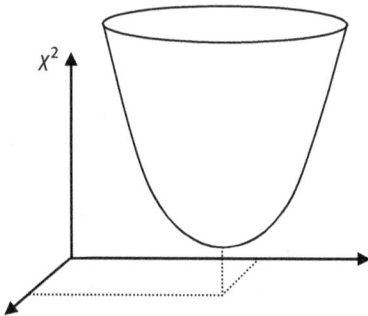

Fig. 6.4: A quadratic objective function.

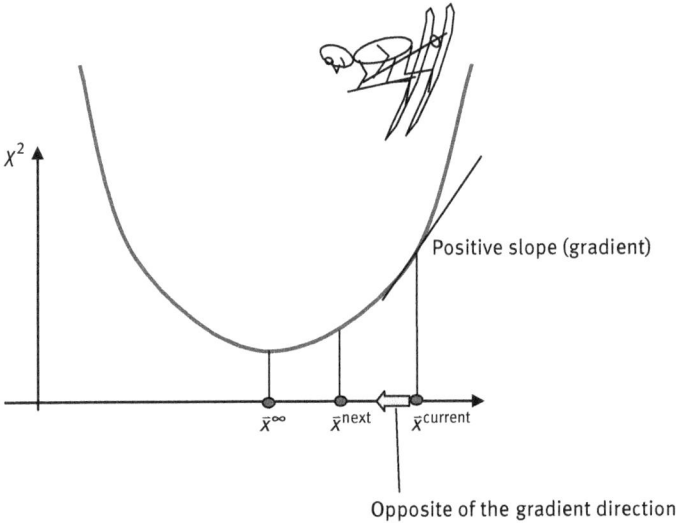

Fig. 6.5: The opposite direction of the gradient is the downhill direction.

where $\vec{\Delta}$ is the gradient of the objective function χ^2 at $\vec{x}^{\text{current}}$ and contains both projection and backprojection at all rays. In fact,

$$\vec{\Delta} = \nabla \|AX - P\|^2 = 2A^T(AX - P),\qquad(6.27)$$

where ∇ is the notation for the gradient operator, the projection AX is the multiplication of X by matrix A, and the backprojection is multiplication by matrix A^T. The data discrepancy is $(AX - P)$. The algorithm converges when $AX = P$ and X does not change any more. If the system is inconsistent, the algorithm converges when $A^T(AX - P) = \vec{0}$. In our notation, X and $\vec{x}$ are the same thing.

In the gradient descent algorithm, the step size α_{current} is calculated at each descent direction so that the quadratic objective function reaches its minimum along the chosen path. Most likely this minimum is not the global minimum of the objective function. At this point, the algorithm finds the current negative gradient direction and travels downhill to reach the next minimum along the new path. This procedure repeats over and over again until the algorithm terminates.

If the system is underdetermined, this least-squares problem does not have a unique solution, and the objective function χ^2 has a long valley (see Figure 6.6). The solution will depend upon the initial solution. If the initial solution is zero (i.e., $\vec{x}^0 = \vec{0}$), then the algorithm will converge to a minimum norm solution.

Due to noise, we seldom ever have $AX = P$ at convergence. Instead, we get a very noisy image when the iteration number is large. We apply an iterative reconstruction algorithm to noisy data generated with the phantom in Figure 2.10. As shown in Figure 6.7,

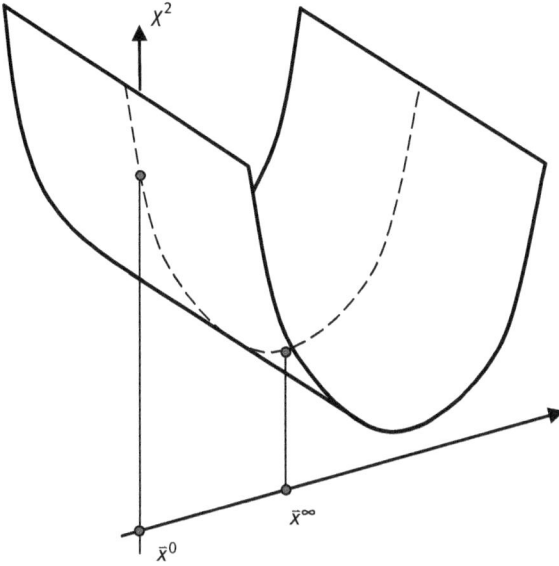

Fig. 6.6: For a degenerated system, the iterative algorithm solution depends on the initial value.

at the early iterations, the images only contain low-frequency components; at higher iterations, high-frequency components are recovered and noise comes into effect, too.

6.3.2 The Landweber algorithm

The Landweber algorithm is a simplified gradient descent algorithm, in which the step size is a fixed value determined by the user. Unlike the gradient decent algorithm, the Landweber algorithm does not necessarily reach the minimum in the decent direction at each iteration. However, the algorithm can converge if the step size is chosen small enough. How small is small enough? The step size α should satisfy

$$0 < \alpha < \frac{2}{\sigma_1^2}, \tag{6.28}$$

where σ_1 is the largest singular value of matrix A. Sometimes, it is advantageous to make the step size α variable, as a function of, for example, iteration number, noise weighting, and/or regions of interest.

6.3.3 The conjugate gradient algorithm

The gradient direction is easy to compute in a practical imaging problem using $\vec{\Delta} = \nabla$ $\|AX - P\|^2 = 2A^T(AX - P)$, but the negative gradient direction may not be the optimal direction to use in finding the optimum image. Let us look at the contour lines of a typical objective function χ^2 in Figure 6.8, where the contour lines are ellipses. The gradient direction at any point is perpendicular to the tangent of the ellipse. The searching directions of two consequential steps, $\vec{u}^{current}$ and $\vec{u}^{next}$, are orthogonal to each other (i.e., $\vec{u}^{current} \cdot \vec{u}^{next} = 0$). The searching path is zigzagging and not optimal.

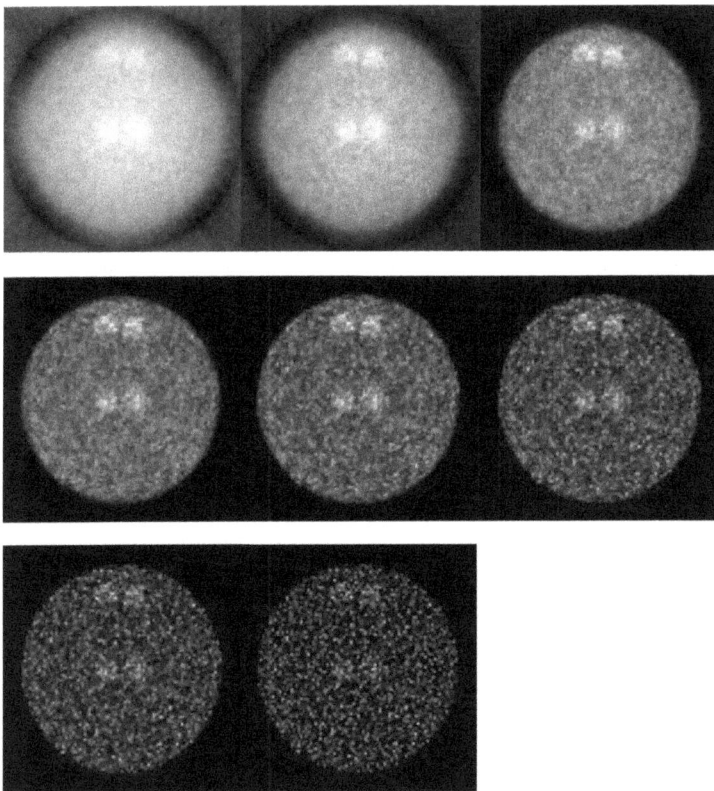

Fig. 6.7: The image gets noisier as the iteration number gets larger.

A better searching direction is to use the concept of *conjugate directions*. The conjugate directions are defined by $\vec{u}^{current} (A^TA) \vec{u}^{next} = 0$ (see Figure 6.9). The shape of the objective function χ^2 is characterized by (A^TA). When we use the conjugate directions, we actually first deform the contour lines into circles and then find the orthogonal directions. The conjugate directions make the algorithm converge faster.

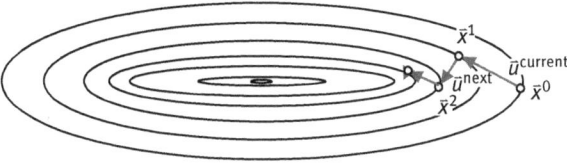

Fig. 6.8: The negative gradient direction may not be the most efficient way to find the function's minimum.

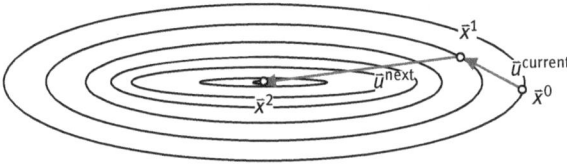

Fig. 6.9: The conjugate gradient direction is more effective than the gradient direction.

6.4 ML-EM algorithms

We do not have to use a least-squares objective function. There are many different ways to set up an objective function. If we use the Poisson noise model or simply use the nonnegativity constraint, we will get a special objective function. By minimizing that special objective function, a multiplicative updating algorithm known as the ML-EM algorithm is derived and can be symbolically expressed as

$$\vec{x}^{\,\text{next}} = \vec{x}^{\,\text{current}} \frac{\text{Backproject}\left\{\text{Measurement}/\text{Project}\left(\vec{x}^{\,\text{current}}\right)\right\}}{\text{Backproject}\left\{\vec{1}\right\}}, \tag{6.29}$$

where $\{\vec{1}\}$ is a vector with elements of 1s. The size of the vector is that of the projection data vector. In this algorithm, the data discrepancy is calculated as a ratio instead of a difference. The distinguishing feature of this algorithm is its nonnegativity. If the initial image $\vec{x}^{\,0}$ does not contain any negative pixels or voxels, the image values will never become negative.

Now let us explain the name of this algorithm: ML-EM. The objective function of this algorithm can be a likelihood function, which is the joint probability density function of Poisson random variables. We are looking for a solution (i.e., the reconstructed image) that can maximize this likelihood function. Therefore, this is a maximum likelihood algorithm.

When we try to maximize or minimize a function (e.g., an objective function or a likelihood function), we usually take the partial derivatives with respect to all of its unknowns (i.e., the pixel or voxel values), set these derivatives to zero, and solve for the unknowns. It turns out that our Poisson likelihood function is too complicated for us to optimize. We take the expectation value (or the statistical mean value) of the likelihood function. This is the "E" step, and it simplifies the problem significantly. We then find the maximum of the expected likelihood function. This is the "M" step. Therefore, we have the name "EM," the expectation-maximization.

This ML-EM algorithm is also called the Richardson–Lucy algorithm or Lucy–Richardson algorithm because Richardson and Lucy developed this algorithm for image deblurring applications in 1972 and 1974. There are many EM algorithms in different fields of research. The usual ML-EM algorithm is derived and used particularly for the emission data reconstruction. We also have transmission data ML-EM algorithms, too, but they are not as popular.

6.5 OS-EM algorithm

In ART, the image is updated after each projection ray is considered. On the other hand, in gradient descent methods and in the ML-EM algorithm, the image is updated only when all projection rays are considered. One way to speed up the convergence rate of an iterative algorithm is to make more frequent image updates.

In an OS-EM algorithm, the projection views are grouped in different sets (called subsets), the algorithm goes through the subsets in a specified order, and the image is updated after each subset is considered. Figure 6.10 shows an example of how the projection views are divided into subsets. There are many strategies for dividing the views into subsets.

Increasing the number of subsets accelerates the convergence rate but may increase the noise as well. Roughly speaking, if you have N subsets, you may accelerate the ML-EM algorithm about N times. Modest acceleration of approximately 10 times is possible with very little increase in noise.

6.6 Noise handling

Nowadays, many people choose an iterative image reconstruction algorithm over an analytical algorithm simply because the iterative algorithm in general can provide images containing less noise with the same or better resolution. We will investigate how the analytical and iterative algorithms handle the noise in this section.

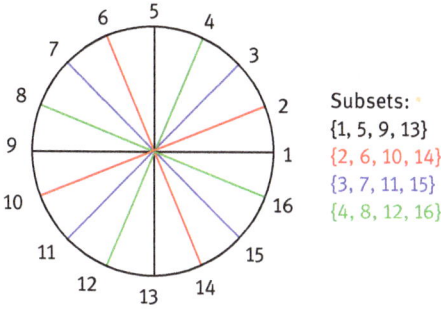

Fig. 6.10: The total projection views are divided into subsets.

Subsets:
{1, 5, 9, 13}
{2, 6, 10, 14}
{3, 7, 11, 15}
{4, 8, 12, 16}

6.6.1 Analytical methods: windowing

In an analytical algorithm, noise regulation is achieved via the application of a window function when the projection data are filtered. The filter in an image reconstruction algorithm is always a high-pass filter (e.g., the ramp filter) in which the high-frequency components are enhanced more than the low-frequency components. In order to suppress the high-frequency noise, a window function is always applied to the ramp filter (see Figure 6.11). Basically, the noise regulation strategy in an analytical algorithm is to control the bandwidth. Thus, both high-frequency noise and high-frequency signal are discarded.

6.6.2 Iterative methods: stopping early

There are many ways to control the noise in an iterative algorithm. We can first consider a rough noise propagation model of a linear iterative algorithm:

$$\text{Error}_{\text{image}} \sim \lambda_n(\omega) \times \text{Error}_{\text{projections}}, \tag{6.30}$$

where $\text{Error}_{\text{image}}$ is the error magnitude in the reconstructed image, $\text{Error}_{\text{projections}}$ is the error magnitude in the projections, and $\lambda_n(\omega)$ is the algorithm transfer function which depends on the frequency ω and the iteration number n.

We can compare an iterative reconstruction algorithm with an SVD matrix pseudo-inverse solution. You may imagine that $\lambda_n(\omega)$ contains the information of both the singular values and singular vectors of the imaging matrix A. The frequency components are in the singular vectors. As the iteration number n increases, more singular vectors join $\lambda_n(\omega)$. The iteration number is somehow related to the cutoff index in an SVD pseudo-inversion expression. With a larger iteration number n, $\lambda_n(\omega)$ contains components with higher frequencies. In some linear algorithms, this relationship can be simplified to

$$\text{Error}_{\text{image}} \sim \kappa \times \text{Error}_{\text{projections}}, \tag{6.31}$$

where κ is similar to the condition number of matrix A, and κ is defined as the ratio of the largest singular value σ_1 over the cutoff singular value σ_n. This simplification is reasonable because the "worst" noise influence comes from the frequency components (i.e., the singular vector) corresponding to the current smallest singular value σ_n. In the SVD pseudo-inverse method, the reconstructed image is a sum of many terms. Each term is a product of a component (i.e., the singular vector) and the reciprocal of its corresponding singular value $1/\sigma$. The largest gain comes from $1/\sigma_n$, which corresponds to a singular vector with very high frequencies.

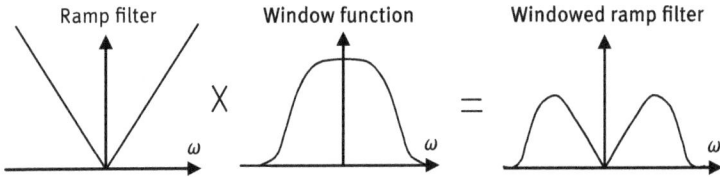

Fig. 6.11: Application of a window function to the ramp filter.

In your mind, you can imagine that $\lambda_n(\omega)$ is a windowed ramp filter (see Figure 6.11). The width of the window increases as the iteration number increases. We immediately see that one way to control noise is to control the iteration number.

This analogy between an iterative reconstruction and an SVD pseudo-inverse is not mathematically correct, but it is a way of showing the similarity of these two approaches. This analogy can give us some insight of an iterative algorithm.

Stopping early is the simplest way to regulate the noise. However, iterative algorithms do not have a uniform convergence rate throughout the image. After an iterative algorithm is stopped, the resultant image will have nonuniform resolution. If you would like your reconstructed image to have uniform resolution, one remedy is to over-iterate (i.e., not to stop early) and then apply a post-filter to suppress the noise.

6.6.3 Iterative methods: choosing pixels

The second way to reduce image noise is to reduce the errors in the data. This approach of regularization is a unique feature for the iterative algorithm. The errors between the projections P and the model AX, $\text{Error}_{\text{data}}$, consist of two parts: deterministic errors and random errors. The deterministic errors are generated from the nonideal system model AX. First of all, discretizing the continuous object into pixels (or voxels) may cause errors. One must consider the trade-offs when deciding pixel size. Smaller pixels give a more accurate model but increase the number of unknowns to be solved. Larger pixels make the image model less accurate, but fewer unknowns can make the inverse problem more stable.

Using nonoverlapping uniform pixels or voxels to model an image is not an ideal approach because this image model contains a lot of discontinuity in the image and introduces too many artificial high-frequency components into the image. Some people have tried to use overlapping nonuniform pixels (or voxels) such as blobs (see Figure 6.12), which results in improved image quality. This gives a more realistic band-limited image model.

One drawback of using blobs as image voxels is the increased computational complexity. An alternative approach has been investigated to achieve the same effect but with better computational efficiency. This strategy uses the traditional nonoverlapping voxels in the image, but a low-pass filter is applied to the backprojected image. The kernel function of the low-pass filter is chosen as the 3D "profile" of the blobs. In other words, the backprojected image is 3D convolved with the blob (see Figure 6.13).

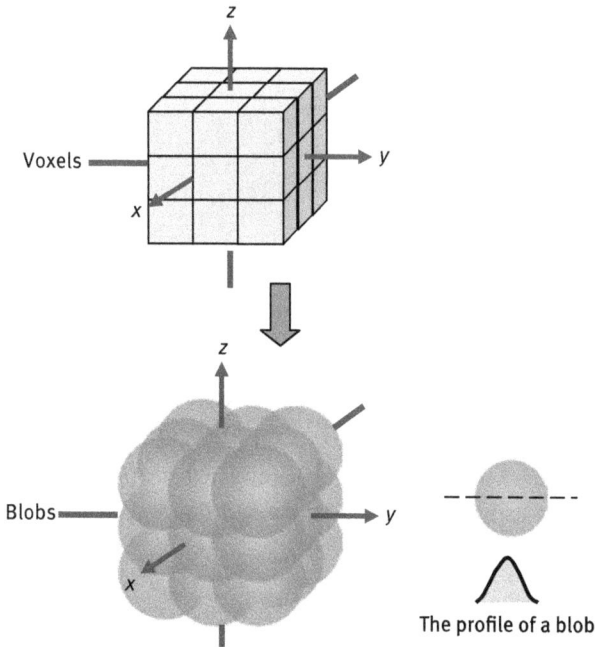

The profile of a blob

Fig. 6.12: Using overlapping blobs to replace the traditional voxels can better model the image.

To make the inverse problem more stable, as a rule of thumb, we would select the pixel size larger than the detector bin size; this makes the number of image pixels smaller than the number of detector bins (see Figure 6.14). In practice, it is advantageous to choose a large array size (with a small bin dimension) on the detector during data acquisition. This makes a big difference in noise control in an iterative algorithm, especially when the system resolution is modeled in the projector/backprojector. A balanced selection of the detector bin size is half the size of the image pixel. If the

3D covolution

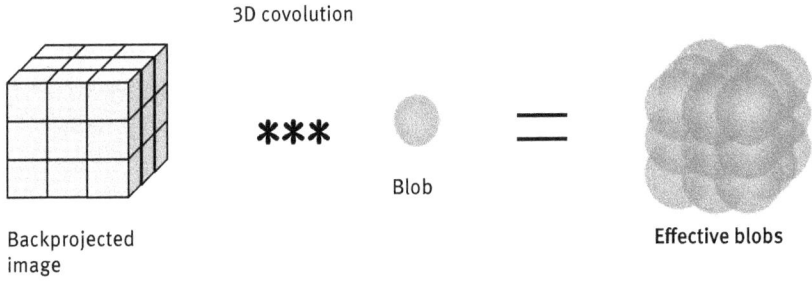

Backprojected
image

Blob

Effective blobs

Fig. 6.13: An alternative approach to get the blob effect.

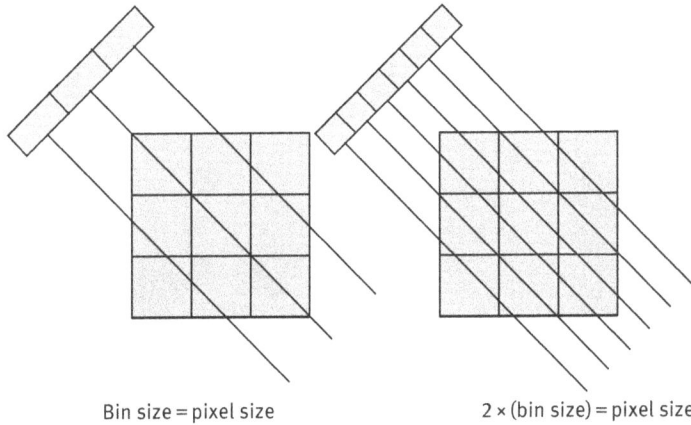

Bin size = pixel size

2 × (bin size) = pixel size

Fig. 6.14: It is advantageous to use a detector bin size that is smaller than the image pixel size.

image size is $256 \times 256 \times 256$ and there is no option on the scanner to acquire 512×512 projections, you can acquire data using the 256×256 mode and interpolate the data into 512×512 arrays during iterative image reconstruction.

6.6.4 Iterative methods: accurate modeling

Modeling the system's point spread function (see Figure 6.15) and patient-induced attenuation and scattering in the matrix A will significantly reduce the deterministic errors between the projections P and the model AX. If you are not ready to model the imaging physics in matrix A, you still have a choice of line-length weighting or area weighting of the image path within the pixel to be used in calculating the elements in matrix A (see Figure 6.16). The freedom to model the imaging system with various geometries and physics effects is the main advantage of using an iterative reconstruction algorithm. We are able to control the errors between the projection data and the model to some degree; we can at least control the deterministic part of them with good system modeling. Smaller data modeling errors result in smaller errors in the image. With reduced data modeling error, we can increase the iteration number to get better image resolution with the same or reduced image noise.

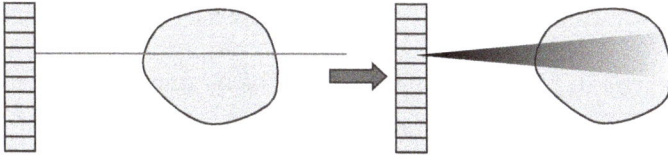

Fig. 6.15: Modeling the system distance-dependent resolution and sensitivity in the projector.

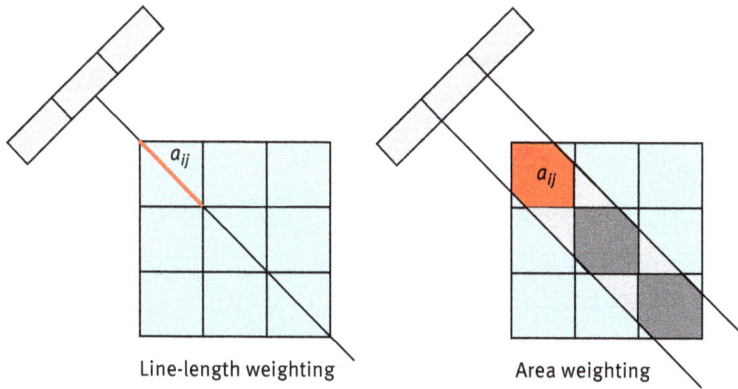

Line-length weighting   Area weighting

Fig. 6.16: Area weighting is a better model than the line-length weighting but has a higher computation cost.

There is a fourth way to control the noise in an iterative algorithm: the random part. We can model the noise distribution in the objective function. Section 6.7 will be dedicated to this topic.

The fifth way is to use the prior knowledge about the image that we are looking for in addition to using the projection data alone. This topic will be covered in Section 6.8.

6.7 Noise modeling as a likelihood function

In order for the noise model to work, we must have redundant measurements; otherwise, the noise model has no effect on the solution. In Figure 6.17, there are two lines, L_1 and L_2, representing two independent measurements described by two independent linear equations. These measurements can be noisy or noiseless. The solution is the intersection of these two lines, regardless of the presence of noise or if you trust L_1 more or L_2 more. Due to noise, this intersection may not be the true solution at all. There is nothing we can do to improve upon the solution if we only have two measurements.

What if we have three measurements L_1, L_2, and L_3 (see Figure 6.18)? Because of noise influence the three lines do not intersect at one point. How should we pick a reasonable solution? A wise decision would depend on the noisiness of each measurement. We should trust the measurement with less noise more, and trust measurement with more noise less.

If we use the variance σ^2 to characterize the noisiness of a measurement, we can assign a weighting factor $1/\sigma^2$ to that measurement. Thus, we can use a variance-based weighting scheme to select a solution. Our old least-squares objective function

$$\chi^2 = \|AX - P\|^2 = (AX - P)^T(AX - P) \tag{6.32}$$

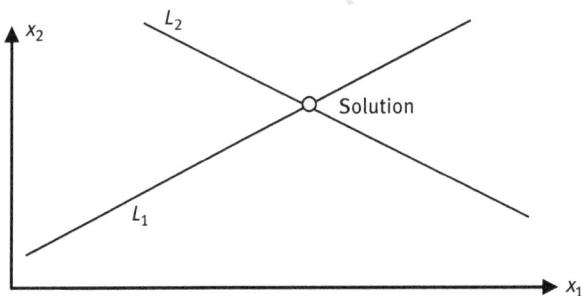

Fig. 6.17: Two lines intersect at one point, which is the solution of the corresponding two equations.

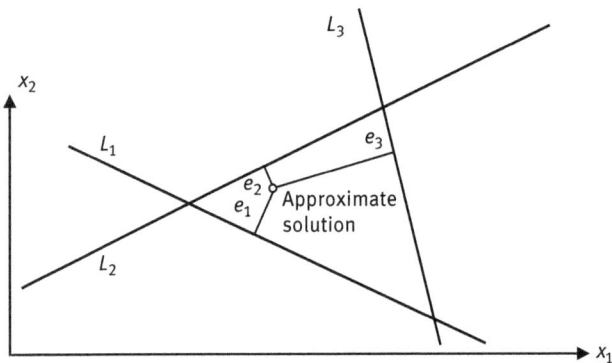

Fig. 6.18: If the system has redundant measurements and is not consistent, one can use the noisiness to weight each measurement and find an approximate solution. In this example, we assume that L_3 is noisier than L_1, and L_2 is less noisy than L_1.

becomes a weighted least-squares objective function

$$\chi_W^2 = (AX - P)^T W(AX - P), \tag{6.33}$$

where W is the diagonal matrix, $W = \text{diag}\ \{1/\sigma_1^2, 1/\sigma_2^2, \ldots, 1/\sigma_N^2\}$, and N is the number of projections. In many cases, it is advantageous to make the definition of the weighting matrix W more general. For example, we can let $W = \text{diag}\ \{1/\sigma_1^\gamma, 1/\sigma_2^\gamma, \ldots, 1/\sigma_N^\gamma\}$, where the value of γ can be different from 2. The weighting W sometimes causes low-frequency shadowing artifacts in the reconstruction. A careful selection of this parameter γ may reduce the artifacts.

This weighted, least-squares objective function can also be obtained through the likelihood function by assuming Gaussian noise in the projections. The ith projection measurement p_i is a Gaussian random variable whose mean value is $\mu_i = \sum_j a_{ij} x_j = A_i X$ and variance is σ_i^2. Here, A_i is the ith row of the matrix A. The Gaussian distribution density function gives

$$\text{Prob}(p_i) = \frac{1}{\sqrt{2\pi}\sigma_i} \exp\left(-\frac{(A_i X - p_i)^2}{2\sigma_i^2}\right). \tag{6.34}$$

We assume that all projections are statistically independent. The likelihood function is the joint probability density function by considering all projections together:

$$\text{Prob}(P) = \prod_i \frac{1}{\sqrt{2\pi}\sigma_i} \exp\left(-\frac{(A_i X - p_i)^2}{2\sigma_i^2}\right). \tag{6.35}$$

Our goal is to find an image X that maximizes the above likelihood function, hence, the term "maximum likelihood solution." Taking the logarithm of the likelihood function, we have

$$\ln(\text{Prob}(P)) = \frac{1}{2}\sum_i \frac{(A_i X - p_i)^2}{\sigma_i^2} + \sum_i \ln\left(\frac{1}{\sqrt{2\pi}\sigma_i}\right). \tag{6.36}$$

The second term in the above equation is a constant. Therefore, maximizing the likelihood function is equivalent to minimizing the following weighted least-squares objective function:

$$\chi_W^2 = \sum_i \frac{(A_i X - p_i)^2}{\sigma_i^2} = (AX - P)^T W(AX - P). \tag{6.37}$$

If the data noise is not Gaussian, the above approach of setting up a likelihood function and an objective function still applies. However, the resultant objective function is different.

6.8 Including prior knowledge (Bayesian)

We sometimes know more about the image that we are looking for – in addition to the measurements. We can enforce this prior knowledge into the image by adding an extra term to the objective function.

For example, if we know in advance that the image X is very smooth, we can add a penalty term $\| \nabla X \|^2$ to suppress sharp jumps and encourage the smoothness:

$$\text{New objective function } (X) = \text{Old objective function } (X) + \beta \| \nabla X \|^2, \quad (6.38)$$

where β is the user-specified controlling parameter. Using the squared norm of the gradient $\| \nabla X \|^2$ as a penalty term can be generalized as using an "energy" function $U(X)$ as penalty term. The energy function $U(X)$ is defined as

$$U(X) = \sum_{i,j} w_{ij} V\left(x_j - x_j\right), \quad (6.39)$$

where the summation is over a neighborhood (clique), and V is a convex function, which may or may not be quadratic (see Figure 6.19). If V is a quadratic function, this energy function encourages smoothness and penalizes jumps. If the function V increases more slowly than a quadratic function (say, V increases linearly), then it can preserve edges and smooth out the noise. How does the algorithm know which is the edge that you want to keep and which is noise that you want to smooth out? Let us answer this question by comparing a linear function $|x|$ and a quadratic function x^2. We assume that the noise consists of small jumps, while the edges separate the large jumps. When $|x|$ is small, we have $|x| > x^2$, and the linear function gives relatively heavier penalty. Thus the noise is suppressed. When $|x|$ is large, we have $|x| < x^2$, and the linear function gives a relatively much lighter penalty. Thus the edges are preserved.

Algorithms that include the prior information carry many names, such as Bayesian methods or MAP (maximum a posteriori) algorithms.

The new objective function is, in fact, the conditional probability $\text{Prob}(X|P)$. Bayes' law states

$$\text{Prob}(X|P) = \frac{\text{Prob}(P|X)\text{Prob}(X)}{\text{Prob}(P)}. \quad (6.40)$$

Taking the logarithm yields

$$\ln(\text{Prob}(X|P)) = \ln(\text{Prob}(P|X)) + \ln(\text{Prob}(X)) - \ln(\text{Prob}(P)), \quad (6.41)$$

where the third term has nothing to do with our unknowns X and can be eliminated from the objective function. The Bayesian objective function then becomes

$$L(X) = \ln(\text{Prob}(P|X)) + \ln(\text{Prob}(X)), \quad (6.42)$$

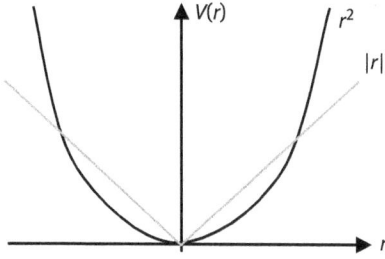

Fig. 6.19: The penalty function can have different function forms, depending on the application.

or symbolically,

$$(\text{Posterior function}) = (\text{Likelihood function}) + \beta(\text{Prior function}). \qquad (6.43)$$

The first term on the right-hand side is the old objective function for ML algorithm, and the second term contains the prior information term about the image X. This justifies our expression of the new objective function at the beginning of this section.

6.9 Mathematical expressions

This section gives the mathematical expressions for the ART algorithm, the MLEM algorithm, the OS-EM algorithm, Green's one-step late algorithm, and ML-TV (total variation) algorithm. Computer implementation steps are given for the conjugate gradient (CG) algorithm. The derivation of the ML-EM algorithm is also presented.

6.9.1 ART

The ART algorithm is a row-action algorithm. It considers one ray sum at a time and can be expressed as

$$X^{\text{next}} = X^{\text{current}} - \frac{A_i X^{\text{current}} - p_i}{\|A_i\|^2} A_i^T, \qquad (6.44)$$

where $A_i X$ performs the forward projection along the ith projection ray, p_i is the measured projection from the ith projection bin, $\|A_i\|^2 = \sum_j a_{ij}^2$ is the sum of the squared "contribution factors" along the ith ray, and $c A_i^T$ backprojects the value c along the ith ray. If we rewrite the above algorithm in the following form:

$$X^{\text{next}} = X^{\text{current}} - \left(\frac{A_i X^{\text{current}}}{\|A_i\|} - \frac{p_i}{\|A_i\|} \right) \frac{A_i^T}{\|A_i\|}, \qquad (6.45)$$

the geometric meaning of this algorithm can be easily explained as in Figure 6.20.

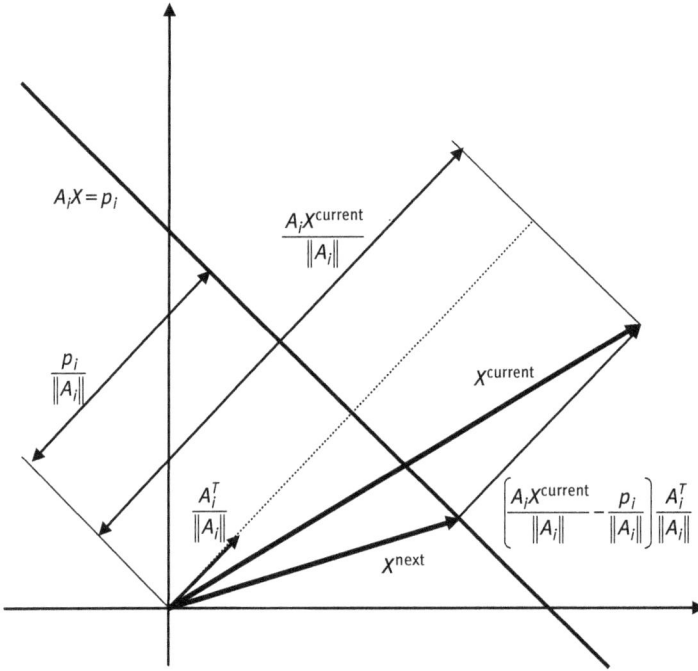

Fig. 6.20: An illustration of the ART algorithm.

There are many versions of the ART algorithms. The simultaneous iterative reconstruction technique does not update the image ray by ray, but update the image once per iteration. Another version is to update the image angle by angle. Another version uses a relaxation (or damping) factor to reduce the step size and stabilize the algorithm. Still another version is called the MART (multiplicative ART) algorithm, in which updating is multiplicative, instead of additive. One advantage of the MART algorithm is that the resultant image is always nonnegative provided the initial guess of the image is nonnegative.

6.9.2 The Landweber algorithm

Each iteration of the Landweber algorithm can be expressed as

$$X^{\text{next}} = X^{\text{current}} - \alpha A^T \left(A X^{\text{current}} - P\right), \tag{6.46}$$

which involves one projection operation (i.e., multiplying by matrix A) and one back-projection (i.e., multiplying by matrix A^T). If the objective function considers noise weighting, that is,

$$\chi_W^2 = (AX - P)^T W (AX - P), \tag{6.47}$$

where $W = \text{diag}\{w_i\}$, the Landweber algorithm becomes

$$X^{\text{next}} = X^{\text{current}} - \alpha A^T W \left(A X^{\text{current}} - P \right).$$ (6.48)

The upper bound of the step size α is determined by the square root of the largest eigenvalue of matrix $A^T WA$. In reality, this eigenvalue is unknown. The following method can narrow down the range of α to $0 < \alpha < 1$:

$$x_i^{\text{next}} = x_i^{\text{current}} - \alpha \frac{\sum_j a_{ji} w_j \left(\sum_n a_{jn} x_n^{\text{current}} - p_j \right)}{\sum_j a_{ji} w_j \sum_n a_{jn}}.$$ (6.49)

In the above expression, $\sum_j a_{ji} w_j \sum_n a_{jn}$ is a scalar for each pixel i. It can be understood as follows. Let us form a constant image volume, and each voxel has a value of 1. Then we perform forward projection on this constant image along the jth path, which is what $\sum_n a_{jn}$ does. Next, we multiply this projection value by a scalar w_j, which is the meaning of $w_j \sum_n a_{jn}$. Finally, we backproject the scalar value $w_j \sum_n a_{jn}$ into the image domain along the jth path (which is the same path as we perform the forward projection) to the ith image voxel. The expression $\sum_j a_{ji} w_j \sum_n a_{jn}$ generates an image that is used to normalize the step size in the Landweber algorithm; this $\sum_j a_{ji} w_j \sum_n a_{jn}$ normalization image can be precalculated.

6.9.3 CG algorithm

This CG algorithm solves the normal equations

$$A^T AX = A^T P.$$ (6.50)

Let $M = A^T A$ and $B = A^T P$. The normal equations become

$$MY = B.$$ (6.51)

Matrix M is real, symmetric, and positive definite (or semidefinite). The implementation steps of the CG algorithm are given as follows.

Set up initial conditions:

$$X^{(0)} = \vec{0}, \quad R_0 = B, \quad \text{and} \quad \Delta_0 = R_0.$$ (6.52)

These three image domain arrays are represented as three 1D column vectors.

Iterations (for $n = 1, 2, 3, \ldots$ do the following):

Update the step size (a scalar) a_n as a ratio of two scalars:

$$a_n = \left(R_{n-1}^T R_{n-1} \right) / \left(\Delta_{n-1}^T M \Delta_{n-1} \right).$$ (6.53)

Update the image with

$$X^{(n)} = X^{(n-1)} + a_n \Delta_{n-1}. \tag{6.54}$$

Calculate the residual image

$$R_n = R_{n-1} - a_n M \Delta_{n-1}. \tag{6.55}$$

Calculate the factor (a scalar) β_n used to find the searching direction

$$\beta_n = \left(R_n^T R_n \right) / \left(R_{n-1}^T R_{n-1} \right). \tag{6.56}$$

Calculate the new searching direction for the next iteration

$$\Delta_n = R_n + \beta_n \Delta_{n-1}. \tag{6.57}$$

endfor

This CG algorithm has some properties:
(1) A Krylov subspace, κ_n, is expanding as the iteration number increases

$$
\begin{aligned}
k_n &= \langle B, MB, M^2 B, \dots, M^{n-1} B \rangle \\
&= \langle X^{(1)}, X^{(2)}, X^{(3)}, \dots, X^{(n)} \rangle \\
&= \langle R_0, R_1, R_2, \dots, R_{n-1} \rangle \\
&= \langle \Delta_0, \Delta_1, \Delta_2, \dots, \Delta_{n-1} \rangle
\end{aligned}
\tag{6.58}
$$

where $\langle R_0, R_1, R_2, \dots, R_{n-1} \rangle$ denotes a space spanned by $R_0, R_1, R_2, \dots, R_{n-1}$.
(2) The residuals are orthogonal to each other, that is,

$$R_n^T R_j = 0, \quad j < n. \tag{6.59}$$

(3) The search directions are M-conjugate with each other, that is,

$$\Delta_n^T M \Delta_j = 0, \quad j < n. \tag{6.60}$$

(4) At each iteration, $X^{(n)}$ minimizes the objective function over the Krylov subspace κ_n. Therefore, if M is $m \times m$, the algorithm converse in at most m iterations.

What makes the iterative CG algorithm remarkable is the choice of the search direction Δ_{n-1}, which has the special property that minimizing the objective function over $X^{(n)} + \langle \Delta_{n-1} \rangle$ actually minimizes it over all of κ_n.

6.9.4 ML-EM

The emission data ML-EM algorithm is the most popular iterative algorithm in emission tomography and is expressed as

$$x_j^{\text{next}} = \frac{x_j^{\text{current}}}{\sum_i a_{ij}} \sum_i a_{ij} \frac{p_i}{\sum_{\hat{j}} a_{i\hat{j}} x_{\hat{j}}^{\text{current}}}, \tag{6.61}$$

where the summation over $\hat{j}$ is the projector identical to A_iX in Section 6.9.1, and the summations over i are the backprojectors. This algorithm compares the measured projection p_i with the forward projection of the current estimate A_iX as a ratio. This ratio is backprojected to the image domain. The summation $\sum_i a_{ij}$ is the backprojection of constant 1 to the image domain. The ratio of these two backprojected images determines a modification factor to update the current estimate of the image.

The following is the derivation of the emission-projection ML-EM algorithm. If p is a random variable with the Poisson distribution, then its probability mass function is given as

$$\text{Prob}\,(p|\lambda) = e^{-\lambda} \frac{\lambda^p}{p!}, \tag{6.62}$$

where λ is the expected value of this random variable. For an imaging problem $AX = P$, the number of photons emitted from each image pixel is a Poisson random variable, and each measurement p_i can be treated as the summation of these Poisson variables. We write

$$p_i = \sum_j c_{ij}, \tag{6.63}$$

where c_{ij} is the Poisson random variable and

$$\lambda_{ij} = E(c_{ij}) = a_{ij}x_j. \tag{6.64}$$

Note that X is not random.

We can set up the likelihood function as the joint probability mass function of all Poisson distributed random variables c_{ij}:

$$\text{Prob} = \prod_{i,j} e^{-\lambda_{ij}} \frac{\lambda_{ij}^{c_{ij}}}{c_{ij}!} = \prod_{i,j} e^{-a_{ij}x_j} \frac{\left(a_{ij}x_j\right)^{c_{ij}}}{c_{ij}!}. \tag{6.65}$$

Taking the logarithm of this likelihood function yields

$$\ln(\text{prob}) = \sum_{i,j} \left(c_{ij} \ln\left(a_{ij}x_j\right) - a_{ij}x_j\right) - \sum_{i,j} \ln\left(c_{ij}!\right). \tag{6.66}$$

The second summation term $\sum_{i,j} \ln(C_{ij}!)$ does not contain the parameters x_j to be estimated; therefore, it can be discarded without changing the ML problem. To find the ML solution of x_j, we will maximize the following objective function:

$$L = \sum_{i,j} \left(c_{ij} \ln\left(a_{ij}x_j\right) - a_{ij}x_j \right). \tag{6.67}$$

The "E" Step:
The above objective function contains random variables c_{ij}. The "E" (expectation) step is to replace it by the expected value using the measurement p_i and the current estimate of the parameters x_j. That is, c_{ij} is replaced by

$$E(c_{ij}|p_i, X^{\text{current}}) = \frac{a_{ij}x_j^{\text{current}}}{\sum_k a_{ik}x_k^{\text{current}}} p_i. \tag{6.68}$$

Thus, after the "E" step, the objective function becomes

$$E(L|P, X^{\text{current}}) = \sum_{i,j} \left(\frac{a_{ij}x_j^{\text{current}}}{\sum_k a_{ik}x_k^{\text{current}}} p_i \ln\left(a_{ij}x_j\right) - a_{ij}x_j \right). \tag{6.69}$$

The "M" Step:
To maximize the new objective function $E(L \mid P, X^{\text{current}})$, we will take the derivative of it with respect to estimation parameters x_j and set the derivatives to zero, that is,

$$\begin{aligned}
\frac{\partial E(L|P, X^{\text{current}})}{\partial x_j} &= \sum_i \left(\frac{a_{ij}x_j^{\text{current}}}{\sum_k a_{ik}x_k^{\text{current}}} p_i \frac{a_{ij}}{a_{ij}x_j} - a_{ij} \right) \\
&= \frac{1}{x_j} \sum_i \frac{a_{ij}x_j^{\text{current}}}{\sum_k a_{ik}x_k^{\text{current}}} p_i - \sum_i a_{ij} \\
&= 0.
\end{aligned} \tag{6.70}$$

Solving for x_j, we finally have the ML-EM algorithm:

$$x_j = \frac{x_j^{\text{current}}}{\sum_i a_{ij}} \sum_i a_{ij} \frac{p_i}{\sum_j a_{ij}x_j^{\text{current}}}. \tag{6.71}$$

In fact, this multiplicative ML-EM algorithm can also be written in an additive form so that it appears like a gradient descent algorithm as

$$X^{\text{next}} = X^{\text{current}} + S^{\text{current}} A^T \Lambda^{\text{current}} \left(P - AX^{\text{current}} \right), \tag{6.72}$$

where the step size is

$$S^{\text{current}} = \text{diag}\left\{ x_j^{\text{current}} / \sum_i a_{ij} \right\}, \tag{6.73}$$

and the noise variance weighting is

$$\Lambda^{\text{current}} = \text{diag}\left\{ 1/\sum_k a_{ik} x_k^{\text{current}} \right\} \approx \text{diag}\{1/p_i\}. \tag{6.74}$$

In reality, the measurement noise may not be exactly Gaussian distributed or Poisson distributed. The author personally believes that it is not very critical what the noise distribution is, while it is very important to know the variance of the measurement noise, because it is the variance that is used in measurement weighting. For example, in an imaging system, the measurement p_i is noisy and the noise variance is related to p_i as

$$(\text{Variance of } p_i) = s_i \bar{p}_i, \tag{6.75}$$

where s_i can be a system scaling factor, p_i may not be Poisson distributed, and $\bar{p}_i$ is the expected value of p_i. Realizing that the imaging model is

$$\sum_k a_{ik} x_k = p_i, \tag{6.76}$$

we can modify the ML-EM algorithm and obtain a general image reconstruction algorithm as

$$x_j^{\text{next}} = \frac{x_j^{\text{current}}}{\sum_i a_{ij} \frac{1}{s_i}} \sum_i a_{ij} \frac{p_i}{s_i \sum_{\hat{j}} a_{i\hat{j}} x_{\hat{j}}^{\text{current}}}. \tag{6.77}$$

If we rewrite this modified algorithm in the additive form, the noise variance weighting is

$$\Lambda^{\text{current}} = \text{diag}\left\{ 1/\left(s_i \sum_k a_{ik} x_k^{\text{current}} \right) \right\} \approx \text{diag}\{1/(s_i p_i)\}. \tag{6.78}$$

6.9.5 OS-EM

With minor changes to the ML-EM algorithm, the ordered subset version of it can be readily obtained as the OS-EM algorithm:

$$x_j^{\text{next}} = \frac{x_j^{\text{current}}}{\sum_{i \in S_k} a_{ij}} \sum_{i \in S_k} a_{ij} \frac{p_i}{\sum_{\hat{j}} a_{i\hat{j}} x_{\hat{j}}^{\text{current}}}, \tag{6.79}$$

where S_k represents the kth subset of the projections.

6.9.6 MAP (Green's one-step-late algorithm)

MAP stands for maximum a posteriori. The ML-EM algorithm can be changed into a MAP (i.e., Bayesian) algorithm by adding a penalty term in the denominator:

$$x_j^{\text{next}} = \frac{x_j^{\text{current}}}{\sum_i a_{ij} + \beta \frac{\partial U(X^{\text{current}})}{\partial x_j^{\text{current}}}} \sum_i a_{ij} \frac{p_i}{\sum_{\hat{j}} a_{i\hat{j}} x_{\hat{j}}^{\text{current}}}, \tag{6.80}$$

where $U(X)$ is the energy function defined in Section 6.8, and β is the control parameter. This is not a true MAP algorithm because the energy function U is supposed to be evaluated using the next estimate X^{next}, which is not yet available, hence, the term "one step late."

This one-step-late MAP algorithm (6.80) can be modified for many applications. In emission tomography, if the projection matrix models the attenuation effects, the regularization in eq. (6.80) varies with the attenuation effects and is not uniform throughout the reconstructed image. Stronger regularization is found in one region, and weaker regularization is found in a different region. To overcome this drawback, we can replace $\sum_i a_{ij} + V$ by $(1 + V) \sum_i a_{ij}$, where

$$V = \beta \frac{\partial U(X^{\text{current}})}{\partial x_j^{\text{current}}}. \tag{6.81}$$

The modified version of eq. (6.80) is given as

$$x_j^{\text{next}} = \frac{1}{1 + V} \frac{x_j^{\text{current}}}{\sum_i a_{ij}} \sum_i \frac{\sum_i a_{ij} p_i}{\sum_{\hat{j}} a_{i\hat{j}} x_{\hat{j}}^{\text{current}}}, \tag{6.82}$$

which is simply the ML-EM algorithm (6.61) divided by $(1 + V)$. As $V \to 0$, the modified Green's one-step-late algorithm reduces to the ML-EM algorithm.

There is an alternative version of eq. (6.82) by using the approximation

$$\frac{1}{1+V} \approx 1 - V, \text{ when } |V| \ll 1. \tag{6.83}$$

Then an alternative version of (6.82) is

$$x_j^{\text{next}} = (1 - V) \frac{x_j^{\text{current}}}{\sum_i a_{ij}} \sum_i \frac{\sum_i a_{ij} p_i}{\sum_{\hat{j}} a_{i\hat{j}} x_{\hat{j}}^{\text{current}}}, \tag{6.84}$$

If we are creative, we can develop many alternative versions of eq. (6.82). For example, to enforce $|V| < 1$, we can replace V by $\frac{V}{\sqrt{1+V^2}}$ in (6.82) and (6.84) if we want, obtaining eqs. (6.85) and (6.86), respectively.

$$x_j^{\text{next}} = \frac{1}{1 + \frac{V}{\sqrt{1+V^2}}\sum_i a_{ij}} \frac{x_j^{\text{current}}}{} \sum_i \frac{\sum_i a_{ij}p_i}{\sum_j a_{\hat{ij}}x_j^{\text{current}}}, \tag{6.85}$$

$$x_j^{\text{next}} = \left(1 - \frac{V}{\sqrt{1+V^2}}\right) \frac{x_j^{\text{current}}}{\sum_i a_{ij}} \sum_i \frac{\sum_i a_{ij}p_i}{\sum_j a_{\hat{ij}}x_j^{\text{current}}}. \tag{6.86}$$

You can see that there are many flexibilities in developing an iterative algorithm. We are free to choose the measurement noise model in this algorithm. The noise model is a simple model of the measurement noise variance as expressed in eq. (6.75). Section 6.9.4 covers the case of the noise variance being Poisson distributed.

If we remove the Poisson noise model and assume the noise variance to be 1, we let

$$s_i = \frac{1}{\bar{p}_i} \approx \frac{1}{\sum_{\hat{j}} a_{\hat{ij}}x_{\hat{j}}^{\text{current}}}. \tag{6.87}$$

Then, eq. (6.77) becomes a modified ML-EM algorithm with a uniform noise variance model. After dividing by $(1 + V)$, a new MAP algorithm is obtained:

$$x_j^{\text{next}} = \frac{1}{1+V} \frac{x_j^{\text{current}} \sum_i a_{ij}p_i}{\sum_i (a_{ij} \sum_{\hat{j}} a_{\hat{ij}}x_{\hat{j}}^{\text{current}})}. \tag{6.88}$$

If we want to modify the MAP algorithm (6.82) for transmission data, we need a noise model for the transmission sinogram (i.e., after taking logarithm). The variance of the transmission tomography sinogram is proportional to the exponential function of the sinogram's mean value:

$$(\text{Variance of } p_i) \propto \exp(\bar{p}_i) \approx \exp\left(\sum_{\hat{j}} a_{\hat{ij}}x_{\hat{j}}^{\text{current}}\right). \tag{6.89}$$

According to eq. (6.75), we have

$$s_i = \frac{\exp\left(\sum_{\hat{j}} a_{\hat{ij}}x_{\hat{j}}^{\text{current}}\right)}{\bar{p}_i} \approx \frac{\exp\left(\sum_{\hat{j}} a_{\hat{ij}}x_{\hat{j}}^{\text{current}}\right)}{\sum_{\hat{j}} a_{\hat{ij}}x_{\hat{j}}^{\text{current}}}. \tag{6.90}$$

Then, eq. (6.77) becomes a modified ML-EM algorithm with a transmission noise model. After dividing by $(1 + V)$, a new transmission data MAP algorithm is obtained:

$$x_j^{\text{next}} = \frac{1}{1+V} \frac{x_j^{\text{current}} \sum_i \left[a_{ij}p_i \exp\left(-\sum_{\hat{j}} a_{\hat{ij}}x_{\hat{j}}^{\text{current}}\right)\right]}{\sum_i \left[a_{ij} \sum_{\hat{j}} a_{\hat{ij}}x_{\hat{j}}^{\text{current}} \exp\left(-\sum_{\hat{j}} a_{\hat{ij}}x_{\hat{j}}^{\text{current}}\right)\right]}. \tag{6.91}$$

Of course, in eqs. (6.88) and (6.91), you can replace $1/(1+V)$ by other versions such as $(1-V), 1/(1+(V/\sqrt{1+V^2})),$ and $(1-(V/\sqrt{1+V^2}))$.

6.9.7 Matched and unmatched projector/backprojector pairs

This topic is quite controversial. It is almost like a religion in which people have their own opinions and carry out their practice accordingly. By projector/backprojector pair, we normally mean matrices A and A^T as described earlier in this chapter. If the backprojection matrix is the transpose of the projection matrix, then this pair is called matched. Otherwise, the pair is called unmatched.

In an analytic reconstruction algorithm, the image pixel is a point. The image is not discretized. The projector, even not used in the reconstruction algorithm, is an integral of the continuous image. Its matched backprojector treats an image pixel as a point, too. Sometimes this backprojector is referred to as the pixel-driven backprojector. When implementing a pixel-driven backprojector, we start with an image pixel location, and find the location on the detector to get the backprojection value.

In an iterative reconstruction algorithm, the image pixel is no longer a point but an area. The projector draws a ray from each detector bin, and determines the contribution of each pixel according to the overlap of this ray with the pixel of interest. This projector is sometimes referred to as the ray-driven projector. Its matched backprojector is the ray-driven backprojector.

We all agree that one should make the projector (i.e., the matrix A) to model the imaging geometry and imaging physics as accurately as possible, significantly reducing the deterministic modeling errors. The question is whether one is allowed to use unmatched projector/backprojector pairs. We have seen many cases from the practical implementations. In some cases, the unmatched pair gives better results than the matched pair, in terms of artifact removal and speedup in computation time. In other cases, the unmatched pair gives more artifacts than the matched pair. A common reason one would use a backprojector other than A^T is to save computation time. We need to be cautioned that the solution can be different if an unmatched pair is used to replace the matched pair in an iterative reconstruction algorithm.

First of all, one cannot just arbitrarily pick up a backprojector and use it in an iterative algorithm. For example, one cannot use a fan-beam backprojector to reconstruct a parallel-beam image. The minimum requirement of choosing a backprojector is that the projection-then-backprojection operator can only blur the image; it cannot cause image distortion in shape and other motions such as rotation and translation. If a projector/backprojector pair is applied to a point source, the result can only be a blurred point source at the same location.

Second of all, we will investigate how a solution can be changed if a different backprojector is used to solve the system $AX = P$. Let us consider a modified Landweber iterative algorithm with a backprojector B^T, where B^T may not be the same as A^T, as

$$X^{\text{next}} = X^{\text{current}} + SB^T \left(P - AX^{\text{current}} \right), \tag{6.92}$$

where $S = \text{diag}\{s_1, s_2, \ldots, s_n\}$ with positive diagonal elements controlling the step size of each iteration for each unknown pixel x_j. Let matrix T be

$$T = I - SB^T A, \tag{6.93}$$

with I being the identity matrix. Then this modified Landweber algorithm has a general expression for each iteration as

$$X^{(k+1)} = T^{k+1} X^{(0)} + \left(T^k + \cdots + T + I\right) SB^T P. \tag{6.94}$$

where k is the iteration index, $X^{(k+1)}$ means X^{next}, $X^{(k)}$ means X^{current}, $X^{(0)}$ means the initial condition, and so on. This algorithm converges if and only if $\max_i\{|\lambda_i|\} < 1$, $i = 1$, $2, \ldots, n$ where λ_i is the eigenvalue of the square matrix T. If the eigenvalues of the matrix $B^T A$ are all positive, then these convergence conditions can be met by choosing small enough step sizes. After the algorithm converges, the final solution is given by

$$X^{(\infty)} = \left(B^T A\right)^{-1} B^T P, \tag{6.95}$$

which has a strong dependency on the choice of the backprojector B^T. Only in a special case when A^{-1} and B^{-1} exist, the solution

$$X^{(\infty)} = A^{-1} P \tag{6.96}$$

is backprojector B^T independent. If any one of the eigenvalues of the matrix $B^T A$ is negative, the corresponding eigenvalue of matrix T will be greater than 1 and the algorithm will diverge.

In general, when an unmatched projector/backprojector pair is used in an iterative algorithm, the final solution and the intermediate solutions are backprojector dependent. The matched pair solves the system

$$A^T AX = A^T P, \tag{6.97}$$

while the unmatched pair solves a different system

$$B^T AX = B^T P. \tag{6.98}$$

In order to speed up the convergence rate of the algorithm, some people include the ramp filter in the backprojector. They treat the combined ramp filter and the backprojector as a new backprojector. The new backprojector first performs ramp filtering and then backprojection.

The unmatched pair enforces a different weighting scheme and has a different noise effect than the original problem. The unmatched pair also has a different sampling and data interpolation properties than the original problem. How much these differences can influence the reconstructed image is problem dependent and the users should exercise their judgment to choose a backprojector in their particular problem.

6.10 Reconstruction using highly undersampled data

It has been reported that it is possible to exactly reconstruct a 256×256 image with only 10 views. It seems magic to us, because more than 100 views are normally required in such an imaging problem. Let us explore how they do it. The setup is quite simple and the goal is to minimize the following Bayesian objective function:

$$F = \|AX - P\|_2^2 + \beta\|\psi X\|_0, \tag{6.99}$$

where $\|\bullet\|_2$ is the l_2 norm, $\|\bullet\|_0$ is the l_0 quasi-norm, and ψ is the sparsifying transform. The zero quasi-norm $\|\bullet\|_0$ is easy to understand; it simply counts the nonzero elements in a vector. For example, $\vec{v} = [3, 0, 0, 1, 7]$, in which there are three nonzero elements; therefore, $\|\vec{v}\|_0 = 3$.

The trick of this method is to design the sparsifying transform ψ. This transform can be anything as long as it transforms your regular image into a sparse image in which most (say, 97%) pixels are zero.

If your regular image is a piecewise-constant image, you can use $\|\nabla\|$ (i.e., the magnitude of the gradient) as your sparsifying transform ψ (see Figure 6.21).

A practical medical image is not a piecewise-constant image. If you take the gradient, the resultant image is not very sparse. Thus, it is tricky to find the sparsifying transform ψ for your application.

Another problem is that the l_0-minimization is difficult to do. To be honest, we do not have an effective and efficient way to find its minimum. We normally set up an objective function using the l_2 norm, which basically gives you a user-friendly least-squares function such as $\sum_i v_i^2$, and we can perform many fun things with it (e.g., taking a derivative). However, the l_2 norm does not work well when data are highly undersampled.

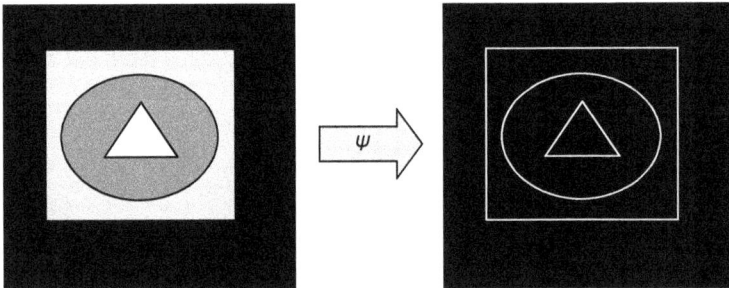

Fig. 6.21: A sparsifying transform extracts some essential information from the original image to produce a sparse image.

The l_1 norm $\{\sum_i |v_i|\}$ is a popular alternative. It does not perform well as the l_0 quasi-norm in selecting the optimum image, but it is better than the l_2 norm. On the other hand, it is not as user friendly as the l_2 norm but is much better than the l_0 quasi-norm.

Let us illustrate the differences among the l_2 norm, l_1 norm, and l_0 quasi-norm. A norm is a sort of measure of the distance or length. Let us consider a point in the 2D x–y coordinate system with coordinates of (3, 4). We want to find its "distance" from the origin (0, 0). If we use the l_2 norm, the l_2 distance is given as $\sqrt{3^2 + 4^2} = 5$. If we use the l_1 norm, the l_1 distance is given as $|3| + |4| = 7$. If we use the l_0 quasi-norm, the l_0 distance is given as $1 + 1 = 2$.

Let us consider a different point in the 2D x–y coordinate system which has coordinates of (0, 4). We want to find its "distance" from the origin (0, 0). If we use the l_2 norm, the l_2 distance is given as $\sqrt{0^2 + 4^2} = 4$. If we use the l_1 norm, the l_1 distance is given as $|0| + |4| = 4$. If we use the l_0 quasi-norm, the l_0 distance is given as $0 + 1 = 1$.

Figure 6.22 shows the unit circles using the l_2 norm, l_1 norm, and l_0 quasi-norm. Except for the l_2 norm unit circle, the other two unit circles do not look like circles at all. For a "measure" to be qualified as a norm, it needs to satisfy a set of axioms. A quasi-norm is almost a norm except that it does not satisfy an axiom called the triangle inequality.

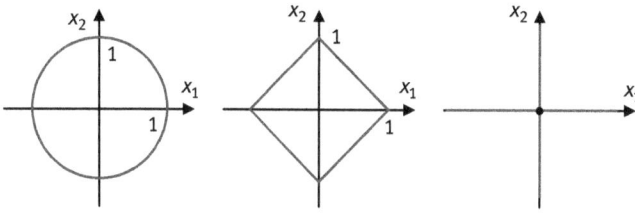

Fig. 6.22: A unit circle is a trajectory of the points that have a distance 1 from the origin. Left: l_2 norm's unit circle. Middle: l_1 norm's unit circle. Right: l_0 quasi-norm's unit circle.

In theory, the l_0 quasi-norm is the best measure to promote the sparsity of ψX and should be used. However, the optimization procedure is not tractable for any l_p norm with $0 \le p < 1$ because its associated objective function is not convex any more. The closest norm that produces a convex objective function is the l_1 norm. The l_1 norm minimization is nearly optimal, in the sense that its solution is not far from the solution with the much more complicated l_0 quasi-norm minimization.

If the derivative of the Bayesian term exists, we can use the one-step-late algorithm introduced in Section 6.9.6 to reconstruct the image. Unfortunately, the derivative of the Bayesian term in the objective function (6.99) does not exist. Rather complicated numerical methods are needed to minimize the objective function (6.99), even if the l_0 quasi-norm is replaced by the l_1 norm.

There is hope. Instead of using l_0 quasi-norm and l_1 norm, you may have also heard of the TV minimization. The TV norm of an image $v(x, y)$ is defined as

$$\text{TV}(v) = \iint \|\nabla v(x,y)\|_2 dx dy = \iint \sqrt{\left(\frac{\partial v}{\partial x}\right)^2 + \left(\frac{\partial v}{\partial y}\right)^2} dx dy. \tag{6.100}$$

The TV norm minimization enforces a flat image with the gradient being zero in most places. The resultant image tends to be piecewise constant. In practice, a small number ε is introduced to calculate the TV norm as

$$\text{TV}(v) \approx \iint \sqrt{\left(\frac{\partial v}{\partial x}\right)^2 + \left(\frac{\partial v}{\partial y}\right)^2 + \varepsilon^2} dx dy \tag{6.101}$$

so that the TV norm is differentiable. In order to get some intuition about the modified TV norm (6.101), let us assume $\varepsilon^2 = 1$. Then the above expression is nothing but the surface area formula!

We see that minimizing the modified TV norm is almost equivalent to minimizing the surface area. If a solution has a minimum surface area, it has the least oscillations. TV norm minimization avoids the l_0 quasi-norm and l_1 norm minimization.

Using Green's one-step-late method presented in Section 6.9.6, a TV-regulated EM (TV-EM) algorithm can be obtained as

$$x_i^{\text{next}} = \frac{x_j^{\text{current}}}{\sum_i a_{ij} + \beta \frac{\partial \text{TV}(x^{\text{current}})}{\partial x_j^{\text{current}}}} \sum_i a_{ij} \frac{p_i}{\sum_j a_{ij} x_j^{\text{current}}}. \tag{6.102}$$

For a 2D image X, if we express each pixel with double indices as $x_{k,l}$, then the partial derivative in the above TV-EM algorithm can be given as

$$\begin{aligned}
\frac{\partial \text{TV}(X)}{\partial x_{k,l}} &= \frac{x_{k,l} - x_{k-1,l}}{\sqrt{\left(x_{k,l} - x_{k-1,l}\right)^2 + \left(x_{k-1,l+1} - x_{k-1,l}\right)^2 + \varepsilon^2}} \\
&+ \frac{x_{k,l} - x_{k,l-1}}{\sqrt{\left(x_{k+1,l-1} - x_{k,l-1}\right)^2 + \left(x_{k,l} - x_{k,l-1}\right)^2 + \varepsilon^2}} \\
&- \frac{x_{k+1,l} + x_{k,l+1} - 2x_{k,l}}{\sqrt{\left(x_{k+1,l} - x_{k,l}\right)^2 + \left(x_{k,l+1} - x_{k,l}\right)^2 + \varepsilon^2}}.
\end{aligned} \tag{6.103}$$

Reconstruction with highly undersampled data is a Bayesian reconstruction problem. Performing the sparsifying transform ψX is one way to extract the prior knowledge from the image X. If other prior information about the image X is available, it should be included in the objective function as well.

6.11 Worked examples

Example 1: For an arbitrary matrix A, its generalized inverse matrix A^+ must satisfy the following four properties:

$$AA^+A = A, \tag{6.104}$$

$$A^+AA^+ = A, \tag{6.105}$$

$$\left(A^+A\right)^* = A^+A, \tag{6.106}$$

$$\left(AA^+\right)^* = AA^+. \tag{6.107}$$

Here M^* is the Hermitian transpose (also called conjugate transpose) of a matrix M. For a real matrix, it is simply the transpose. In the following, we assume that matrix A is real.

Please verify that if $(A^TA)^{-1}$ exists, then $A^+ = (A^TA)^{-1}A^T$ is a generalized inverse matrix of A. Please also verify that if $(AA^T)^{-1}$ exists, then $A^+ = A^T(AA^T)^{-1}$ is a generalized inverse matrix of A.

Proof.

Case 1: If $(A^TA)^{-1}$ exists, $A^+ = (A^TA)^{-1}A^T$.

Property 1:

$$\text{Left} = AA^+A = A\left(A^TA\right)^{-1}A^TA = A\left(A^TA\right)^{-1}\left(A^TA\right) = A = \text{Right} \tag{6.108}$$

Property 2:

$$\text{Left} = A^+AA^+ = \left(A^TA\right)^{-1}A^TA\left(A^TA\right)^{-1}A^T = \left(A^TA\right)^{-1}\left(A^TA\right)\left(A^TA\right)^{-1}A^T$$

$$= \left(A^TA\right)^{-1}A^T = A^+ = \text{Right} \tag{6.109}$$

Property 3:

$$\text{Left} = (A^+A)^T = \left(\left(A^TA\right)^{-1}A^TA\right)^T = \left(\left(A^TA\right)^{-1}\left(A^TA\right)\right)^T = I \tag{6.110}$$

$$\text{Right} = A^+A = \left(A^TA\right)^{-1}A^TA = \left(A^TA\right)^{-1}\left(A^TA\right) = I \tag{6.111}$$

Property 4:

$$\text{Left} = (AA^+)^T = \left(A\left(A^TA\right)^{-1}A^T\right)^T = A\left(\left(A^TA\right)^{-1}\right)^TA^T$$

$$= A\left(\left(A^TA\right)^T\right)^{-1}A^T = A\left(A^TA\right)^{-1}A^T = AA^+ = \text{Right} \tag{6.112}$$

Case 2: If $(AA^T)^{-1}$ exists, $A^+ = A^T(AA^T)^{-1}$.

Property 1:

$$\text{Left} = AA^+ A = AA^T \left(AA^T\right)^{-1} A = \left(AA^T\right) \left(AA^T\right) A = A = \text{Right} \tag{6.113}$$

Property 2:

$$\text{Left} = A^+ AA = A^T \left(AA^T\right)^{-1} AA^T \left(AA^T\right) = A^T \left(AA^T\right)^{-1} \left(AA^T\right) \left(AA^T\right)^{-1}$$
$$= A^T \left(AA^T\right)^{-1} = A^+ = \text{Right} \tag{6.114}$$

Property 3:

$$\text{Left} = \left(A^+ A\right)^T = \left(A^T \left(AA^T\right)^{-1} A\right)^T = A^T \left(\left(AA^T\right)^{1}\right)^T A$$
$$= A^T \left(\left(AA^T\right)^T\right)^{-1} A = A^T \left(AA^T\right) A = A^+ A = \text{Right} \tag{6.115}$$

Property 4:

$$\text{Left} = \left(AA^+\right)^T = \left(AA^T \left(AA^T\right)^{-1}\right)^T = \left(\left(AA^T\right) \left(AA^T\right)^{-1}\right)^T = I \tag{6.116}$$

$$\text{Right} = AA^+ = AA^T \left(A^T A\right)^{-1} = \left(AA^T\right) \left(A^T A\right)^{-1} = I \tag{6.117}$$

Example 2: In Section 6.1, we used SVD to decompose matrix $A_{m \times n}$ into $A_{m \times n} = U_{m \times m} \Sigma V^T{}_{n \times n}$ and defined $A^+ = V \Sigma^+ U^T$ (assuming $\sigma_r > 0$) with $\Sigma^+_{n \times m} = \begin{bmatrix} D_r & 0 \\ 0 & 0 \end{bmatrix}$ and D_r, = diag $\{1/\sigma_1,$ $1/\sigma_2, \ldots, 1/\sigma_r, 0, \ldots, 0\}$. Is $A^+ = V\Sigma^+ U^T$ a generalized inverse of matrix A according to the four properties stated in Example 1? Find the condition under which $A^+ = V\Sigma^+ U^T$ is a generalized inverse of matrix A.

Solution
Let us apply the first property to $A^+ = V\Sigma^+ U^T$ and $A_{m \times n} = U_{m \times m} \Sigma V^T_{n \times n}$:

$$AA^+ A = U\Sigma V^T V\Sigma^+ U^T U\Sigma V^T = U\Sigma \left(V^T V\right) \Sigma^+ \left(U^T U\right) \Sigma V^T = U\Sigma\Sigma^+ \Sigma V^T. \tag{6.118}$$

In order to have $AA^+ A = A^+$, we must have

$$\Sigma\Sigma^+ \Sigma = \Sigma, \tag{6.119}$$

which is equivalent to

$$\text{diag}\{ \sigma_1, \sigma_2, \ldots, \sigma_r, \ldots, \sigma_m\} \text{diag}\left\{\tfrac{1}{\sigma_1}, \tfrac{1}{\sigma_2}, \ldots, \tfrac{1}{\sigma_r}, 0, \ldots, 0\right\} \text{diag}\{ \sigma_1, \sigma_2, \ldots, \sigma_r, \ldots, \sigma_m\}$$
$$= \text{diag}\{ \sigma_1, \sigma_2, \ldots, \sigma_r, \ldots, \sigma_m\}, \tag{6.120}$$

or

$$\sigma_{r+1} = \cdots = \sigma_m = 0. \tag{6.121}$$

Since

$$\sigma_1 \geq \sigma_2 \geq \cdots \geq \sigma_l \geq \cdots \geq 0, \tag{6.122}$$

Property 1 is equivalent to

$$\sigma_{r+1} = 0 \quad \text{and} \quad \sigma_r > 0. \tag{6.123}$$

Now let us look at Property 2:

$$A^+ A A^+ = A^+, \tag{6.124}$$

which is equivalent to

$$\text{diag}\left\{\frac{1}{\sigma_1}, \frac{1}{\sigma_2}, \ldots, \frac{1}{\sigma_r}, 0, \ldots, 0\right\} = \text{diag}\left\{\frac{1}{\sigma_1}, \frac{1}{\sigma_2}, \ldots, \frac{1}{\sigma_r}, 0, \ldots, 0\right\}. \tag{6.125}$$

This property is always satisfied.

Property 3 $A^+ A^* = A^* A$ and Property 4 $(AA^+)^* = AA^+$ both imply that

$$\text{diag}\{1, 1, \ldots, 1, 0, \ldots, 0\} = \text{diag}\{1, 1, \ldots, 1, 0, \ldots, 0\}, \tag{6.126}$$

which is always satisfied.

To summarize the above discussion, the condition under which $A^+ = V\Sigma^+ U^T$ is a generalized inverse of matrix A is

$$\sigma_{r+1} = 0 \quad \text{and} \quad \sigma_r > 0. \tag{6.127}$$

However, in practice, a cutoff index much smaller than this r is used to obtain a stable solution.

Example 3: None of the following images match the measured projections. Which one is the best solution among the three solutions in Figure 6.23? (Compare χ^2)

Solution
i) $\chi^2 = 0^2 + 0^2 + 1^2 + 1^2 = 2,$ ($\leftarrow$ Best)

ii) $\chi^2 = 0^2 + 0^2 + 2^2 + 0^2 = 4,$

iii) $\chi^2 = 1^2 + 0^2 + 1^2 + 2^2 = 6.$

Example 4: Find a least-squares reconstruction to the imaging problem shown in Figure 6.24.

Solution
The imaging matrix A for this problem $AX = P$ is given as

$$A = \begin{bmatrix} 1 & 0 & 1 & 0 \\ 0 & 1 & 0 & 1 \\ 1 & 1 & 0 & 0 \\ 0 & 0 & 1 & 1 \end{bmatrix}. \tag{6.128}$$

The rank of A is 3. The system $AX = P$ is also not consistent. Therefore, there is no solution for this problem. We can use SVD-based method to find a pseudo-solution. Using Matlab with $X = pinv(A) * P$, we get $x_1 = 2.25$, $x_2 = 1.75$, $x_3 = 1.75$, and $x_4 = 1.25$.

Example 5: Let $A = \begin{bmatrix} 2 & 0 \\ 0 & 1 \end{bmatrix}$. Find the conjugate direction $\vec{u}_1$ of $\vec{u}_0 = \begin{bmatrix} 1 \\ 1 \end{bmatrix}$.

Solution

The conjugate direction is defined by this relationship: $\vec{u}_0 \cdot (ATA)\, \vec{u}_1 = 0$, where

$$A^T A = \begin{bmatrix} 2 & 0 \\ 0 & 1 \end{bmatrix}\begin{bmatrix} 2 & 0 \\ 0 & 1 \end{bmatrix} = \begin{bmatrix} 4 & 0 \\ 0 & 1 \end{bmatrix}. \tag{6.129}$$

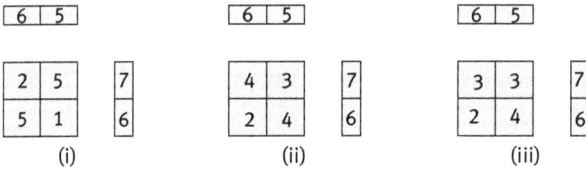

Fig. 6.23: Three images try to match the projections.

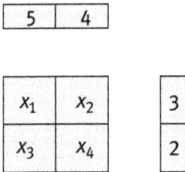

Fig. 6.24: A 2 × 2 image reconstruction problem.

We also know

$$\begin{bmatrix} 1 \\ 1 \end{bmatrix} \cdot \begin{bmatrix} 1 \\ -1 \end{bmatrix} = 0. \tag{6.130}$$

Thus $\begin{bmatrix} 4 & 0 \\ 0 & 1 \end{bmatrix} \vec{u}_1 = \begin{bmatrix} 1 \\ -1 \end{bmatrix}$, which leads to

$$\vec{u}_1 = \begin{bmatrix} 1/4 & 0 \\ 0 & 1 \end{bmatrix}\begin{bmatrix} 1 \\ -1 \end{bmatrix} = \begin{bmatrix} 1/4 \\ -1 \end{bmatrix}. \tag{6.131}$$

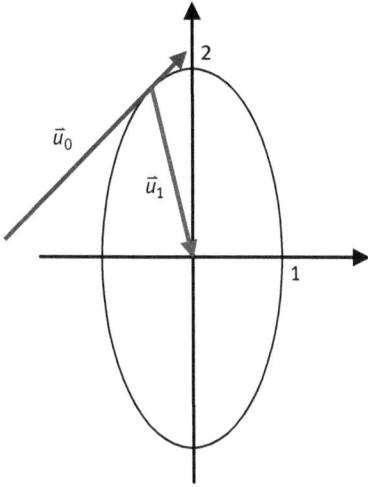

Fig. 6.25: Conjugate directions.

(If you want, you could normalize $\vec{u}_1$ and make it a unit vector.)

The matrix $A^T A$ defines a quadratic form, which is an ellipse in our case. The point of this example is as follows: For any initial direction $\vec{u}_0$, draw an ellipse specified by the quadratic form $A^T A$ such that $\vec{u}_0$ is a tangential direction. If you travel along the conjugate direction $\vec{u}_1$, you will reach the center of the ellipse (see Figure 6.25).

Example 6: Compute one iteration of the ML-EM algorithm with the initial image and projections given in Figure 6.26.

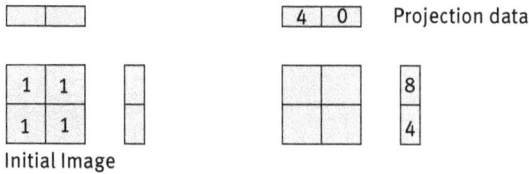

Fig. 6.26: Initial image and projection data for Example 6.

Solution

Step 1: Find the forward projections:

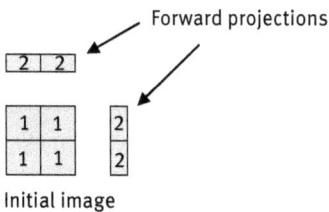

Step 2: Find the ratios of given (i.e., measured) projections and the forward projections of the current image:

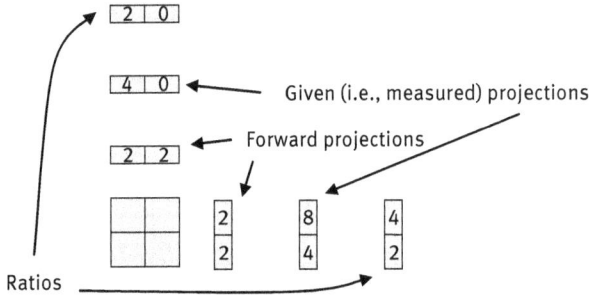

Step 3: Backproject the ratios:

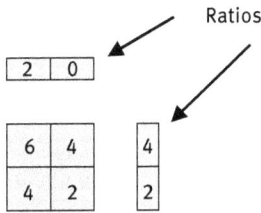

Step 4: Backproject a constant 1 from all rays:

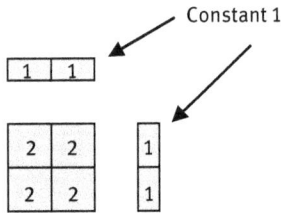

Step 5: Find the pixel-by-pixel ratio of the backprojected image obtained from Step 3 and the backprojected image from Step 4:

Step 6: Update the current image pixel by pixel (using point-by-point multiplication, not matrix multiplication):

$$
\begin{array}{|c|c|}
\hline
1 & 1 \\
\hline
1 & 1 \\
\hline
\end{array}
\;\times\;
\begin{array}{|c|c|}
\hline
3 & 2 \\
\hline
2 & 1 \\
\hline
\end{array}
\;=\;
\begin{array}{|c|c|}
\hline
3 & 2 \\
\hline
2 & 1 \\
\hline
\end{array}
$$

Example 7: Use a Bayesian method to find a stable solution of the system

$$
\begin{cases}
x_1 + 0.01x_2 = 1.2 \\
x_1 + 0.001x_2 = 1.
\end{cases}
\tag{6.132}
$$

Solution
If we consider the potential measurement errors, the system of equations can be written as

$$
\begin{cases}
x_1 + 0.01x_2 = 1.2 + \delta_1 \\
x_2 + 0.001x_2 = 1 + \delta_2.
\end{cases}
\tag{6.133}
$$

The solution of this modified system is given as

$$
\begin{cases}
x_1 = 0.978 - 0.111\delta_1 + 1.11\delta_2 \\
x_2 = 22.22 + 111.1\delta_1 - 111.1\delta_2.
\end{cases}
\tag{6.134}
$$

It is seen that x_2 is sensitive to measurement noise. Let us assume a priori that "x_1 and x_2 are close" and solve this problem using the Bayesian method. Here, a prior is a probability distribution representing knowledge or belief about an unknown quantity a priori, that is, before any data have been observed. First, we set up an objective function and use the prior knowledge as a penalty term:

$$
F(x_1, x_2) = (x_1 + 0.01x_2 - 1.2 - \delta_1)^2 + (x_1 + 0.001x_2 - 1 - \delta_2)^2 + \beta(x_1 + x_2)^2.
\tag{6.135}
$$

Note that β in this expression is not a Lagrange multiplier, but a preassigned constant. To minimize $F(x_1, x_2)$, we set $\partial F/\partial x_1 = 0$ and $\partial F/\partial x_2 = 0$. This results in a *different* problem:

$$
\begin{bmatrix}
2+\beta & 0.011-\beta \\
0.011-\beta & 0.01^2 + 0.001^2 + \beta
\end{bmatrix}
\begin{bmatrix}
x_1 \\
x_2
\end{bmatrix}
=
\begin{bmatrix}
(1.2+\delta_1) + (1+\delta_2) \\
(0.01)(1.2+\delta_1) + (0.001)(1+\delta_2)
\end{bmatrix}.
\tag{6.136}
$$

The solution of this system depends on the value of β, as well as the values of δ_1 and δ_2. Some MAP solutions are listed in the following two tables:

Case 1: Noiseless ($\delta_1 = \delta_2 = 0$).

β	0	0.01	0.1	1	10	100
Condition #	50,000	200	22	5.7	22	200
x_1	0.978	1.094	1.094	1.094	1.094	1.094
x_2	22.222	1.178	1.103	1.095	1.094	1.094

Case 2: Noisy ($\delta_1 = -\delta_2 = 0.2$).

β	0	0.01	0.1	1	10	100
Condition #	50,000	200	22	5.7	22	200
x_1	0.733	1.094	1.095	1.095	1.095	1.095
x_2	66.667	1.357	1.122	1.098	1.096	1.095

The system becomes more stable with an additional prior term in the objective function. The stability is reflected by much smaller condition numbers. This example also tells us that using a prior changes the original problem. Even for noiseless data, the MAP solutions may be different from the true solution. The Bayesian constraint $\beta(x_1 - x_2)^2$ in eq. (6.135) is not a good one, and a better constraint should have been used. When you plan to use the Bayesian method, be careful and be sure that your prior knowledge is reasonable.

Example 8: Show that for the ML-EM algorithm at each iteration, the total number of the counts of the forward projections is the same as the total number of the counts of the original projection data:

$$\sum_i \left(\sum_j a_{ij} x_j \right) = \sum_j p_i. \tag{6.137}$$

Proof.

$$\sum_i \sum_j a_{ij} x_j = \sum_i \sum_j a_{ij} \left(\frac{x_j^{\text{old}}}{\sum_n a_{nj}} \sum_n a_{nj} \frac{p_n}{\sum_k a_{nk} x_k^{\text{old}}} \right)$$

$$= \sum_j x_j^{\text{old}} \frac{\sum_i a_{ij}}{\sum_n a_{nj}} \sum_n a_{nj} \frac{p_n}{\sum_k a_{nk} x_k^{\text{old}}} \quad \text{[Change the oder of summations.]}$$

$$= \sum_j x_j^{\text{old}} \sum_n a_{nj} \frac{p_n}{\sum_k a_{nk} x_k^{\text{old}}} \quad \left[\because \sum_i a_{ij} = \sum_n a_{nj} \right]$$

$$= \sum_n p_n \frac{\sum_j a_{nj} x_j^{old}}{\sum_k a_{nk} x_k^{old}} \text{ [Change the oder of summations.]}$$

$$= \sum_n p_n. \tag{6.138}$$

Example 9: Use a computer simulation to demonstrate that the resolution recovery rate is location dependent in the ML-EM algorithm.

Solution

A 2D computer-generated phantom shown in Figure 6.27 is used to generate projection data with and without Poisson noise. In data generation, attenuation, system blurring, and photon scattering are not included. That is, the projections are ideal line integrals of the object. The image size is 256×256, and 256 views uniformly distributed over $360°$ are used for projection data generation.

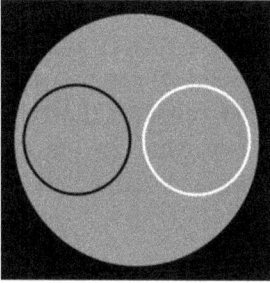

Fig. 6.27: The true image (a mathematical phantom).

The iterative ML-EM algorithm is used for image reconstruction. Two reconstructed images are shown in Figure 6.28: one of which is obtained after 25 iterations and the other is obtained after 250 iterations.

After 250 iterations, the algorithm has almost converged, and the resolution throughout the image is uniform. When the algorithm is stopped early at the 25th iteration, the resolution is not uniformly recovered. Higher resolution can be observed at the edge of the object. The resolution recovery rate is slower at the center. The motivation of early stopping is to regulate the noise. This is demonstrated with the noisy data reconstructions as shown in Figure 6.29, where two images are displayed. One of the images is obtained after 25 iterations and the other is obtained after 250 iterations.

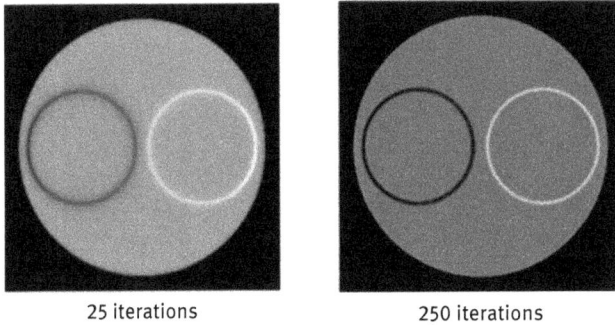

25 iterations   250 iterations

Fig. 6.28: Iterative ML-EM reconstructions.

The image with 250 iterations is noisier than that with 25 iterations. If we apply a low-pass filter to the noisy image obtained with 250 iterations, the image becomes less noisy and still maintains the uniform resolution throughout the image (see Figure 6.30). Therefore, it is a good strategy to over-iterate and then to apply a low-pass filter to control noise.

On the other hand, if the projections are ideal line integrals, the 2D parallel-beam FPB algorithm can provide images with uniform resolution (see Figure 6.31). However, if the imaging system has a spatially variant resolution and sensitivity, the conventional FBP algorithms are unable to model the system accurately and cannot provide uniform resolution in their reconstructions.

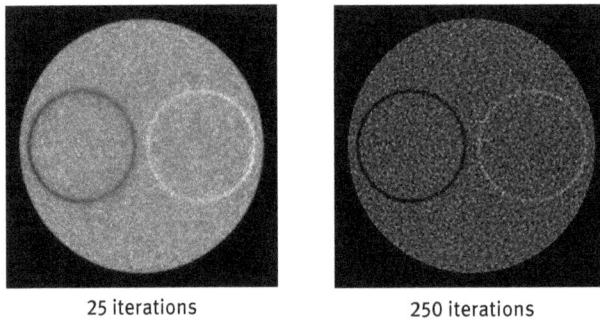

25 iterations   250 iterations

Fig. 6.29: Iterative ML-EM reconstructions with the same noisy projection data.

Example 10: When the system of imaging equations is underdetermined, use 2-pixel (x_1, x_2) or 3-pixel (x_1, x_2, x_3) examples to illustrate solutions with the minimum l_2 norm, l_1 norm, and l_0 quasi-norm, respectively.

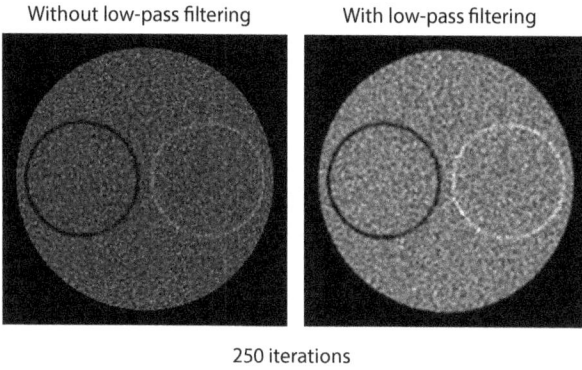

Fig. 6.30: Applying a low-pass filter to the image reconstructed with 250 iterations of ML-EM results in uniform resolution and reduced noise.

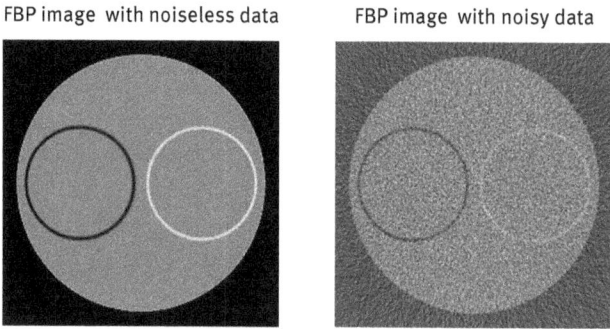

Fig. 6.31: Filtered backprojection reconstructions with noiseless and noisy data.

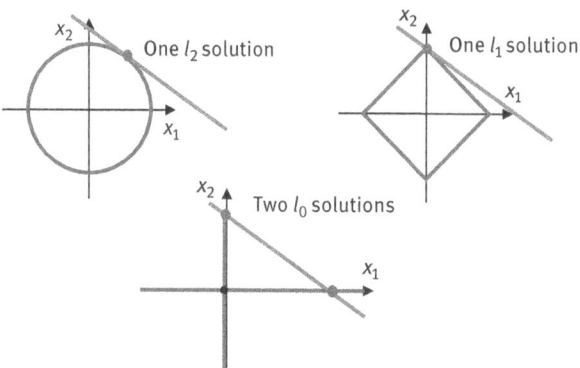

Fig. 6.32: Illustration of 2-pixel solutions with the minimum l_2 norm, l_1 norm, and l_0 quasi-norm.

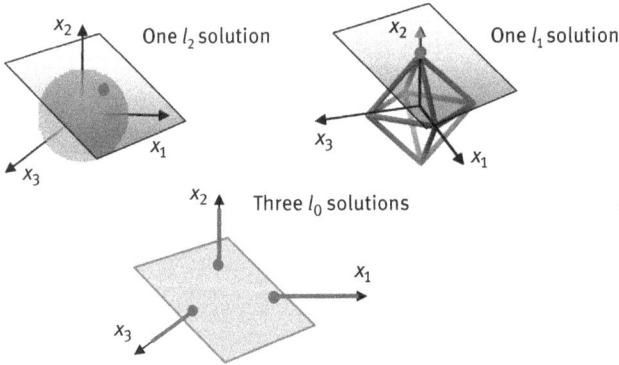

Fig. 6.33: Illustration of 3-pixel 1-measurement solutions with the minimum l_2 norm, l_1 norm, and l_0 quasi-norm.

Solution

Let us consider a 2-pixel (x_1, x_2) case, and there is one measurement. This measurement can be represented as a straight line in the (x_1, x_2) coordinate system. Any point on the line is a valid solution. The solutions with the minimum l_2 norm, l_1 norm, and l_0 quasi-norm are shown in Figure 6.32.

In the 3-pixel (x_1, x_2, x_3) case, there can be one measurement or two measurements. If there is one measurement, this measurement can be represented as a plane in the (x_1, x_2, x_3) coordinate system. Any point on the plane is a valid solution. The solutions with the minimum l_2 norm, l_1 norm, and l_0 quasi-norm are shown in Figure 6.33.

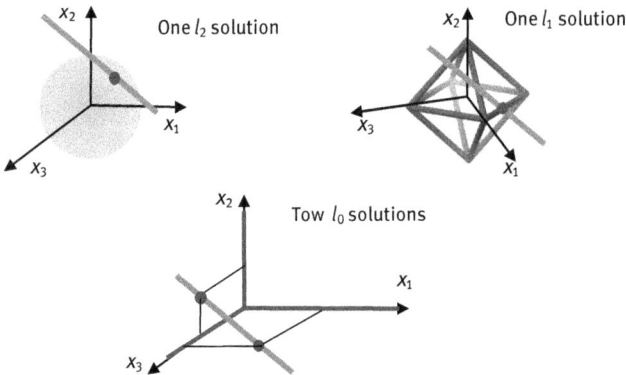

Fig. 6.34: Illustration of 3-pixel 2-measurement solutions with the minimum l_2 norm, l_1 norm, and l_0 quasi-norm.

In the 3-pixel (x_1, x_2, x_3) case, let us consider the case of two measurements. Each measurement can be represented as a plane in the (x_1, x_2, x_3) coordinate system. Two measurements correspond to two planes. These two planes intersect and form a straight line. Any point on the line is a valid solution. The solutions with the minimum l_2 norm, l_1 norm, and l_0 quasi-norm are shown in Figure 6.34, respectively.

6.12 Summary

- The main difference between an analytic image reconstruction algorithm and an iterative image reconstruction algorithm is in image modeling. In an analytic algorithm, the image is assumed to be continuous, and each image pixel is a point. The set of discrete pixels is for display purpose. We can make those display points any way we want. However, in an iterative algorithm, a pixel is an area, which is used to form the projections of the current estimate of the image. The pixel model can significantly affect the quality of the reconstructed image.
- Another difference between an analytic image reconstruction algorithm and an iterative image reconstruction algorithm is that the analytic algorithm tries to solve an integral equation, while the iterative algorithm tries to solve a system of linear equations.
- A system of linear equations is easier to solve than an integral equation. This allows the linear equations to model more realistic and more complex imaging geometry and imaging physics. In other words, the iterative algorithm can solve a more realistic imaging problem than an analytic algorithm. As a result, the iterative algorithm usually provides a more accurate reconstruction.
- Iterative algorithms are used to minimize an objective function. This objective function can effectively incorporate the noise in the measurement. Currently, analytic algorithms control noise by frequency windowing.
- The iterative ML-EM and OS-EM algorithms are most popular in emission tomography image reconstruction. They assume Poisson's noise statistics. The transmission data counterparts exist.
- Even though noise is modeled in the objective function, the reconstructed image is still noisy. There are five methods to control noise.
- The first method is to stop the iterative algorithm early before it converges. This simple method has a drawback that it may result in an image with nonuniform resolution. One remedy is to iterate till convergence and apply a post low-pass filter to suppress the noise.
- The second method is to replace the flat, nonoverlapping pixels by smooth, overlapping pixels to represent the image. The method can remove the artificially introduced high-frequency components by the flat, nonoverlapping pixels in the image. A drawback of using smooth, overlapping pixels is the increased computational complexity. One remedy is to use the traditional flat, nonoverlapping pixels and apply a low-pass filter to backprojected images.

- The third method is to model more accurate imaging geometry and physics in the projector/backprojector pair. The aim of this method is to reduce the deterministic discrepancy between the projection model and the measured data.
- The fourth method is to use the correct noise model to set up an objective function. The author's personal belief is that the noise model that is based on the joint probability density function is not very critical. You can have a not-so-accurate noise model (i.e., a not-so-accurate joint probability density function) but you must have the accurate variance. The important part is to incorporate a correct measurement noise variance to weigh the data. It is not as important to worry about whether the noise is strictly Gaussian, strictly Poisson, and so on.
- The fifth method is the use of prior knowledge. If the projection data do not carry enough information about the object, due partially to insufficient measurements and partially to noise, prior knowledge about the object can supplement more information about the object and make the iterative algorithm more stable. Be careful that if the prior knowledge is not really true, it can mislead the algorithm to converge to a wrong image.
- The readers are expected to understand how the system of imaging equations is set up and how the iterative ML-EM algorithm works in this chapter.

Problems

Problem 6.1 Some iterative algorithms, for example, the ART and OS-EM algorithms, update the image very frequently. For those algorithms, the processing order of the data subsets is important. In this problem, we use the ART algorithm to graphically solve a system of linear equations $\{L_1, L_2, L_3, L_4\}$ with two variables as shown in the figure below. The initial estimated solution is X^0. Solve the system with two different orders: (a) $L_1 \rightarrow L_2 \rightarrow L_3 \rightarrow L_4$ and (b) $L_1 \rightarrow L_3 \rightarrow L_2 \rightarrow L_4$, respectively. Compare their performance in terms of convergence rate.

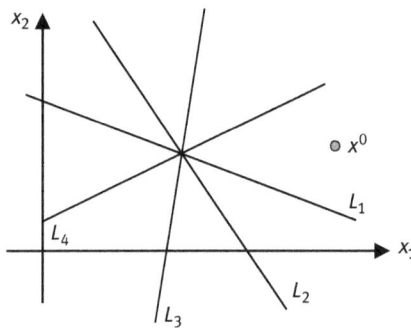

Problem 6.2 The iterative ML-EM algorithm $x_j^{(k+1)} = x_j^{(k)} \dfrac{\sum_i a_{ij} \frac{p_i}{\sum_j a_{ij} x_j^{(k)}}}{\sum_i a_{ij}}$, or the iterative OS-EM algorithm, has many modified versions. One of the versions introduces a new parameter h:

$$x_j^{(k+1)} = x_j^{(k)} \left(\frac{\sum_i a_{ij} \frac{p_i}{\sum_j a_{ij} x_j^{(k)}}}{\sum_i a_{ij}} \right)^h.$$

This parameter h usually takes a real value in the interval between 1 and 5. The purpose of using this parameter h is to increase the iteration step size and make the algorithm converge faster. If the parameter h is chosen in the interval between 0 and 1, it reduces the iteration step size. Does this new algorithm satisfy the property of total count conservation as studied in Worked Problem 8 in this chapter? If that property is not satisfied any more, you can always scale the newly updated image with a factor. Find this scaling factor so that the total count is conserved for each iteration.

Problem 6.3 The modified algorithm discussed in Problem 6.2 can only be used in emission imaging applications because the linear equations are weighted by the Poisson variance of the emission measurements. One can develop a similar algorithm for the transmission measurements to find the attenuation map of the object as

$$\mu_j^{(k+1)} = \mu_j^{(k)} \left(\frac{\sum_i a_{ij} \frac{N_0 e^{-\sum_j a_{ij} \mu_j^{(k)}}}{N_i}}{\sum_i a_{ij}} \right)^h.$$

where the measurements are modeled as $N_i = N_0 e^{-\sum_j a_{ij}\mu_j}$. Here, we do not take the logarithm and convert these nonlinear equations into linear equations. Instead, we go ahead and solve this system of nonlinear equations $N_i = N_0 e^{-\sum_j a_{ij}\mu_j}$ directly. Use the Taylor expansion to convert this algorithm to its corresponding additive updating form. Discuss how the parameter h controls the iteration step size, how the nonlinear equations are weighted, and what the weighting quantity is.

Problem 6.4 We have learned that in an iterative image reconstruction algorithm the projector should model the imaging system as accurately as possible. For a set of practical data acquired from an actual imaging system, there is a simple way to verify whether the modeling in the projector is accurate enough. This method is described as follows. Run an iterative algorithm to reconstruct the image. After the algorithm is converged, calculate and display the data discrepancy images at every view. A discrepancy image is the difference between the projection of the reconstructed image and the measured projection data, or is the ratio of the projection of the reconstructed image to the measured projection data. If the projector models the imaging system accurately, you do not see the shadow of the object in the discrepancy images and you only see the random noise in the discrepancy images. Verify this method with a computer simulation.

Problem 6.5   In Example 7, a bad Bayesian constraint $\beta(x_1 - x_2)^2$ in eq. (6.135) was chosen. Redo Example 7 with a different Bayesian constraint of your choice. For example, you could choose a minimum norm constraint.

Bibliography

[1] Bai C, Zeng GL, Gullberg GT, DiFilippo F, Miller S (1998) Slab-by-slab blurring model for geometric point response and attenuation correction using iterative reconstruction algorithms. IEEE Trans Nucl Sci 45:2168–2173.

[2] Barrett HH, Wilson DW, Tsui BMW (1994) Noise properties of the EM algorithm, I. Theory Phys Med Biol 39:833–846.

[3] Beekman F, Kamphuis C, Viergever M (1996) Improved SPECT quantitation using fully three-dimensional iterative spatially variable scatter response compensation. IEEE Trans Med Imaging 15:491–499.

[4] Byrne CL (1996) block-iterative methods for image reconstruction from projections. IEEE Trans Imaging Process 5:792–794.

[5] Candès E, Romberg J, Tao T (2006) Robust uncertainty principles: Exact signal reconstruction from highly incomplete frequency information. IEEE Trans Inf Theory 52:489–509.

[6] Censor Y, Eggermont PPB, Fordon D (1983) Strong underrelaxation in Kaczmarz's method for inconsistent system. Numer Math 41:83–92.

[7] Chang L (1979) Attenuation and incomplete projection in SPECT. IEEE Trans Nucl Sci 26:2780–2789.

[8] Chen GH, Tang J, Hsieh YL (2009) Temporal resolution improvement using PICCS in MDCT cardiac imaging. Med Phys 36:2130–2135.

[9] Chiao PC, Rogers WL, Fessler JA, Clinthorne NH, Hero AO (1994) Model-based estimation with boundary side information or boundary regularization. IEEE Trans Med Imaging 13:227–234.

[10] Dempster AP, Laird NM, Rubin DB (1977) Maximum likelihood from incomplete data via the EM algorithm. J Royal Stat Soc B 39:1–38.

[11] DiBella E, Barclay A, Eisner R, Schaefer R (1996) A comparison of rotation-based methods for iterative reconstruction algorithms. IEEE Trans Nucl Sci 43:3370–3376.

[12] Donoho D (2006) Compressed sensing. IEEE Trans Inf Theory 52:1289–1306.

[13] Elbakri A, Fessler JA (2002) Statistical image reconstruction for polyenergetic x-ray computed tomography. IEEE Trans Med Imaging 21:89–99.

[14] Fessler J (1996) Mean and variance of implicitly defined biased estimators (such as penalized maximum likelihood): Applications to tomography. IEEE Trans Image Process 5:493–506.

[15] Fessler J, Hero A (1994) Space-alternating generalized expectation maximization algorithm. IEEE Trans Signal Process 42:2664–2677.

[16] Fessler JA, Rogers WL (1996) Spatial resolution properties of penalized-likelihood image reconstruction: Space-invariant tomography. IEEE Trans Imaging Process 5:1346–1358.

[17]   Frey EC, Ju ZW, Tsui BMW (1993) A fast projector-backprojector pair for modeling the asymmetric spatially-varying scatter response in SPECT imaging. IEEE Trans Nucl Sci 40:1192–1197.

[18]   Geman S, Geman D (1984) Stochastic relaxation, Gibbs distributions, and Bayesian restoration of images. IEEE Trans Pattern Anal Mach Intell 6:721–741.

[19]   Gilbert P (1972) Iterative methods for the reconstruction of three dimensional objects from their projections. J Theor Biol 36:105–117.

[20]   Gindi G, Lee M, Rangarajan A, Zubal IG (1991) Bayesian reconstruction of functional images using registered anatomical images as priors. In: Colchester ACF, Hawkes DJ (eds) Information Processing in Medical Imaging, Springer-Verlag, New York, 121–131.

[21]   Gordon R, Bender R, Herman GT (1970) Algebraic reconstruction techniques (ART) for three-dimensional electron microscopy and X-ray photography. J Theor Biol 29:471–481.

[22]   Hebert TJ, Leahy R (1992) Statistic-based MAP image reconstruction from Poisson data using Gibbs priors. IEEE Trans Signal Process 40:2290–2303.

[23]   Herman GT (1980) Image Reconstruction from Projections: The Fundamentals of Computerized Tomography, Academic Press, New York.

[24]   Hsieh YL, Zeng GL, Gullberg GT (1998) Projection space image reconstruction using strip functions to calculate pixels more "natural" for modeling the geometric response of the SPECT collimator. IEEE Trans Med Imaging 17:24–44.

[25]   Hudson HM, Larkin RS (1994) Accelerated image reconstruction using ordered subsets of projection data. IEEE Trans Med Imaging 13:601–609.

[26]   Hwang DS, Zeng GL (2005) A new simple iterative reconstruction algorithm for SPECT transmission measurement. Med Phys 32:2312–2319.

[27]   Hwang DS, Zeng GL (2005) Reduction of noise amplification in SPECT using smaller detector bin size. IEEE Trans Nucl Sci 52:1417–1427.

[28]   Hwang DS, Zeng GL (2006) Convergence study of an accelerated ML-EM algorithm using bigger step size. Phys Med Biol 51:237–252.

[29]   Qi J (2003) A unified noise analysis for iterative image estimation. Phys Med Biol 48:3505–3519.

[30]   Kacmarz S (1937) Angenaherte Aufosung von Systemen linearer Gleichungen. Bull Acad Polon Sci A 35:355–357.

[31]   King MA, Miller TR (1985) Use of a nonstationary temporal Wiener filter in nuclear medicine. Eur J Nucl Med 10:458–461.

[32]   Lalush DS, Tsui BMW (1993) A generalized Gibbs prior for maximum a posteriori reconstruction in SPECT. Phys Med Biol 38:729–741.

[33]   Lalush DS, Tsui BMW (1994) Improving the convergence of iterative filtered backprojection algorithms. Med Phys 21:1283–1286.

[34]   Lange K (1990) Convergence of EM image reconstruction algorithms with Gibbs smoothing. IEEE Trans Med Imaging 9:439–446.

[35]   Lange K, Carson R (1984) EM reconstruction algorithms for emission and transmission tomography. J Comput Assist Tomogr 8:306–316.

[36]   Leng S, Tang J, Zambelli J, Nett B, Tolakanahalli R, Chen GH (2008) High temporal resolution and streak-free four-dimensional cone-beam computed tomography. Phys Med Biol 53:5653–5673.

[37]   Lewitt R (1992) Alternatives to voxels for image representation in iterative reconstruction algorithms. Phys Med Biol 37:705–716.

[38]   Lewitt RM, Muehllehner G (1986) Accelerated iterative reconstruction for positron emission tomography based on the EM algorithm for maximum likelihood estimation. IEEE Trans Med Imaging 5:16–22.

[39]   Liang Z (1993) Compensation for attenuation, scatter, and detector response in SPECT reconstruction via iterative FBP methods. Med Phys 20:1097–1106.

[40]   Liang Z, Jazczak R, Greer K (1989) On Bayesian image reconstruction from projections: Uniform and nonuniform *a priori* source information. IEEE Trans Med Imaging 8:227–235.
[41]   Liow J, Strother SC (1993) The convergence of object dependent resolution in maximum likelihood based tomographic image reconstruction. Phys Med Biol 38:55–70.
[42]   Lucy LB (1974) An iterative technique for the rectification of observed distribution. Astrophys J 79:745–754.
[43]   Naraynan VM, King MA, Soare E, Byrne C, Pretorius H, Wernick MN (1999) Application of the Karhunen-Loeve transform to 4D reconstruction of gated cardiac SPECT images. IEEE Trans Nucl Sci 46:1001–1008.
[44]   Nuyts J, Michel C, Dupont P (2001) Maximum-likelihood expectation-maximization reconstruction of sinograms with arbitrary noise distribution using NEC-transformations. IEEE Trans Med Imaging 20:365–375.
[45]   Pan TS, Yagle AE (1991) Numerical study of multigrid implementations of some iterative image reconstruction algorithms. IEEE Trans Med Imaging 10:572–588.
[46]   Pan X, Wong WH, Chen CT, Jun L (1993) Correction for photon attenuation in SPECT: Analytical framework, average attenuation factors, and a new hybrid approach. Phys Med Biol 38:1219–1234.
[47]   Panin VY, Zeng GL, Gullberg GT (1999) Total variation regulated EM algorithm. IEEE Trans Nucl Sci 46:2202–2210.
[48]   Qi J, Leahy RM (1999) Fast computation of the covariance of MAP reconstructions of PET images. Proc SPIE 3661:344–355.
[49]   Rucgardson WH (1972) Bayesian-based iterative method of image restoration. J Opt Soc Am A 62:55–59.
[50]   Shepp LA, Vardi Y (1982) Maximum likelihood estimation for emission tomography. IEEE Trans Med Imaging 1:113–121.
[51]   Snyder DL, Miller MI (1985) The use of sieves of stabilize images produced with the EM algorithm for emission tomography. IEEE Trans Nucl Sci 32:3864–3872.
[52]   Tanaka E (1987) A fast reconstruction algorithm for stationary positron emission tomography based on a modified EM algorithm. IEEE Trans Med Imaging 6:98–105.
[53]   Trzasko J, Manduca A (2009) Highly undersampled magnetic resonance image reconstruction via homotopic l_0-minimization. IEEE Trans Med Imaging 28:106–121.
[54]   Wallis JW, Miller TR (1993) Rapidly converging iterative reconstruction algorithms in single-photon emission computed tomography. J Nucl Med 34:1793–1800.
[55]   Wilson DW, Tsui BMW, Barrett HH (1994) Noise properties of the EM algorithm, II. Monte Carlo Simulations. Phys Med Biol 39:847–871.
[56]   Yu H, Wang G (2009) Compressed sensing based interior tomography. Phys Med Biol 54:2791–2805.
[57]   Zeng GL, Gullberg GT (2000) Unmatched projector/backprojector pairs in an iterative reconstruction algorithm. IEEE Trans Med Imaging 19:548–555.
[58]   Zeng GL, Hsieh YL, Gullberg GT (1994) A rotating and warping projector-backprojector for fan-beam and cone-beam iterative algorithm. IEEE Trans Nucl Sci 41:2807–2811.
[59]   Zeng GL, Gullberg GT, Tsui BMW, Terry JA (1991) Three-dimensional iterative reconstruction algorithms with attenuation and geometric point response correction. IEEE Trans Nucl Sci 38:693–702.
[60]   Zhang B, Zeng GL (2006) An immediate after-backprojection filtering method with blob-shaped window functions for voxel-based iterative reconstruction. Phys Med Biol 51:5825–5842.
[61]   Zeng GL (2018) Emission expectation-maximization look-alike algorithms for x-ray CT and other applications. Med Phys 45:3721–3727.
[62]   Zeng GL (2019) Modification of Green's one-step-late algorithm for attenuated emission data. Biomed Phys Eng Exp 5:037001.
[63]   Zeng GL, Li Y (2019) Extension of emission expectation maximization lookalike algorithms to Bayesian algorithms. Vis Comput Ind Biomed Art 2:14.

7 MRI reconstruction

The magnetic resonance imaging (MRI) physics is quite different from that of transmission or emission imaging as we discussed in the previous chapters. This chapter first introduces the imaging physics of MRI, showing how the MRI signals are formed. We will see that the MRI signals are in the Fourier domain and the image reconstruction is achieved via the inverse Fourier transform.

7.1 The "$\vec{M}$"

The MRI working principle is quite different from that of emission and transmission tomography. MRI is an image of "proton density" in a cross section of the patient. The data in MRI can be simplified as weighted plane integrals of the proton density function in that plane with "frequency"-dependent weighting functions.

In this chapter, we present a watered-down version of the MRI principle. The goal of regular MRI is to get a picture of the distribution of hydrogen atoms (H^+) within the patient body.

The hydrogen atom is simply a proton that carries a positive charge and is continuously spinning. A spinning charge generates a magnetic field around it, as if it were a tiny magnet (see Figure 7.1). We call this tiny magnet a "magnetic moment," $\vec{\mu}$, which is a vector. The orientation is important for a vector.

In the absence of a strong external magnetic field, the proton magnetic moments $\vec{\mu}$ are randomly oriented inside the body. Thus, the *net* magnetic moment is zero $\vec{0}$ (see Figure 7.2).

When a strong external magnetic field is applied, the net magnetic moment is no longer zero. We denote this net magnetic moment as a vector $\vec{M}$ (see Figure 7.3) must point out that even in this case, not all $\vec{\mu}$ are pointing in the same direction. About half of them point in the direction of the magnetic field, and the other half point in the opposite direction. Therefore, the magnitude of $\vec{M}$ is very small and is proportional to

$$\frac{\gamma \hbar B_o}{2KT}, \tag{7.1}$$

where T is the absolute temperature of the patient (in K), K is the Boltzmann constant (8.62×10^{-11} MeV/K), $\hbar$ is the Planck constant (6.6252×10^{-27} ergs), B_0 is the magnetic strength (in Tesla), and γ is an atom-dependent constant (42.58 MHz/T for H^+) called the gyromagnetic ratio. We can see that a stronger magnetic field helps generate a stronger signal.

Another concept that we need to explain MRI physics is *precession*. Let us take a look at a spinning toy top (or a gyroscope). Besides spinning, there is another motion:

https://doi.org/10.1515/9783111055404-007

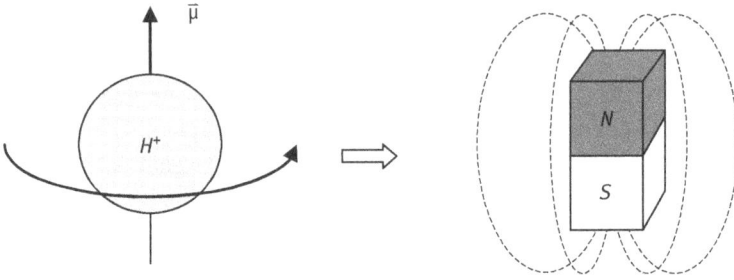

Fig. 7.1: A spinning proton acts like a tiny magnet.

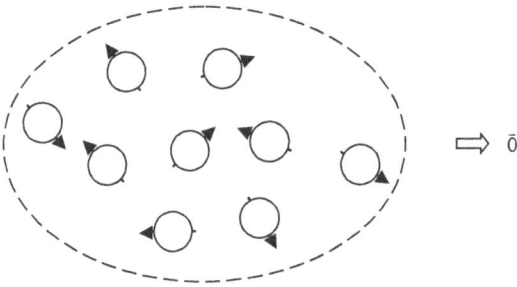

Fig. 7.2: The net magnetic moment is zero without a strong external magnetic field.

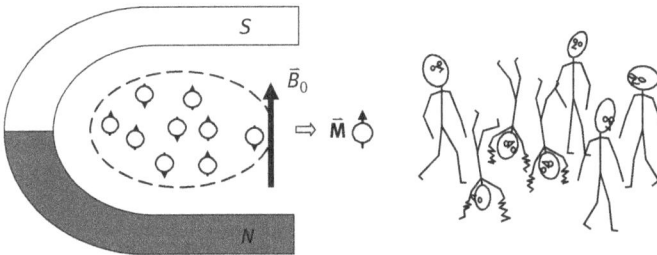

Fig. 7.3: In the strong external magnetic field, a small nonzero net magnetization can be observed.

the spin axis rotates around the direction of gravity (an external force). This motion of the spin axis rotating about the direction of gravity is referred to as precession (see Figure 7.4). If the top does not spin, precession would not happen; it just falls.

We must point out that it is not accurate to use classic physics to explain MRI physics. A more accurate explanation should use quantum mechanics. A neutron with no net electric charge can also act like a tiny magnet and has a magnetic moment. Therefore, besides the hydrogen atoms, MRI can also, in principle, image a nucleus with either an odd atomic number or an odd mass number, for example, ^{13}C, ^{14}N, ^{19}F, ^{23}Na, and ^{31}P. On the other hand, MRI cannot image atomic nuclei consisting of an

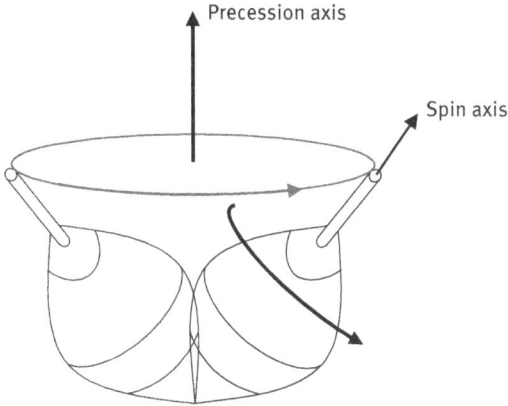

Fig. 7.4: A spinning toy top precesses about a vertical axis due to the gravity.

even number of protons and even number of neutrons because they do not have un-paired protons or neutrons. Paired protons or neutrons have zero magnetic moment.

7.2 The "R"

We have a vector $\vec{M}$, called net magnetic moment, which is spinning by itself. Nor-mally, the vector $\vec{M}$ points at the same direction of the external magnetic field $\vec{B}_0$. If we somehow knock the vector $\vec{M}$ off balance, and $\vec{M}$ is not in the direction of $\vec{B}_0$ any-more, then the vector $\vec{M}$ will precess about the direction of $\vec{B}_0$ just as the toy top pre-cesses about the direction of gravity (see Figure 7.5). The precession frequency is called the Larmor frequency and is given as

$$\omega_0 = \gamma B_0, \tag{7.2}$$

where B_0 is the external magnetic field strength, and γ is the gyromagnetic ratio. For a proton, $\gamma = 42.58$ MHz/T. If the MRI machine has a magnetic field of 1.5 T, then the Larmor frequency is approximately 64 MHz, which is close to the frequency range of an FM radio.

The MRI signal is nothing but this 64 MHz radio frequency (RF) electromagnetic wave sent out from the patient's body after the net magnetic moment $\vec{M}$ is somehow knocked off balance. Thus, the MRI signal is also called the RF signal. The strength of the signal is proportional to the proton density inside the patient.

To knock the net magnetic moment $\vec{M}$ off balance is not an easy task. In order to move the vector $\vec{M}$, we need to create a virtual situation so that the vector $\vec{M}$ does not feel the existence of the external $\vec{B}_0$ field.

We will first put the vector $\vec{M}$ on a virtual rotating merry-go-round or a rotating platform (see Figure 7.6), which is rotating at the Larmor frequency. We will later

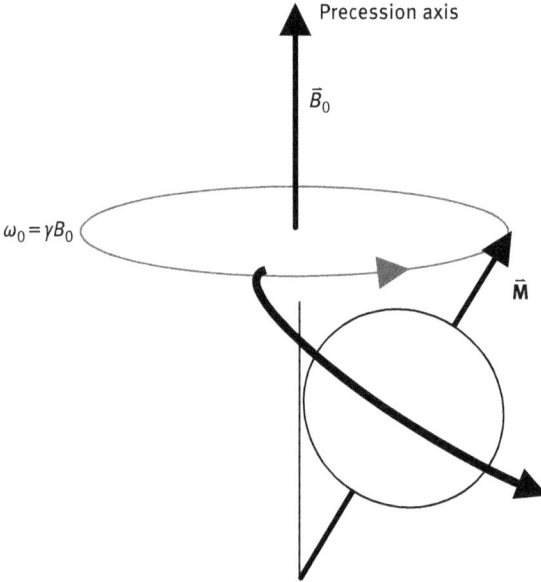

Fig. 7.5: Protons precess at the Larmor frequency ω_0.

find a way to create precession while the vector $\vec{M}$ is on that rotating platform. The vector $\vec{M}$ now is currently standing on the rotating platform upright and is spinning on its own axis; there is no motion relative to the platform. Even if the vector $\vec{M}$ is not standing upright and it has a nonzero angle with the vertical line, vector $\vec{M}$ will stay with that angle and will not have any motion relative to the platform.

Next, we step on the rotating platform and apply a new magnetic field $\vec{B}_1$, orthogonal to the main magnetic field $\vec{B}_0$. On the rotating platform, the vector $\vec{M}$ does not feel the existence of the main field $\vec{B}_0$; it only feels the push from the new field $\vec{B}_1$. Since $\vec{M}$ is not aligned with $\vec{B}_1$, it will precess about the direction of $\vec{B}_1$ (see Figure 7.7) at a precession frequency $\omega_1 = \gamma B_1$. Once $\vec{M}$ reaches the platform floor, we turn the new field $\vec{B}_1$ off. Thus, the mission of knocking off $\vec{M}$ is accomplished. The $\vec{B}_1$ field is only turned on very briefly; it is called a 90° RF pulse if it is turned off as soon as $\vec{M}$ touches the platform floor.

What is this $\vec{B}_1$ field anyway? The magnetic field $\vec{B}_1$ is applied on a virtual rotating platform, which rotates at the Larmor frequency, say 64 MHz for a proton in a 1.5 T MRI machine. Therefore, $\vec{B}_1$ is an alternating electromagnetic RF field with the *same* frequency as the Larmor frequency. Another term for the same frequency is the *resonance* frequency. The $\vec{B}_1$ RF signal is sent to the patient through an RF coil, which is basically an antenna. The procedure of turning the $\vec{B}_1$ RF field on and knocking $\vec{M}$ off balance is called RF excitation.

After RF excitation, we turn off $\vec{B}_1$. Now the net magnetic moment $\vec{M}$ is not in the equilibrium state, but in the excited state. The vector $\vec{M}$ is precessing about the main

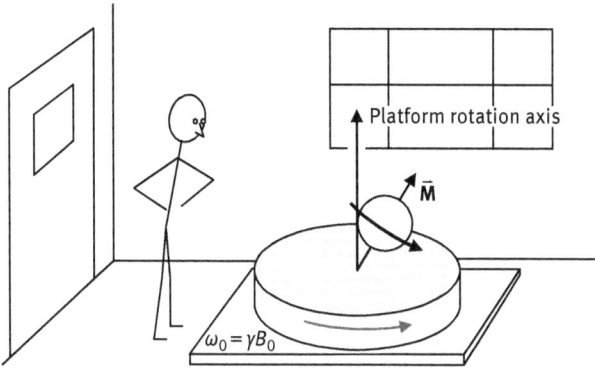

Fig. 7.6: The magnetization vector does not precess relative to the rotating platform.

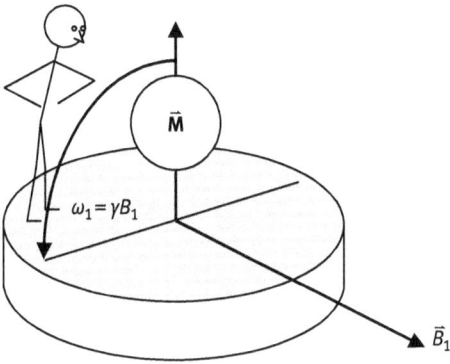

Fig. 7.7: On the rotating platform, the effect of the main field can be ignored. Only the new B_1 field is effective.

field direction $\vec{B}_0$, and RF signals that contain the patient's proton density information are emitted. The excited state is unstable. After excitation, the vector $\vec{\mathbf{M}}$ then goes through the relaxation period and eventually returns to the original equilibrium state, where the vector $\vec{\mathbf{M}}$ points to the $\vec{B}_0$ direction (see Figure 7.8). In a Cartesian system, the $\vec{B}_0$ direction is the z-direction, and the "platform floor" is the x–y plane.

The x- and y-components of the vector $\vec{\mathbf{M}}$ make up the MRI signal. After $\vec{\mathbf{M}}$ relaxes back to its equilibrium position, both its x- and y-components are zero; hence, no more signals can be detected. Another RF pulse excitation is needed for further data acquisition. This procedure is repeated over and over again until enough data are acquired for imaging.

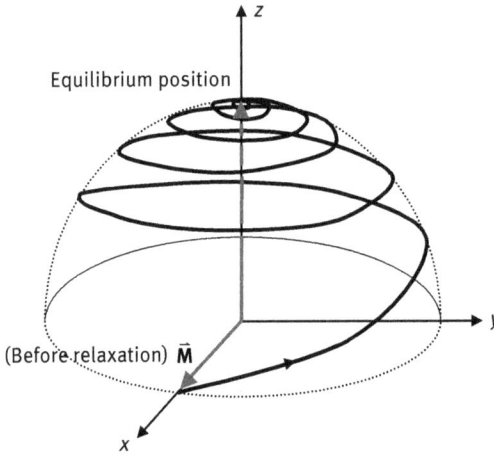

Fig. 7.8: Relaxation of the magnetization vector.

7.3 The "I"

The RF signal emitted by the patient as described in Section 7.2 cannot be used to form an image because it is a combined signal from everywhere. We need a way to code the location information. This is achieved by the gradient coils.

In the MRI machine, there are many coils. The main large superconducting coil immersed in liquid helium is used to generate the strong static $\vec{B_0}$ field. There are RF coils, which are used to emit RF pulses for excitation and to receive RF imaging signals. The other coils in the machine to create gradients are x-gradient coils, y-gradient coils, and z-gradient coils.

7.3.1 To obtain z-information: slice selection

The z-gradient coils are shown in Figure 7.9. The currents in the two coils are running in the opposite directions and generate the local magnetic fields to enhance and reduce the main field $\vec{B_0}$, respectively. The resultant magnetic field is still pointing in the z-direction; however, the field strength is stronger at locations with larger z-values and is weaker at locations with smaller z-values. These coils create a gradient of the magnetic field strength in the z-direction.

This z-gradient makes a nonuniform magnetic field, which results in different Larmor frequencies ω for a different z-position. The Larmor frequency is ω_0 only at one z-slice. We turn on and off the z-gradient and the $\vec{B_1}$ RF pulse at the same time. The $\vec{B_1}$ RF pulse is at the frequency of ω_0. According to the resonance frequency principle, only

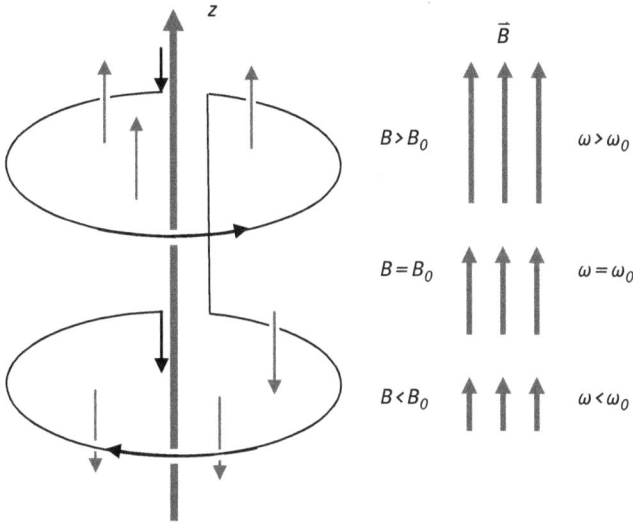

Fig. 7.9: The z-gradient coils create a nonuniform field in the z-direction.

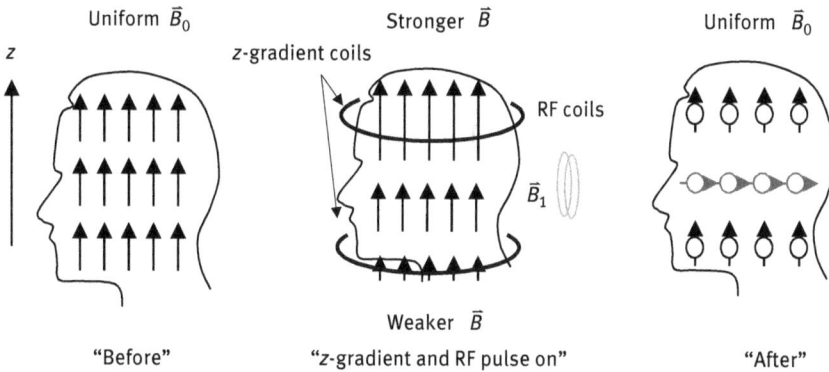

Fig. 7.10: Slice selection is done by the z-gradient and the B_1-field.

one z-slice of the patient body is affected by the RF pulse. That is, only the protons in this particular z-slice get excited and send out the RF signal (see Figure 7.10).

7.3.2 To obtain x-information: frequency encoding

We use the x-gradient to provide the x-position information. When we are ready to receive the RF signal from the patient, we turn on the x-gradient. The x-gradient is generated by the x-gradient coils, and it makes the strength of the main magnetic field to vary in the x-direction. The principle of using x-gradient to code x-position is illustrated in Figure 7.11. Using the relation $\omega = \gamma B$, the stronger magnetic field B gives the higher frequency. Thus, the x-location can be determined by the received frequency.

The x-gradient coils are depicted in Figure 7.12. We assume that the x-direction is the direction from the patient's right ear to left ear. The x-gradient is turned on only when the RF signal is received, and this gradient is also called the readout gradient. Do not turn it on during slice selection. When you turn two gradients on at the same time, they will combine and form a gradient in the third direction.

7.3.3 To obtain *y*-information: phase encoding

After slice selection and before RF signal readout, we turn on the y-gradient for a short time. Before a gradient is turned on, $\overrightarrow{\mathbf{M}}$ at all locations precess at the same frequency (i.e., same speed). However, when the y-gradient is turned on, the field strength at a different y-position is different. As a result, $\overrightarrow{\mathbf{M}}$ at a different y-position precesses at a different Larmor frequency (e.g., precession at a faster rate for a position with a larger y value, Figure 7.13). After the y-gradient is on for a short while, it is turned off. At this moment, $\overrightarrow{\mathbf{M}}$ at a different y-position will arrive at a different phase (i.e., angle). This phase carries the information of the y-position.

The y-gradient coils look exactly like the x-gradient coils, except for a 90° rotation. The effective gradient is confined to the gap between two pairs of the coils (see Figure 7.14).

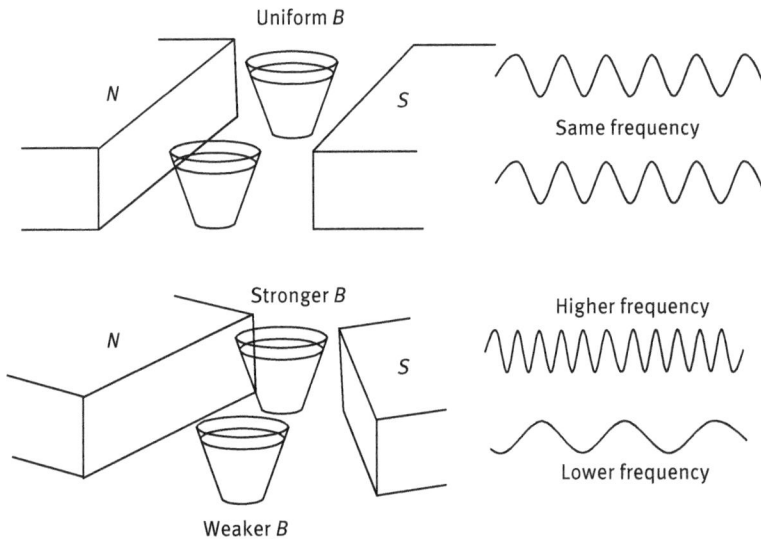

Fig. 7.11: Stronger field produces higher frequency.

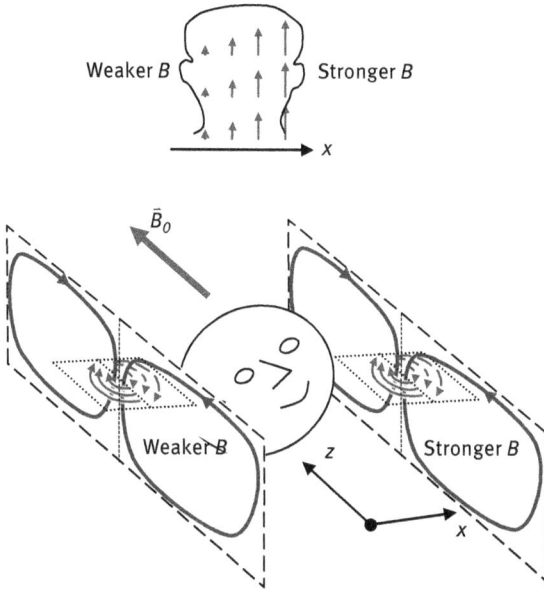

Fig. 7.12: The x-gradient coils generate a nonuniform magnetic field in the readout direction.

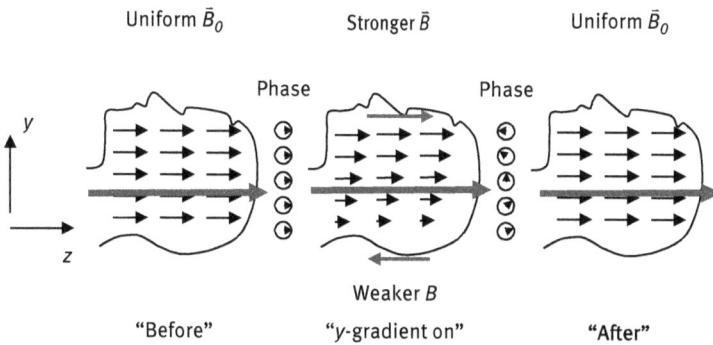

Uniform $\vec{B}_0$ · Stronger $\vec{B}$ · Uniform $\vec{B}_0$

Phase · Phase

"Before" · "y-gradient on" · "After"

Weaker B

Fig. 7.13: The y-gradient causes the phase displacement as a function of y-location.

The time diagram shown in Figure 7.15 summarizes this basic MRI data acquisition procedure. This procedure is repeated many times. At each time, a different value of the y-gradient is used. The next section will show that the acquired data is nothing but the 2D Fourier transform of the image $f(x, y)$, which is closely related to the proton density distribution inside the patient body. A 2D inverse Fourier transform is used to reconstruct the image. A typical MRI image is shown in Figure 7.16. There is a small negative pulse in front of the x-gradient readout pulse. The purpose of it is to create an "echo" to make the strongest signal at the center.

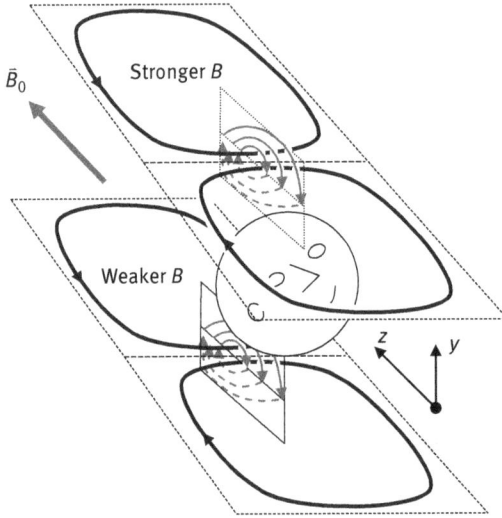

Fig. 7.14: The y-gradient coils make the magnetic field nonuniform in the y-direction.

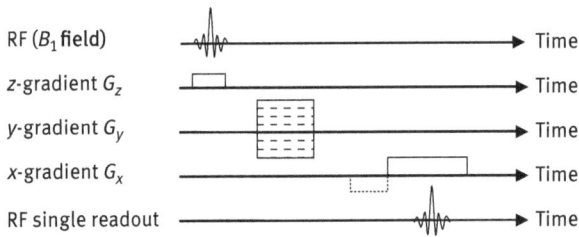

Fig. 7.15: The timing diagram for an MRI pulse sequence.

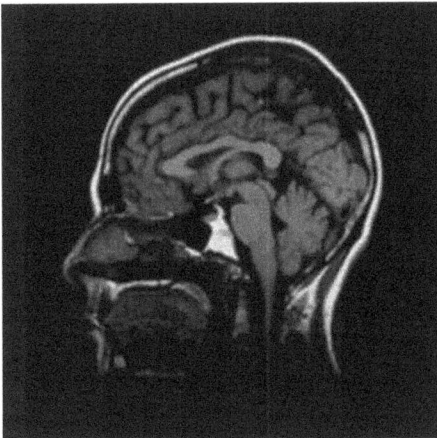

Fig. 7.16: An MRI of the head.

7.4 Mathematical expressions

In this section, we assume that a slice selection has been done and $\vec{M}(x, y)$ is a function of x and y. The vector $\vec{M}(x, y)$ can be decomposed into the x-component $M_x(x, y)$, the y-component $M_y(x, y)$, and the z-component $M_z(x, y)$. We define a complex function $f(x, y)$ as

$$f(x,y) = M_x(x,y) + iM_y(x,y). \tag{7.3}$$

The goal of MRI is to obtain this function $f(x, y)$ and display its magnitude $|f(x, y)|$ as the final output for the radiologists to read.

Let us first consider the effect of the readout (x) gradient. When the x-gradient is turned on, the magnetic field strength is a function of x as

$$B(x) = B_0 + xG_x, \tag{7.4}$$

and the associated Larmor frequency is calculated as

$$\omega(x) = \gamma(B_0 + xG_x). \tag{7.5}$$

At readout, the function $f(x, y)$ will be encoded as

$$f(x,y)\cos(2\pi\omega(x)t) = f(x,y)\cos(2\pi\gamma(B_0 + xG_x)t), \tag{7.6}$$

where γB_0 is the carrier frequency and has no contribution for image reconstruction. In the MRI receiver, there is a demodulator that can remove this carrier frequency. After the removal of the carrier frequency, the leftover baseband signal is given as

$$f(x,y)\cos(2\pi\gamma xG_x t). \tag{7.7}$$

Since the signal comes from the entire x–y plane, the received baseband signal is the summation of the signals from each location (x, y):

$$\int_{-\infty}^{\infty}\int_{-\infty}^{\infty} f(x,y)\cos(2\pi\gamma xG_x t)dxdy. \tag{7.8}$$

Second, let us consider the effect of the phase-encoding (y) gradient. When the y-gradient is turned on for a period of time T, the magnetic field strength is a function of y as

$$B(y) = B_0 + yG_y, \tag{7.9}$$

and the associated Larmor frequency is calculated as

$$\omega(y) = \gamma\left(B_0 + yG_y\right). \tag{7.10}$$

After the period of T, the phase change is a function of y as

$$\phi(y) = T\omega(y) = \gamma B_0 T + \gamma y G_y T. \tag{7.11}$$

Recall that the function $f(x, y)$ is complex with a magnitude and a phase and can be expressed as

$$f(x,y) = |f(x,y)|e^{i\varphi(x,y)}. \tag{7.12}$$

After a phase change of $\phi(y)$, $f(x, y)$ becomes

$$|f(x,y)|e^{i\varphi(x,y)}e^{-i\phi(y)} = f(x,y)e^{-i\left(\gamma B_0 T + \gamma y G_y T\right)}. \tag{7.13}$$

We can ignore the first term $\gamma B_0 T$ in the exponent because it introduces the same phase change to all y-positions and carries no information of the image.

The function $f(x, y)$ is now encoded by the phase-changing factor as

$$f(x,y)e^{-i2\pi\gamma y G_y T}, \tag{7.14}$$

which is the signal that we try to read out. Therefore, the readout signal is

$$\int_{-\infty}^{\infty}\int_{-\infty}^{\infty} \left[f(x,y)e^{-i2\pi\gamma y G_y T}\right] \cos(2\pi\gamma x G_x t) dx dy. \tag{7.15}$$

The MRI machines use quadrature data acquisition, which has two outputs at 90° out of phase. One output gives

$$\int_{-\infty}^{\infty}\int_{-\infty}^{\infty} \left[f(x,y)e^{-i2\pi\gamma y G_y T}\right] \cos(2\pi\gamma x G_x t) dx dy, \tag{7.16}$$

and the other gives

$$\int_{-\infty}^{\infty}\int_{-\infty}^{\infty} \left[f(x,y)e^{-i\pi\gamma y G_y T}\right] \sin(2\pi\gamma x G_x t) dx dy. \tag{7.17}$$

We can combine them into a complex signal, with one output as the real part and the other output as the imaginary part,

$$\text{signal}(t) = \int_{-\infty}^{\infty}\int_{-\infty}^{\infty} f(x,y)e^{-i2\pi\gamma y G_y T}e^{-i2\pi\gamma y G_x t} dx dy. \tag{7.18}$$

If we rewrite it as

$$\text{signal}_{G_y}(t) = \int_{-\infty}^{\infty}\int_{-\infty}^{\infty} f(x,y)e^{-i2\pi\left[x(\gamma G_x t) + y(\gamma G_y T)\right]} dx dy, \tag{7.19}$$

we immediately recognize it as the 2D Fourier transform of $f(x, y)$:

$$F(k_x, k_y) = \int\limits_{-\infty}^{\infty} \int\limits_{-\infty}^{\infty} f(x,y) e^{-i2\pi \left[x(\gamma G_x t) + y(\gamma G_y T) \right]} \, dx \, dy, \tag{7.20}$$

with $k_x = \gamma G_x t$ and $k_y = \gamma G_y T$. When we sample the time signal over time t, we get the samples of the x-direction frequencies, k_x. When we repeat the scan with a different value of G_y, we get the samples of the y-direction frequencies, k_y. For this reason, people often call the MRI signal space the k-space (see Figure 7.17), which is the Fourier space. During data acquisition, the k-space is filled out one line at a time according to $k_x = \gamma G_x t$ and $k_y = \gamma G_y T$.

Finally, we will consider a polar k-space scanning strategy in which the x-gradient and the y-gradient are turned on and off simultaneously (see Figure 7.18). In this case, we do not have the phase-encoding step; we only have the readout gradient, which is determined by both the x- and y-gradients.

The signal readout is given as

$$\text{signal}_{G_x, G_y}(t) = F(k_x, k_y) = \int\limits_{-\infty}^{\infty} \int\limits_{-\infty}^{\infty} f(x,y) e^{-i2\pi \left[x(\gamma G_x t) + y(\gamma G_y t) \right]} \, dx \, dy, \tag{7.21}$$

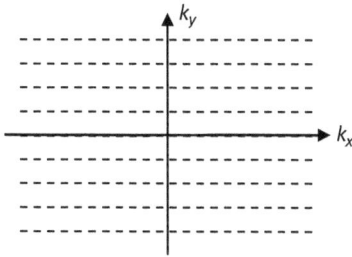

Fig. 7.17: The k-space.

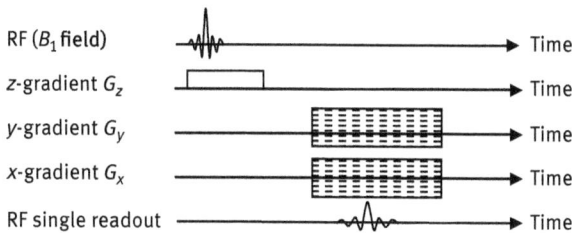

Fig. 7.18: The timing diagram for polar scanning.

with $k_x = \gamma G_x t$ and $k_y = \gamma G_y t$. The ratio $k_y/k_x = G_y/G_x$ tells us that each RF excitation cycle measures a line in the k-space with a slope of G_y/G_x (see Figure 7.19). For a different RF excitation, a new set of G_x and G_y is used, and a new line in the k-space is

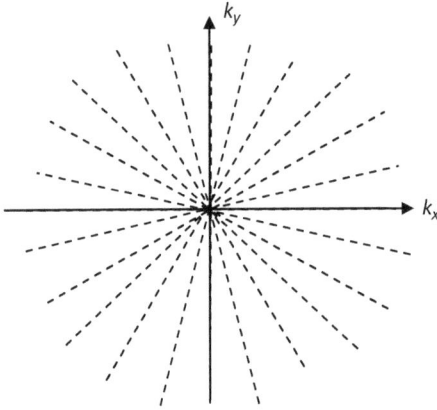

Fig. 7.19: The *k*-space sampling for the polar scan.

obtained. Figure 7.19 reminds us of the central slice theorem. We therefore can use the filtered backprojection algorithm (Section 2.3) to reconstruct this MRI.

7.5 Image reconstruction for MRI

7.5.1 Fourier reconstruction

If the *k*-space is sufficiently sampled in Cartesian grids, $F(k_x, k_y)$ is known in a square region in the Fourier domain. The image $f(x, y)$ can be readily obtained using a 2D inverse fast Fourier transform (2D IFFT).

In fact, the result of this 2D IFFT is not really the original image (but really close). The result is the true image convolved with a point spread function (PSF) *h*. At each dimension, *h* is the ratio of a sine function over another sine function, and *h* is periodic. The PSF *h* is characterized by the number of samples and the sampling interval in the *k*-space. Each period of *h* looks similar to a sinc function that has a main lobe and ringing side lobes. The *k*-space sampling interval determines the period of *h*. If the period is smaller than the object size, we will see aliasing artifacts in the reconstruction, as shown in Figure 7.20, where the nose appears at the back of the head. Some Gibbs ringing artifacts sometimes can be seen around the sharp edges.

For many applications, the MRI data are acquired using non-Cartesian trajectories such as radial and spiral. Regridding is a common technique to convert a non-Cartesian data set into a Cartesian data set. Then the 2D IFFT method is used to reconstruct the image. Regridding is essentially data interpolation, which moves the sampled data value to its close grid neighbors with proper weighting. The proper weighting considers the local sampling density. Nonuniform FFT is basically a regridding method; it is fast and can be used for MRI reconstruction.

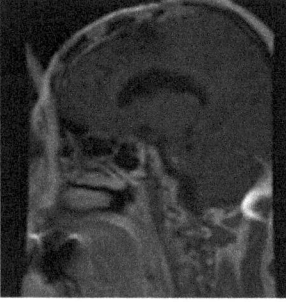

Fig. 7.20: Aliasing artifacts in an MRI.

If the k-space is sampled in a radial format, the conventional filtered backprojection algorithm can be directly used for image reconstruction.

7.5.2 Iterative reconstruction

As in emission and transmission computed tomography, better images can be obtained if more imaging physics aspects are modeled and incorporated into the image reconstruction algorithm. The simple Fourier transform model discussed earlier in this chapter is not adequate for this model-based image reconstruction.

In many situations, the k-space is not fully sampled. Constraints are needed to supply some prior information. A common method is to set up an objective function that includes the imaging physics models and constraints. An iterative algorithm is used to reconstruct the image by minimizing the objective function as discussed in Chapter 6.

A typical objective function for iterative MRI image reconstruction is shown as follows:

$$F = ||AX - P||^2 + \text{Bayesian terms}, \tag{7.22}$$

where AX symbolically represents two actions: 2D Fourier transform of the image X and a binary selecting function to identify measured k-space sample locations. The k-space measurements are denoted by P. The optimization algorithm to minimize the objective function (7.22) produces a complex-valued image, which should be converted to a real and nonnegative image at the end.

7.6 Worked examples

Example 1: The vector $\vec{M}$ has three components: M_x, M_y, and M_z. Does the magnitude $\sqrt{M_x^2 + M_y^2 + M_z^2}$ always remain constant?

Answer
No. During relaxation, M_x and M_y relax to zero faster than M_z relaxes back to its equilibrium maximum value. This makes the magnitude $\sqrt{M_x^2 + M_y^2 + M_z^2}$ time varying.

Example 2: The received MRI signal is converted to a discrete signal via an analog-to-digital converter (ADC). Does the sampling rate of the ADC determine the image resolution?

Answer
No. The image resolution in MRI is determined by how far out the k-space is sampled. The distance from the farthest sample in the k-space to the origin (i.e., the DC point) gives the highest resolution in the image. The sampling time interval in the ADC determines the image field of view. If the sampling rate of the ADC is not high enough, you will see image aliasing artifacts (e.g., the nose appears at the back of the head).

Example 3: Design a pulse sequence that gives a spiral k-space trajectory.

solution
In this case, the x- and y-gradients must be turned on simultaneously. The generic expressions for k_x and k_y are

$$k_x(t) = \gamma \int_0^t G_x(\tau)d\tau \quad \text{and} \quad k_y(t) = \gamma \int_0^t G_y(\tau)d\tau, \qquad (7.23)$$

respectively. On the other hand, a k-space spiral can be expressed as

$$\begin{cases} k_x(t) = a(t)\cos(\beta(t)), \\ k_y(t) = a(t)\sin(\beta(t)), \end{cases} \qquad (7.24)$$

respectively, for some parameters $a(t)$ and $\beta(t)$. Therefore, we can choose

$$\begin{cases} G_x(t) = \frac{1}{\gamma}\frac{d[a(t)\cos(\beta(t))]}{dt}, \\ G_y(t) = \frac{1}{\gamma}\frac{d[a(t)\sin(\beta(t))]}{dt}. \end{cases} \qquad (7.25)$$

The corresponding time diagram and the k-space trajectory are shown in Figure 7.21. Its image reconstruction is normally performed by first regridding the k-space samples on the spiral trajectory into regularly spaced Cartesian coordinates, and then the 2D inverse Fourier transform is taken to obtain the final image.

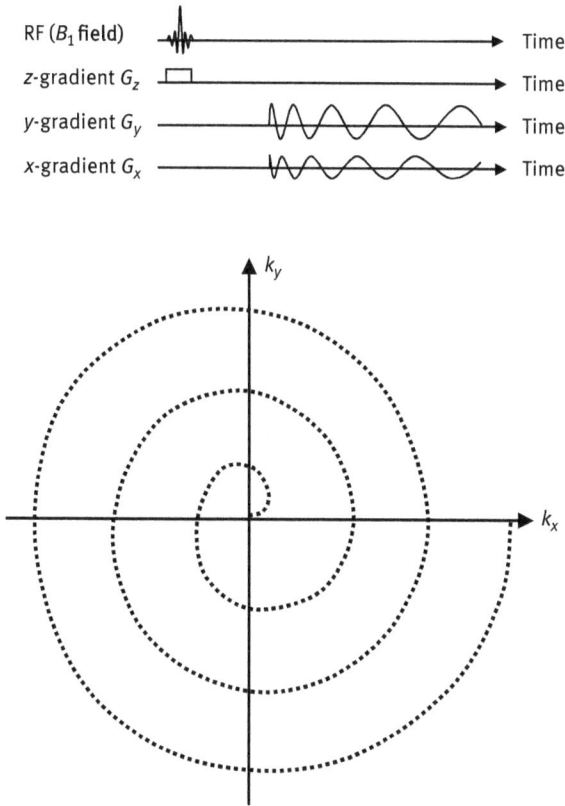

Fig. 7.21: The timing diagram and the k-space representation of a spiral scan.

7.7 Summary

- The working principle of MRI is quite different from that of transmission and emission tomography. The MRI signal is in the form of radio waves and is received by antennas (called coils).
- The "M" part: The patient must be positioned in a strong magnetic field so that the magnetic moments created by the proton spins have a chance to line up.
- The "R" part: A resonant RF signal is required to be emitted toward the patient so that the net magnetic moments can be tipped over and do not align with the main

magnetic field. When the net magnetic moments precess about the direction of the main magnetic field, RF signals are sent out from the patient body.

- The "I" part: Gradient coils are turned on and off to encode the outcoming RF signals so that the signals can carry the position information.
- The received MRI signal by the RF coils is in the Fourier domain (or spatial frequency space, or k-space). The image is reconstructed by performing a 2D inverse Fourier transform.
- The readers are expected to understand how the MRI signal is encoded to carry position information and why the MRI signal in the k-space is the 2D Fourier transform of the object.

Problems

Problem 7.1 During slice selection in the z-direction, the slice thickness is not zero, but is a positive value Δz. Therefore, the RF pulse that generates the alternative magnetic field B_1 should have a proper bandwidth. How is this bandwidth determined by the slice thickness Δz?

Problem 7.2 If you plan to reconstruct an MRI with an iterative algorithm, how do you handle the complex data? You may want to process the real part of the data and the imaginary part of the data separately. How do you set up an objective function?

Problem 7.3 According to $k_x(t) = \gamma \int_0^t G_x(\tau)d\tau$, $k_y(t) = \gamma \int_0^t G_y(\tau)d\tau$, and the gradient waveforms given in the figure below, sketch the corresponding k-space trajectory.

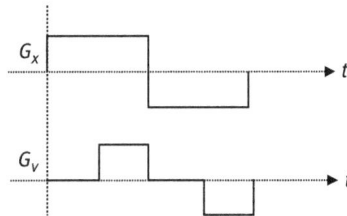

Bibliography

[1]   Abragam A (1961) The Principle of Nuclear Magnetic Resonance, Oxford University Press, Oxford.
[2]   Aiver MN (1997) All You Really Need to Know about MRI Physics, Simply Physics, Baltimore, MD.
[3]   Brown MA, Semelka RC (2003) MRI Basic Principles and Applications, Wiley, Hoboken, NJ.
[4]   Fessler JA, Sutton BP (2003) Nonuniform fast Fourier transforms using min-max interpolation. IEEE Trans Signal Process 51:560–574.

[5]   Fukushima E, Raeder SBW (1981) Experimental Pulse NMR, A Nuts and Bolts Approach, Addison-Wesley, Reading, MA.

[6]   Gerald LW, Carol P (1984) MRI A Primer for Medical Imaging, Slack, Thorofare, JN.

[7]   Haacje EM, Brown RW, Thompson MR, Venkatesan R (1999) Magnetic Resonance Imaging: Physical Principles and Sequence Design, Wiley, New York.

[8]   Liang ZP, Lauterbur PC (2000) Principles of Magnetic Resonance Imaging, A Signal Processing Perspective, IEEE Press, Piscataway, NJ.

[9]   Macovski A (1996) Noise in MRI. Magn Reson Med 36:494–497.

[10]   Stark D, Bradley W (1992) Magnetic Resonance Imaging, Mosby, St. Louis, MO.

[11]   Adluru G, McGann C, Speier P, Kholmovski EG, Shaaban A, DiBella EVR (2009) Acquisition and reconstruction of undersampled radial data for myocardial perfusion magnetic resonance imaging. J Magn Reson Imaging 29(2):466–473.

8 Using FBP to perform iterative reconstruction

The readers should be aware that this chapter is based solely on author's recent research. Many results have not been carefully cross-validated by the imaging community. A Chinese idiom goes like this: To throw a brick to attract jade. It means that the author's attempt is only preliminary and it may trigger further progress.

This chapter is motivated by the fact that the iterative algorithms are much robust than the filtered backprojection (FBP) algorithms and provide better images for the same given projection data set. On the other hand, the FBP algorithms are much faster than the iterative algorithms. It would be nice to develop FBP algorithms that can give almost the same images as those provided by the iterative algorithms. In fact, under some fairly common situations, this is possible.

8.1 The Landweber algorithm: from recursive form to nonrecursive form

As discussed in Sections 6.3.2 and 6.9.2, the kth iteration of the Landweber algorithm can be expressed as

$$
\begin{aligned}
X^{(k)} &= X^{(k-1)} + \alpha A^T \left(P - AX^{(k-1)} \right) \\
&= \alpha A^T P + \left(I - \alpha A^T A \right) X^{(k-1)} \\
&= \alpha A^T P + \left(I - \alpha A^T A \right) \left[\alpha A^T P + \left(I - \alpha A^T A \right) X^{(k-2)} \right] \\
&= \alpha A^T P + \left(I - \alpha A^T A \right) \alpha A^T P + \left(I - \alpha A^T A \right)^2 X^{(k-2)} \\
&= \alpha A^T P + \left(I - \alpha A^T A \right) \alpha A^T P + \left(I - \alpha A^T A \right)^2 \left[\alpha A^T P + \left(I - \alpha A^T A \right) X^{(k-3)} \right] \quad (8.1) \\
&= \alpha A^T P + \left(I - \alpha A^T A \right) \alpha A^T P + \left(I - \alpha A^T A \right)^2 \alpha A^T P + \left(I - \alpha A^T A \right)^3 X^{(k-3)} \\
&= \cdots \\
&= \left[I + \left(I - \alpha A^T A \right) + \cdots + \left(I - \alpha A^T A \right)^{k-1} \right] \alpha A^T P + \left(I - \alpha A^T A \right)^k X^{(0)} \\
&= \left[\sum_{n=0}^{k-1} \left(I - \alpha A^T A \right)^n \right] \alpha A^T P + \left(I - \alpha A^T A \right)^k X^{(0)}.
\end{aligned}
$$

The first line of eq. (8.1) is a recursive expression, in which $X^{(k)}$ depends on $X^{(k-1)}$. The last line of eq. (8.1) in nonrecursive and $X^{(n)}$, $n = 1, 2, \ldots, k-1$ do not appear on the right-hand side.

The Landweber iterative algorithm is used to solve the system $AX = P$. If noise weighting is also considered, the last line of eq. (8.1) becomes

https://doi.org/10.1515/9783111055404-008

$$X^{(k)} = \left[\sum_{n=0}^{k-1} \left(I - \alpha A^T WA \right)^n \right] \alpha A^T WP + \left(I - \alpha A^T WA \right)^k X^{(0)}. \tag{8.2}$$

Our FBP algorithm will be derived based on eq. (8.2). There are two terms in eq. (8.2). We will study them one at a time. Let us define two matrices in eq. (8.2):

$$F^{(k)} \equiv \left[\sum_{n=0}^{k-1} \left(I - \alpha A^T WA \right)^n \right] \alpha A^T W \tag{8.3}$$

and

$$H^{(k)} \equiv \left(I - \alpha A^T WA \right)^k. \tag{8.4}$$

Thus eq. (8.2) can be rewritten as

$$X^{(k)} \equiv F^{(k)} P + H^{(k)} X^{(0)}. \tag{8.5}$$

The first term in eq. (8.5) only depends on the projections P, while the second term only depends on the initial condition $X^{(0)}$. Equation (8.2) is not in a closed form because there is a $\sum_{n=0}^{k-1}$ on the right-hand side. Our next step is to remove this $\sum_{n=0}^{k-1}$ sign.

8.2 The Landweber algorithm: from nonrecursive form to closed form

In order to turn $\sum_{n=0}^{k-1} \left(I - \alpha A^T WA \right)^n$ into a closed form, let us recall the summation formula for the geometric series:

$$\sum_{n=0}^{k-1} r^n = \frac{1 - r^k}{1 - r}. \tag{8.6}$$

A similar summation formula exists for matrices:

$$\sum_{n=0}^{k-1} B^n = (I - B)^{-1} (I - B^K) \tag{8.7}$$

if $(I - B)^{-1}$ exists. Equation (8.7) can be easily verified by left-multiplying $(I - B)$ on both sides. Thus, we have

$$F^{(k)} = \alpha \sum_{n=0}^{k-1} \left(I - \alpha A^T WA \right)^n A^T W = \left(A^T WA \right)^{-1} \left[I - \left(I - \alpha A^T WA \right)^k \right] A^T W. \tag{8.8}$$

If the initial condition is zero, that is, $X^{(0)} = \vec{0}$, the kth iteration of the Landweber algorithm has a closed form of

$$X^{(k)} = F^{(k)} P \qquad (8.9)$$

in the sense that $F^{(k)}$ in the right-hand side of (8.8) does not contain the summation sign $\sum_{n=0}^{k-1}$. Our next step is to convert the closed-form Landweber algorithm into a backprojection-then-filtering analytic algorithm.

8.3 The Landweber algorithm: from closed form to backprojection-then-filtering algorithm

For the sake of simplicity, we drop the noise weighting W in this section. We will put it back in the next section when we derive an FBP algorithm. Let us rewrite eq. (8.9) as follows:

$$X^{(k)} = \left(A^T A\right)^{-1} \left[I - \left(I - \alpha A^T A\right)^k\right] A^T P \qquad (8.10)$$

and make some observations. Here A is the projection matrix and A^T is the backprojection matrix. Hence, $A^T A$ is the matrix for the projection-then-backprojection operation. Equation (8.10) implies that the projections P are first backprojected into the image domain by $A^T P$, which is filtered by an image domain filter $(A^T A)^{-1}[I - (I - \alpha A^T A)^k]$.

8.3.1 Implementation of $(A^T A)^{-1}$ in the Fourier domain

In Section 2.6.7, we had the following two expressions:

$$B_{\text{polar}}(\omega, \theta) = \frac{F_{\text{polar}}(\omega, \theta)}{|\omega|}, \qquad (2.37)$$

$$B(\omega_x, \omega_y) = \frac{F(\omega_x, \omega_y)}{\sqrt{\omega_x^2 + \omega_y^2}}. \qquad (2.38)$$

Equation (2.37) is in the polar coordinate system, and eq. (2.38) is its counterpart in the Cartesian coordinate system. They are both in the Fourier domain. Here F is the 2D Fourier transform of the original object f, and B is the 2D Fourier transform of the projection-then-backprojection of f. Therefore, multiplying an image by $A^T A$ is equivalent to filtering the image with a transfer function $1/|\omega|$, where ω is the radial variable in the polar Fourier space.

Multiplying an image by $(A^T A)^{-1}$ is equivalent to filtering the image with a transfer function $|\omega|$. We will show that $[I - (I - \alpha A^T A)^k]$ is equivalent to a window function in the Fourier domain next.

8.3.2 Implementation of $I - (I - \alpha A^T A)^k$ in the Fourier domain

Multiplying an image with an identity matrix I does nothing to the image. In the Fourier domain, this is equivalent to filtering the image with a transfer function 1 (one).

As pointed out in Section 8.3.1, multiplying an image by $A^T A$ is equivalent to filtering the image with a transfer function $1/|\omega|$. Hence, multiplying an image by $(I - \alpha A^T A)$ is equivalent to filtering the image with a transfer function $(1 - \alpha/|\omega|)$. Multiplying an image by $(I - \alpha A^T A)^k$ is equivalent to filtering the image k times with a transfer function $(1 - \alpha/|\omega|)$, which is the same as filtering the image with a transfer function $(1 - \alpha/|\omega|)^k$. Thus, multiplying an image by $I - (I - \alpha A^T A)^k$ is equivalent to filtering the image with a transfer function $1 - (1 - \alpha/|\omega|)^k$.

8.3.3 Landweber algorithm: backprojection-then-filtering algorithm

Using the results in Sections 8.3.1 and 8.3.2, we are ready to implement the Landweber algorithm using an analytic backprojection-then-filtering algorithm. Let us once again rewrite eq. (8.9) here

$$X^{(k)} = \left(A^T A\right)^{-1}\left[I - \left(I - \alpha A^T A\right)^k\right]A^T P. \tag{8.11}$$

We can do part-by-part translation from the matrix notation to the Fourier domain transfer function and arrive at the following implementation steps:
1. Let $B = A^T P$. This is backprojection of the sinogram P into the image domain.
2. Take the 2D Fourier transform of the backprojected image B.
3. In the Fourier domain, apply the 2D Fourier transform of the backprojected image with a transfer function $|\omega|[1 - (1 - \alpha/|\omega|)^k]$, where $|\omega| = \sqrt{\omega_x^2 + \omega_y^2}$.
4. Take the 2D inverse Fourier transform to obtain the final image.

If we compare this algorithm with the standard backprojection-then-filtering algorithm in Section 2.3.4, they are identical except that here we replace the 2D ramp filter $|\omega|$ by a windowed 2D ramp filter $|\omega|[1 - (1 - \alpha/|\omega|)^k]$. The only thing new is the window function $[1 - (1 - \alpha/|\omega|)^k]$ with parameters α and k, where α corresponds to the step size in the Landweber algorithm and k corresponds to the number of iterations in the Landweber algorithm.

8.3.4 Numerical examples of the window function

In order to see what this window function $W(\omega) = 1 - (1 - \alpha/|\omega|)^k$ looks like, let us plot some cases of the window function in Figure 8.1. Before plotting, discretization is required. The frequency ω is defined in the range of $[-0.5, 0.5]$. This window function is not defined at $\omega = 0$. We define $W(0) = 1$ as a usual practice for a window function. Let the sampling interval on $[-0.5, 0.5]$ be $1/1{,}024$.

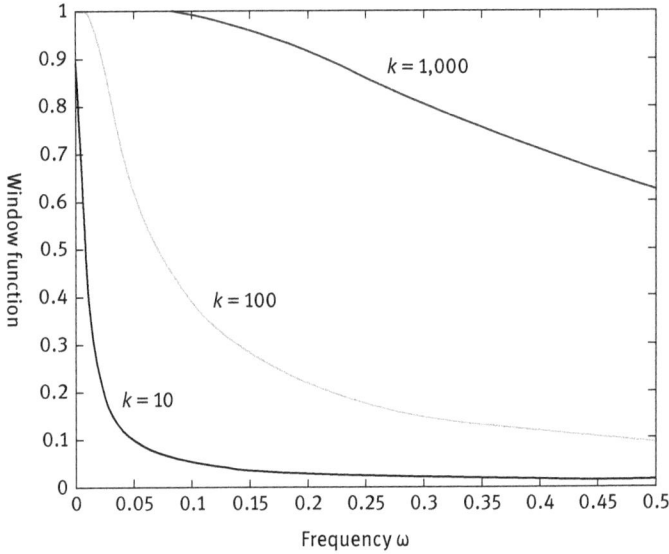

Fig. 8.1: Window functions for $k = 10$, 100, and 1,000, respectively.

The selection of the parameter α can be tricky. If α is not properly chosen $(1 - \alpha/|\omega|)^k$ can be unbounded with $k \to \infty$. We will use the worst case to find a proper α. The smallest positive (discrete) frequency is the sampling interval $1/1{,}024$. We require that

$$|1 - 1{,}024\alpha| < 1, \tag{8.12}$$

which gives

$$0 < \alpha < \frac{2}{1{,}024} = \frac{1}{\text{Number of frequency samples on } [0,\, 0.5]}. \tag{8.13}$$

We pick the median value for α as

$$\alpha = \frac{1}{1{,}024} = \frac{1}{\text{Number of frequency samples on } [-0.5,\, 0.5]}. \tag{8.14}$$

The window function is even and only the positive frequency half is shown in Figure 8.1. We observe that the window function $W(\omega) = 1 - (1 - \alpha/|\omega|)^k$ is like a lowpass filter. For a smaller k, the bandwidth is narrower. As k increases, the bandwidth gets wider.

8.4 The Landweber algorithm: the weighted FBP algorithm

8.4.1 Landweber algorithm: FBP without noise weighting

In Section 8.3, a backprojection-then-filtering algorithm is derived to emulate the iterative Landweber algorithm. This algorithm is almost the same as the conventional backprojection-then-filtering algorithm except that the ramp filter is modified by a window function, which is controlled by the iteration number k and a parameter α.

As presented in Chapter 2, many algorithms are equivalent and they belong to a big FBP family. Figure 2.9 lists 10 algorithms in this big family, and they can be converted from any one of them to another.

To convert a backprojection-then-filtering algorithm to an FBP algorithm, all one has to do is to take the central slice of the 2D ramp filter as the 1D filter. In our case, the 2D ramp filter and the 2D window function are isotropic (that is, radially symmetrical). If presented in the polar coordinate system, they are functions of the radius and independent from the angles. The 1D central sections of them have the same expressions as the 2D radial expression (see Figure 8.2). Therefore, the FBP algorithm that emulates the Landweber algorithm consists of the following steps:

1. Find the 1D Fourier transform of the projections $p(s, \theta)$ with respect to the first variable s, obtaining $P(\omega, \theta)$.
2. Multiply $P(\omega, \theta)$ with a windowed ramp filter $|\omega|[1 - (1 - \alpha/|\omega|)^k]$, obtaining $Q(\omega, \theta)$.
3. Find the 1D inverse Fourier transform of $Q(\omega, \theta)$ with respect to the first variable ω, obtaining $q(s, \theta)$.
4. Backproject $q(s, \theta)$ to obtain the final reconstruction.

This FBP algorithm is almost the same as the conventional FBP algorithm, except that the conventional ramp filter is replaced by a windowed ramp filter and the window function is controlled by the iteration number k and a parameter α. The transition from an iterative algorithm to an FBP algorithm is not so bad, is it?

You are encouraged to implement this Landweber-FBP algorithm and the iterative Landweber algorithm. When you compare their reconstructed images, the images are not 100% identical. These two algorithms are not truly equivalent. The FBP algorithm assumes continuous functions and has a shift-invariant property. On the other hand, the iterative algorithm depends on how the image array pixels are defined. The combined projection and backprojection $A^T A$ in an iterative algorithm do not give a perfect $1/r$ point spread function. The discrepancy between the results of these two algorithms is caused by the discrepancy between $A^T A$ and the $1/r$ point spread function. However, the discrepancies are not large. Figure 8.3 shows a pair of reconstructions by these two algorithms with $k = 20$, where the values of the parameter α are different for two algorithms.

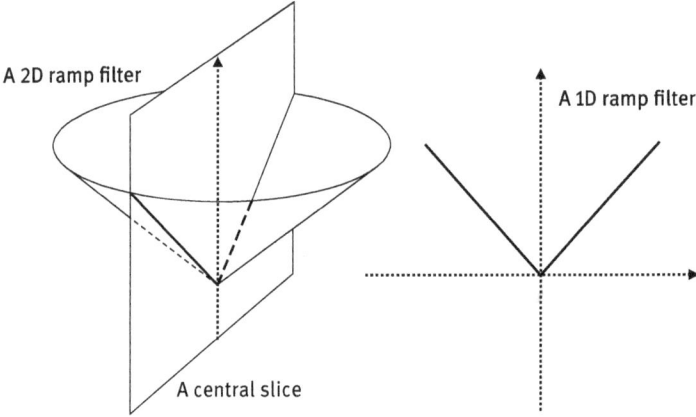

Fig. 8.2: The 1D ramp filter is a central slice of the 2D ramp filter.

8.4.2 Landweber algorithm: FBP with view-based noise weighting

One of the most important reasons to use an iterative algorithm is to incorporate the noise model during image reconstruction. This section considers a simplified case: the noise weighting is view-based. In other words, the weighting factors for all rays at a particular view are the same.

If the initial image is zero, the closed-form Landweber algorithm is

$$X^{(k)} = \left(A^T W A\right)^{-1}\left[I - \left(I - \alpha A^T W A\right)^k\right]A^T WP. \tag{8.15}$$

If the noise weighting is view-based, matrix W is block-diagonal. Each block corresponds to all the rays in a view. Within each block the diagonal elements are identical.

It is not easy to see what matrix W is doing in $A^T W A$. However, a central slice at a particular angle of $A^T W A$ is simply $|\omega|$ multiplied by a scalar w_m, where m is the view index. Thus, the windowed ramp filter at the mth view is given as

$$\frac{|\omega|}{w_m}\left[1 - \left(1 - \frac{\alpha w_m}{|\omega|}\right)^k\right]. \tag{8.16}$$

In eq. (8.15), $A^T WP$ is weighted backprojection of the projections P. At the mth view, all the projections are first multiplied by the scalar w_m and then are backprojected to the image domain. In fact, at the mth view, the projection data multiplier w_m cancels the w_m in the first part $|\omega|/w_m$ of eq. (8.16). The net result is that for view-based noise weighting eq. (8.15) is equivalent to

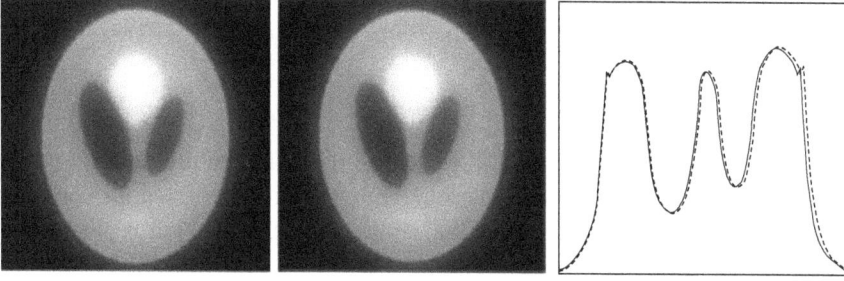

Fig. 8.3: Images reconstructed by the iterative Landweber algorithm (left and the broken line profile) and the Landweber-FBP algorithm (right and the solid line profile). The profiles are drawn across the central horizontal line.

$$X^{(k)} = \left(A^T A\right)^{-1}\left[I - \left(I - \alpha A^T W A\right)^k\right] A^T P. \tag{8.17}$$

The corresponding FBP algorithm for eq. (8.17) is given as follows:

1. Find the 1D Fourier transform of the projections $p(s, \theta)$ with respect to the first variable s, obtaining $P(\omega, \theta)$.
2. Multiply $P(\omega, \theta)$ with a windowed ramp filter $|\omega|[1 - (1 - \alpha w_\theta/|w|)^k]$, obtaining $Q(\omega, \theta)$.
3. Find the 1D inverse Fourier transform of $Q(\omega, \theta)$ with respect to the first variable ω, obtaining $q(s, \theta)$.
4. Backproject $q(s, \theta)$ to obtain the final reconstruction.

The view-based noise weighting only changes the window function of the ramp filter. The window function depends on the iteration number k, a parameter α, and the view-based weighting w_θ.

8.4.3 Landweber algorithm: FBP with ray-based noise weighting

The FBP algorithms without noise weighting and with view-based noise weighting give shift-invariant point spread function in the reconstructed image. Using ray-based noise weighting will make the FBP algorithm have a shift-variant point spread function. The reconstruction time for ray-based noise weighting can be significantly lengthened. The advantage of using an FBP algorithm would be diminished.

In order to make the algorithm fast, let us be creative. A little deviation in the weighting function has almost no effects on the reconstruction. At each view angle, we quantize the ray-based weighting function into $N + 1$ values: $w_0, w_1, \ldots, w_N$, which in turn give $N + 1$ different filters. They are

$$W_{k,a,w_n}(\omega) = 1 - \left(1 - \frac{aw_n}{|\omega|}\right)^k, \text{ when } \omega \neq 0,$$

$$W_{k,a,w_n}(0) = 1, \tag{8.18}$$

for $n = 0, 1, 2, \ldots, N$. Using these $N+1$ filters, $N+1$ sets of filtered projections are obtained. Before backprojection, one of these $N+1$ projections is selected for each ray according to its proper weighting function. Only one backprojection is performed using the selected filtered projections.

As an example, let us assume that N is 10 and the measurements are transmission data. Before the projections data are ready to process, we form 11 Fourier domain filter transfer functions $W_{k,a,w_n}(\omega)$ as defined in eq. (8.18) with, say, $w_n = \exp(-0.1 \times n \times p_{max})$, $n = 0, 1, 2, \ldots, 10$, respectively. Here p_{max} is the maximum value of the measured line integrals. In implementation, ω is a discrete frequency index and takes the discrete values of between -0.5 and 0.5 (and "0.5" corresponds to the highest frequency). Note that $w_n = \exp(-0.1 \times n \times p_{max})$ is for one exemplary noise model. One can use other weighting functions for other noise models. The ray-based FBP algorithm can be implemented as

1. At each view angle θ, find the 1D Fourier transform of $p(s, \theta)$ with respect to s, obtaining $P(\omega, \theta)$.
2. Form $N+1$ versions of $Q_n(\omega, \theta) = P(\omega, \theta) \times W_{k,a,w_n}(\omega)$ with $n = 0, 1, \ldots, N$.
3. Take the 1D inverse Fourier transform of $Q_n(\omega, \theta)$ with respect to ω, obtaining $q_n(s, \theta)$ with $n = 0, 1, \ldots, N$.
4. Construct $q(s, \theta)$ by letting $q(s, \theta) = q_n(s, \theta)$ if $p(s, \theta) \approx n \times p_{max}/N$.
5. Backproject $q(s, \theta)$ to obtain the final image.

This ray-based FBP algorithm is fast because it only performs backprojection once. Backprojection is the most time-consuming part in an FBP algorithm. We now demonstrate the effectiveness of this ray-based noise weighting FBP algorithm using a low-dose X-ray CT patient data set (see Figure 8.4). The severe horizontal streaking artifacts across the body are significantly reduced.

The noise model is $w = \frac{1}{variance}$, where "variance" is referred to as the variance of the sinogram (that is, the line integral data). We can have a more general expression of $w = \frac{1}{variance^\gamma}$, where γ is a parameter. When $\gamma = 0$, the weighting function is a constant without any variation. In this case, the noise weighting has no effect. A larger γ gives a larger variation of the weighting function. When $\gamma = 1$, the weighting function is the so-called "correct" weighting which is widely used among researchers. Since the weighting function is affected by the "iteration number," the "correct" weighting may not yield the best results. In practice, the readers are encouraged to use a γ value that is different from 1. In many cases, a γ value between 0 and 1 gives a better result than using $\gamma = 1$.

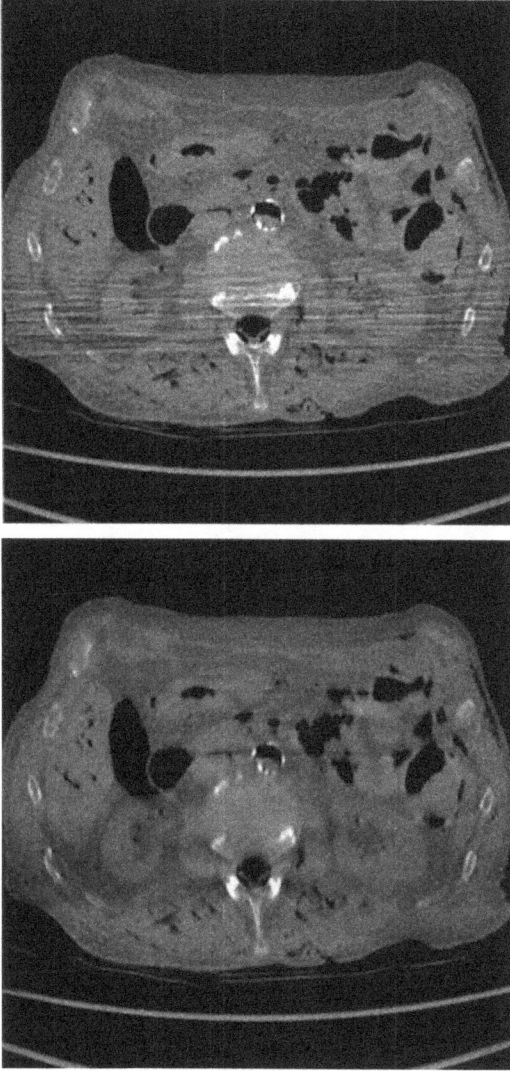

Fig. 8.4: Reconstruction results for the clinical data: (Top) The conventional FBP reconstruction; (Bottom) the ray-based noise-weighted FBP reconstruction. Display window is from −400 to 400 HU. [Thanks Raoul M. S. Joemai of Leiden University Medical Center for the CT scan.].

8.5 FBP algorithm with quadratic constraints

Other than being able to model the data noise, the iterative algorithm can also incorporate constraints as Bayesian optimization. If the constraints are quadratic functions of the image, an FBP algorithm can be developed to incorporate them.

The Bayesian constraints regulate the solutions and effectively control the noise propagation. Therefore, the common iterative Bayesian algorithms do not diverge and, in theory, can iterate till infinity. There is no need to stop the algorithms early.

Many researchers use both noise weighting and Bayesian constraints for iterative image reconstruction.

One can set up a Bayesian objective function

$$
\begin{aligned}
\chi^2 &= \|AX - P\|_w^2 + \beta\|FX - GY\|^2 \\
&= X^T A^T WAX - 2X^T A^T WP + P^T WP \\
&\quad + \beta(X^T F^T FX - 2X^T F^T GY + Y^T G^T GY),
\end{aligned}
\tag{8.19}
$$

where Y is a reference image, and F and G are feature extracting matrices. This objective function pushes the selected feature of X to look like the selected feature of Y. For example, if Y is zero and F is the identity matrix, this objective function is looking for a minimum norm solution. If F and G are edge extracting matrices, this objective function is looking for an X that has edges similar to the edges of Y.

The gradient of the objective function (8.19) is

$$
\nabla\chi^2 = 2\left[A^T WAX - 2A^T WP + \beta\left(F^T FX - F^T GY\right)\right],
\tag{8.20}
$$

and its corresponding iterative Landweber algorithm can be readily obtained as

$$
\begin{aligned}
X^{(k)} &= X^{(k-1)} - \alpha\left[A^T W\left(AX^{(k-1)} - P\right) + \beta F^T\left(FX^{(k-1)} - GY\right)\right] \\
&= \left(I - \alpha A^T WA - \alpha\beta F^T F\right)X^{(k-1)} + \alpha A^T WP + \alpha\beta F^T GY \\
&= \left(\alpha A^T WA + \alpha\beta F^T F\right)^{-1}\left[I - \left(I - \alpha A^T WA - \alpha\beta F^T F\right)^k\right]\left(\alpha A^T WP + \alpha\beta F^T GY\right) \\
&\quad + \left(I - \alpha A^T WA - \alpha\beta F^T F\right)^k X^{(0)} \\
&= \left(A^T WA + \beta F^T F\right)^{-1}\left[I - \left(I - \alpha A^T WA - \alpha\beta F^T F\right)^k\right]\left(A^T WP + \beta F^T GY\right) \\
&\quad + \left(I - \alpha A^T WA - \alpha\beta F^T F\right)^k X^{(0)}.
\end{aligned}
\tag{8.21}
$$

If the initial image is zero, the Landweber Bayesian algorithm becomes

$$
X^{(k)} = \left(A^T WA + \beta F^T F\right)^{-1}\left[I - \left(I - \alpha A^T WA - \alpha\beta F^T F\right)^k\right]\left(A^T WP + \beta F^T GY\right).
\tag{8.22}
$$

The linear iterative algorithm (8.22) can be readily translated into an FBP algorithm.

8.5.1 Example of minimum-norm constrained FBP

In order to convert this iterative Bayesian Landweber algorithm (8.22) to an FBP algorithm, we need to know F, G, and Y. As an example, let Y be zero, and F be the identity matrix. In this case, a minimum norm solution is sought. We also assume that W is view-based noise weighting. Then eq. (8.22) becomes

$$X^{(k)} = \left(A^T W A + \beta I\right)^{-1} \left[I - \left(I - \alpha A^T W A - \alpha \beta I\right)^k\right] A^T W P, \tag{8.23}$$

and the corresponding modified ramp function for the noise-weighted Bayesian FBP algorithm is given as

$$\text{RAMP}_m(\omega) = \frac{w_m}{\frac{w_m}{|\omega|} + \beta}\left[1 - \left(1 - \frac{\alpha w_m}{|\omega|} - \alpha\beta\right)^k\right]$$

$$= |\omega|\frac{1}{1 + \beta|\omega|/w_m}\left[1 - \left(1 - \frac{\alpha w_m}{|\omega|} - \alpha\beta\right)^k\right], \tag{8.24}$$

where m is the angle index and w_m is the noise weighting for the mth angle. In a Bayesian algorithm, the iteration number can be arbitrarily large. Let $k \to \infty$, then the window function associated with eq. (8.24) becomes

$$W_m(\omega) = \frac{1}{1 + \beta|\omega|/w_m}. \tag{8.25}$$

In order to see what this window function $W_m(\omega)$ in eq. (8.25) looks like, let us plot some cases of the window function in Figure 8.5. Let the sampling interval on [0, 0.5] be 0.5/512 = 1/1,024. Let β/α be 1. It is observed from Figure 8.5 that a smaller weight w_m gives a narrower bandwidth and stronger lowpass filtering.

The implementation of a noise-weighted Bayesian-FBP algorithm is almost the same as a noise-weighted Landweber FBP algorithm; the only difference is in the window function expressions.

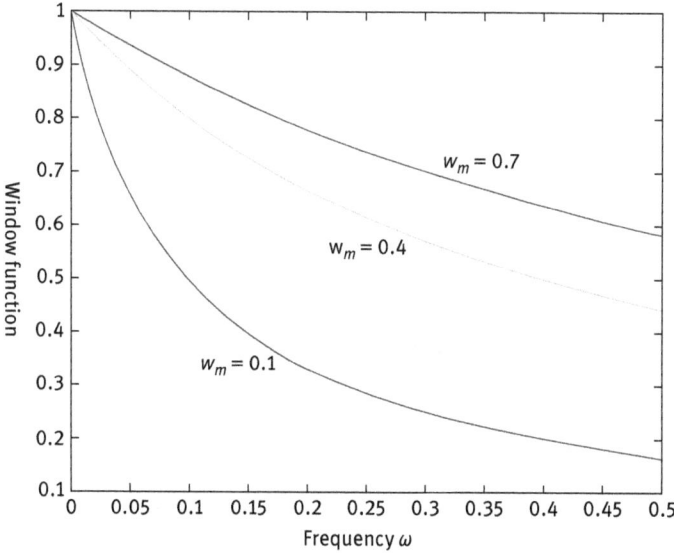

Fig. 8.5: Window functions for the Bayesian-FBP algorithm with $w_m = 0.1$, 0.4, and 0.7, respectively.

8.5.2 Example of reference-image constrained FBP

Let us present another example in which there is no noise weighting, but there is a reference image Y, which can also be represented as a pure backprojection (without ramp filtering) of a certain sinogram Q, i.e., $Y = A^T Q$. Let $F = G = I$. We want X to look like Y. In this example, eq. (8.22) becomes

$$
\begin{aligned}
X^{(k)} &= (A^T A + \beta I)^{-1} [I - (I - \alpha A^T A - \alpha \beta I)^k] (A^T P + \beta Y) \\
&= (A^T A + \beta I)^{-1} \left[I - (I - \alpha A^T A - \alpha \beta I)^k \right] (A^T P + \beta A^T Q).
\end{aligned}
\tag{8.26}
$$

For a Bayesian algorithm, we can let $k \to \infty$. Thus eq. (8.26) is reduced to

$$
X^{(k)} = (A^T A + \beta I)^{-1} (A^T P + \beta A^T Q).
\tag{8.27}
$$

The Fourier domain counterpart of $(A^T A + \beta I)^{-1}$ is

$$
\frac{1}{\frac{1}{|\omega|} + \beta} = \frac{|\omega|}{1 + \beta |\omega|},
\tag{8.28}
$$

which leads to the corresponding window function for the ramp filter as

$$
W(\omega) = \frac{1}{1 + \beta |\omega|}.
\tag{8.29}
$$

This window function is not a stranger to us and is very similar to eq. (8.25). If we plot it with respect to the frequency ω, it looks like a lowpass filter, similar to the curves in Figure 8.5.

Something new to us is in $A^T P + \beta Y$ and in $A^T P + \beta A^T Q$, where P is the projection array and $A^T P$ backprojects P into the image domain. The reference Y is given in the image domain, and $Y = A^T Q$. What is this Q? Well, if we backproject Q to get Y, then Q is the ramp-filtered version of AY.

The Bayesian-FBP version of eq. (8.27) can be implemented as follows:

1. We are given projections $p(s, \theta)$ and a reference image $f_{ref}(x, y)$. At each view angle θ, we forward project $f_{ref}(x, y)$, obtaining $p_{ref}(s, \theta)$. Then apply the ramp filter to $p_{ref}(s, \theta)$, obtaining $q_{ref}(s, \theta)$. Form new projections $p_{new}(s, \theta) = p(s, \theta) + \beta q_{ref}(s, \theta)$.
2. Find the 1D Fourier transform of $P_{new}(\omega, \theta)$ with respect to s, obtaining $P_{new}(\omega, \theta)$
3. Use the window function (8.29) and compute $Q_{new}(\omega, \theta) = P_{new}(\omega, \theta) \times |\omega| \times W(\omega)$.
4. Take the 1D inverse Fourier transform of $Q_{new}(\omega, \theta)$ with respect to ω, obtaining $q_{new}(s, \theta)$.
5. Backproject $q_{new}(s, \theta)$ to obtain the final image.

You may ask: "What is the reference image Y?" We use an MRI example to answer this question. We want to generate a movie of a human beating heart using a radial k-

space data acquisition method. We do not have time to acquire a full *k*-space data set at each time frame. At each time frame, we barely have time to measure data in 24 views. At a different time frame, we measure different 24 views. If we sum up measurements from four consecutive time frames, we would get 96 different views and they can be considered to form a full data set. For each time frame, we would like to reconstruct an image *X*. Before we reconstruct *X*, we use the data from closest five time frames (present + 2 before + 2 after) to reconstruct a reference image *Y*. When we actually reconstruct image *X*, the reference image *Y* comes in to help. Some reconstruction results are shown in Figure 8.6.

8.6 Convolution backprojection

Once we have an FBP algorithm, there should be no problem at all to obtain a convolution backprojection algorithm. All we have to do is to take the inverse Fourier transform of the weighted ramp filter to get a convolution kernel.

This is easier said than done. The problem is that we have a closed-form expression for the frequency domain transfer function for the windowed ramp filter; however, we do not have a closed-form expression for the spatial domain convolution kernel. Numerical evaluation of the convolution kernel costs time.

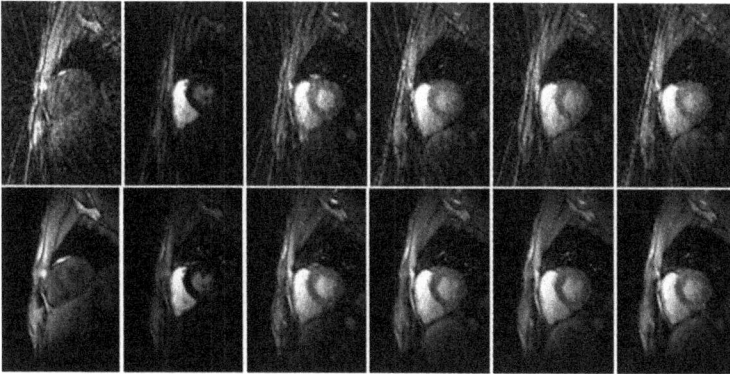

Fig. 8.6: FBP reconstruction results using dynamic MRI data. Top: Reconstructions without using the reference image *Y*. Bottom: Reconstructions with assistance from the reference image *Y*. [Thanks Drs. Adluru and DiBella of the University of Utah for the MRI scan.].

If we do not have a closed-form convolution kernel, the second best thing is to obtain an explicit expression for an approximate convolution kernel. Here we use the example in Section 8.5.1 to illustrate our strategy. The noise-weighted ramp filter transfer function is given as

$$H(\omega) = \frac{|\omega|}{1+\gamma_0|\omega|},$$ (8.30)

where

$$\gamma_0 = \frac{\beta}{w_m}.$$ (8.31)

We do not have an explicit inverse Fourier transform expression for (8.30). We need to find an expansion of $H(\omega) = \frac{|\omega|}{1+\gamma_0|\omega|}$ and each term in the expansion should have a closed-form inverse Fourier transform expression.

Now the question is: What basis functions should we choose? The Fourier series expansion does not work because $H(\omega)$ is not periodic. The Taylor expansion does not work well because the polynomials do not look like the function $H(\omega) = \frac{|\omega|}{1+\gamma_0|\omega|}$, which is monotonically increasing from 0 to 1 in the positive axis of ω.

If we factor the ramp filter $|\omega|$ out, $1/(1+\gamma_0|\omega|)$ is monotonically decreasing and looks like an exponential function $e^{-\gamma_0\omega}$ for $\omega > 0$. Based on this strategy, we propose the following expansion for $\omega > 0$:

$$\frac{1}{1+\gamma_0\cdot\omega} \approx \frac{1}{3}\left(e^{-\gamma_0\omega} + e^{-\gamma_1\omega} + e^{-\gamma_2\omega}\right),$$ (8.32)

where the parameters γ_1 and γ_2 have to be determined. The range of ω is [0, 1/2]. Approximation (8.32) is already exact at $\omega = 0$. We further request that eq. (8.32) to be exact at $\omega = 1/4$ and $\omega = 1/2$. Thus, we have two unknowns (1 and 2) and two equations:

$$\frac{1}{1+\gamma_0/2} = \frac{1}{3}\left(e^{-\gamma_0/2} + e^{-\gamma_1/2} + e^{-\gamma_2/2}\right),$$ (8.33)

$$\frac{1}{1+\gamma_0/4} = \frac{1}{3}\left(e^{-\gamma_0/4} + e^{-\gamma_1/4} + e^{-\gamma_2/4}\right).$$ (8.34)

Solving these two eqs. (8.33) and (8.34) yields

$$\gamma_1 = -4\cdot\ln\left(\frac{A+\sqrt{2B-A^2}}{2}\right)$$ (8.35)

and

$$\gamma_2 = -4\cdot\ln\left(\frac{A-\sqrt{2B-A^2}}{2}\right),$$ (8.36)

where

$$A = \frac{3}{1+\gamma_0/4} - e^{-\gamma_0/4}$$ (8.37)

and

$$B = \frac{3}{1 + \gamma_0/2} - e^{-\gamma_0/2}. \qquad (8.38)$$

Using the above results and an integral table, the closed-form filter kernel $h(n)$ can be obtained by performing 1D Fourier transform and sampling at the integer points:

$$\begin{aligned}
h(n) &= \int_{-1/2}^{1/2} H(\omega) e^{2\pi i n \omega} d\omega \\
&= \frac{2}{3} \int_0^{1/2} \omega (e^{-\gamma_0 \omega} + e^{-\gamma_1 \omega} + e^{-\gamma_2 \omega}) \cos(2\pi n\omega) d\omega \\
&= \frac{-2}{3} \sum_{k=0}^{2} \frac{(-1)^n e^{-\frac{\gamma_k}{2}} \left(\gamma_k^3 + 4\gamma_k \pi^2 n^2 + 2\gamma_k^2 - 8\pi^2 n^2\right)}{\left(\gamma_k^2 + 4\pi^2 n^2\right)^2}.
\end{aligned} \qquad (8.39)$$

Figure 8.7 illustrates that the approximation relationship (8.32) is fairly accurate. There are six curves in Figure 8.7: three cases of true curves and corresponding three cases of approximated curves. We can only see three curves in Figure 8.7. This is because the true and approximated curves match so well and on the top of each other.

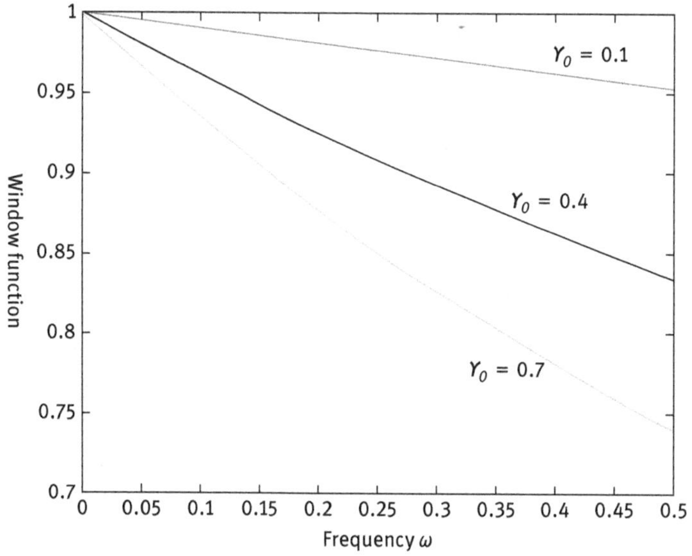

Fig. 8.7: Verification of the approximation expression (8.32) with different values of γ_0.

If there is no noise weighting, that is, $w_m \equiv 1$ in eq. (8.32), expression (8.39) gives a convolution kernel for a Bayesian convolution backprojection algorithm, which can be implemented in the following simple two steps:
1. Convolve the projections $p(n, \theta)$ with the convolution kernel $h(n)$ as defined in eq. (8.39), obtaining $q(n, \theta)$.
2. Backproject $q(n, \theta)$ to get the final reconstruction.

When ray-by-ray noise weighting is required, w_m is a function of ray, and y_0 is no longer a constant. In this case, expression (8.39) still valid, but $h(n)$ is shift variant and ray-dependent. Thus, it is now not proper to refer $h(n)$ to as a convolution kernel because "convolution" assumes the shift invariant property. Other than not using the term "convolution," the implementation steps are still the same as the simple two steps:
1. For each the projection $p(n, \theta)$, determine a kernel $h_{n,\theta}(k)$ as defined in eq. (8.39), obtaining a value $q(n, \theta)$ as

$$q(n,\theta) = \sum_k p(k,\theta) h_{n,\theta}(n-k). \tag{8.40}$$

2. Backproject $q(n, \theta)$ to get the final reconstruction.

Equation (8.40) still looks like the convolution formula. What is the difference? At a fixed view angle θ, the true convolution uses the same kernel $h(n)$ to calculate $q(n, \theta)$ for every index n on the detector. On the other hand, the kernel $h_{n,\theta}(k)$ changes in the calculation of $q(n, \theta)$ for different n, as indicated in eq. (8.40). The computational complexity of eq. (8.40) is the same as that of a real convolution.

8.7 Nonquadratic constraints

Many people choose to use the iterative image reconstruction algorithms because they are so versatile. For example, they work well when the objective function have nonquadratic constraints as long as they are convex. Most edge-preserving constraints are nonquadratic, but are convex. The FBP algorithms are linear algorithms; they can only solve linear equations. If the objective function is quadratic, the gradient of the objective function is linear. By setting the gradients to zero we obtain a set of linear equations. To our knowledge, the FBP algorithms are unable to solve nonlinear equations.

Of course, one can apply nonlinear filters as preprocessing or postprocessing procedures. For example, one can apply a nonlinear median filter to denoise the projections and apply a nonlinear edge-preserving filter to denoise the FBP reconstructed image.

In many situations, the iterative reconstruction outperforms pre- and postfiltering. In this section, we propose a fast method to apply nonlinear filtering during reconstruction.

A unique feature in an iterative algorithm is that the result depends on the initial image. The conventional FBP algorithm does not accept the initial image. We want to change that!

Let us recall the beginning of Section 8.1, where the Landweber algorithm has two parts:

$$X^{(k)} = F^{(k)}P + H^{(k)}X^{(0)}. \tag{8.41}$$

The first part $X^{(k)} = F^{(k)}P$ only depends on the projections P and can be implemented as an FBP algorithm. The second part only depends on the initial image $X^{(0)}$, which has been assumed to be zero in the discussion in the first part of this chapter. From now on the initial image $X^{(0)}$ does not have to be a zero.

According to eq. (8.4), $H^{(k)}$ is defined as

$$H^{(k)} = \left(I - \alpha A^T W A\right)^k, \tag{8.42}$$

which can be implemented in the Fourier domain with a 2D transfer function in the polar coordinate system:

$$H(\omega) = \left(1 - \frac{\alpha w}{|\omega|}\right)^k \text{ when } \omega \neq 0, \tag{8.43}$$

and we define $H(0) = 0$. As $|\omega| \to \infty$, $H(\omega) \to 1$. As k increases, the bandwidth of the high-pass filter gets narrower. A fast implementation of the second term in eq. (8.41) can be

1.   Take 2D fast Fourier transform (FFT) of $X^{(0)}$.
2.   Multiply the Fourier transformed image by $H(\omega)$.
3.   Take 2D inverse FFT.

Thus eq. (8.41) can be implemented in two parts: a windowed FBP algorithm for the projections P and a high-pass filter for the initial image $X^{(0)}$. Let us introduce our strategy with an example.

Say, we have a task of image reconstruction. This task can be achieved by using 2,000 iterations of the Landweber algorithm, which can also be done by using a windowed FBP algorithm with $k = 2,000$. We are now experts for that.

Let us give algorithm (8.41) a name: *the 2-input FBP*, one in put being P and the other input being the initial image $X^{(0)}$. For the first input P, we apply an FBP; for the second input $X^{(0)}$, we apply a high-pass filter. Instead of using the windowed FBP with $k = 2,000$, we can achieve the same goal by applying the 2-input FBP (with $k = 500$) four times. The result of one application of the 2-input FBP algorithm is the initial image of the next application.

It is nice to notice that the first part of the algorithm "$F^{(500)}P$" only needs to be calculated once, even though the entire algorithm needs to be applied four times.

However, the second part "$H^{(500)}X^{(0)}$" needs to be calculated four times because each time the initial image $X^{(0)}$ is different.

You may think that it is silly to break down one application of the "$k = 2,000$" algorithm into four applications of the "$k = 500$" algorithm. Yes, it is silly. It will not be silly if we apply an edge-preserving filter at the end of each application of the 2-input FBP algorithm. There are many edge-preserving denoising filters such as bilateral filter and guided filter. You can use any of your favorite one. The flowchart of our suggested algorithm is shown in Figure 8.8 (still using our example).

Let us symbolically represent the chosen nonlinear filter as G. Thus, the algorithm illustrated in Figure 8.8 can be expressed as

$$Y^{(1)} = G\left[F^{(500)}P + 0\right],$$

$$Y^{(2)} = G\left[F^{(500)}P + H^{(500)}Y^{(1)}\right],$$

$$Y^{(3)} = G\left[F^{(500)}P + H^{(500)}Y^{(2)}\right], \tag{8.44}$$

$$Y^{(4)} = G\left[F^{(500)}P + H^{(500)}Y^{(3)}\right].$$

Here $Y^{(4)}$ is the final reconstruction. We use notation $Y^{(n)}$ (instead of $X^{(n)}$) to avoid the confusion with the result of the algorithm (8.41) which does not use any nonlinear filters. In other words, notation $X^{(k)}$ is for the results of the iterative algorithm (8.41) without any nonlinearity involved; notation $Y^{(n)}$ is for the results of the algorithm (8.45) with a nonlinear filter G:

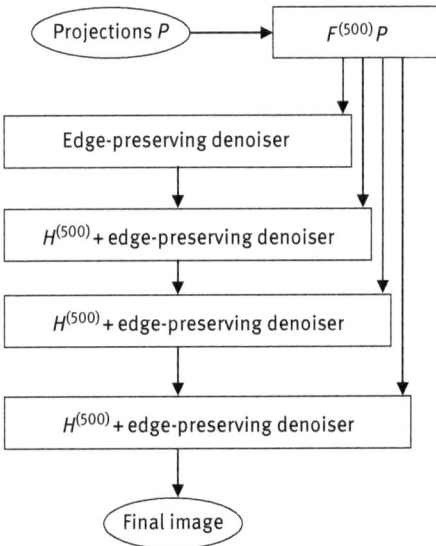

Fig. 8.8: A nonquadratic edge-preserving constraint is incorporated in the 2-input FBP algorithm, where the FBP part only performs once and the image filtering part performs multiple times.

$$Y^{(n)} = G\left[F^{(k)}P + H^{(k)}Y^{(n-1)}\right].\tag{8.45}$$

The nonlinear filter G can be a Huber filter, a guided filter, a bilateral filter, a BM3D filter, and so on. The approach discussed in this section is sometimes referred to as an iterative POCS approach. Here, POCS is an abbreviation for projections onto projections onto convex sets. In layman's language, this approach alternatively uses different methods back-and-forth, and each method has its own goal. The effectiveness of this POCS method is illustrated in Figure 8.9, where an FBP reconstruction is compared with a POCS reconstruction for a low-dose CT patient study.

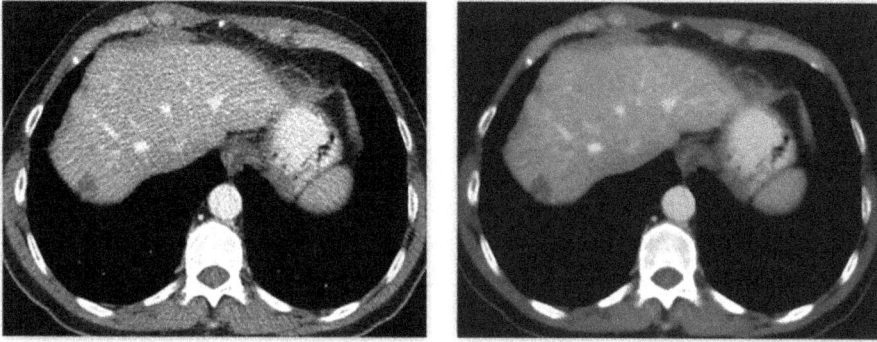

Fig. 8.9: (Left) An FBP reconstruction of low-dose CT image. (Right) A POCS reconstruction of the same image.

8.8 A viewpoint from calculus of variations

The discussion in the previous sections gives an impression that the new FBP algorithms can only be derived from the iterative Landweber algorithm. In fact, these new FBP algorithms can be derived without using the iterative Landweber algorithm. In this section, we give an example of deriving a new FBP algorithm by using calculus of variations.

Let the image to be reconstructed be $f(x, y)$ and its Radon transform be $[Rf](s, \theta)$, defined as

$$[Rf](s, \theta) = \int_{-\infty}^{\infty} \int_{-\infty}^{\infty} f(x,y)\delta(x\cos\theta + y\sin\theta - s)dxdy,\tag{8.46}$$

where δ is the Dirac delta function, θ is the detector rotation angle, and s is the line-integral location on the detector. The Radon transform $[Rf](s, \theta)$ is the line integral of the object $f(x, y)$. Image reconstruction is to solve for the object $f(x, y)$ from its Radon transform $[Rf](s, \theta)$. Let the noisy line-integral measurements be $p(s, \theta)$. The objective function v depends on the function $f(x, y)$ as follows:

$$v(f) = \|[Rf](s,\theta) - p(s,\theta)\|^2 + \beta\|f - g^2\|, \tag{8.47}$$

where the first term enforces data fidelity, the second term imposes prior information about the image f, the parameter $\beta > 0$ controls the level of influence of the prior information to the image f, and g is a reference image. Using integral expressions, v in eq. (8.47) can be written as

$$v(f) = \int_{\theta=0}^{\pi}\int_{s=-\infty}^{\infty}\left[\int_{x=-\infty}^{\infty}\int_{y=-\infty}^{\infty} f(x,y)\delta(\cos\theta + y\sin\theta - s)dydx - p(s,\theta)\right]^2 dsd\theta$$

$$+ \beta \int_{x=-\infty}^{\infty}\int_{y=-\infty}^{\infty} [f(x,y)-g(x,y)]^2 dydx. \tag{8.48}$$

Now we are ready to use the calculus of variations to find the optimal function $f(x, y)$ that minimizes the objective function $v(f)$ in eq. (8.48). The initial step is to replace the function $f(x, y)$ in eq. (8.48) by the sum of two functions: $f(x, y) + \varepsilon\eta(x, y)$, where $\eta(x, y)$ can be any arbitrary function. The next step is to evaluate $\frac{dv}{d\varepsilon}|_{\varepsilon=0}$ and set it to zero. That is,

$$0 = 2\int_{\theta=0}^{\pi}\int_{s=-\infty}^{\infty}\left\{\left[\int_{x=-\infty}^{\infty}\int_{y=-\infty}^{\infty} f(x,y)\delta(x\cos\theta + y\sin\theta - s)dydx - p(s,\theta)\right]\right.$$

$$\times \left.\left[\int_{x=-\infty}^{\infty}\int_{y=-\infty}^{\infty} \eta(x,y)\delta(x\cos\theta + y\sin\theta - s)dydx\right]dsd\theta\right\}$$

$$+ 2\beta \times \int_{x=-\infty}^{\infty}\int_{y=-\infty}^{\infty} [f(x,y)-g(x,y)]n(x,y)dydx. \tag{8.49}$$

In practice, the function $f(x, y)$ is compact, bounded, and continuous almost everywhere. After changing the order of integrals, we have

$$0 = \int_{x=-\infty}^{\infty}\int_{y=-\infty}^{\infty} \eta(x,y) \times \left\{\int_{\theta=0}^{\pi}\int_{s=-\infty}^{\infty}\left[\int_{\hat{x}=-\infty}^{\infty}\int_{\hat{y}=-\infty}^{\infty} f(\hat{x},\hat{y})\delta(\hat{x}\cos\theta + \hat{y}\sin\theta - s)d\hat{y}d\hat{x} - p(s,\theta)\right]\right.$$

$$\left. \delta(x\cos\theta + y\sin\theta - s)dsd\theta + \beta[f(x,y)-g(x,y)] \right\} dydx. \tag{8.50}$$

Equation (8.50) is in the form of $\int \int \eta(x, y)c(x, y)dydx = 0$. Since $\eta(x, y)$ can be any arbitrary function, according to the calculus of variations, one must have $c(x, y) = 0$, which is the Euler–Lagrange equation. The Euler–Lagrange equation in our case is

$$
\int_{\theta=0}^{\pi} \int_{s=-\infty}^{\infty} \left[\int_{\hat{x}=-\infty}^{\infty} \int_{\hat{y}=-\infty}^{\infty} f(\hat{x},\hat{y})\delta(\hat{x}\cos\theta+\hat{y}\sin\theta-s)d\hat{y}d\hat{x} - p(s,\theta) \right]
$$

$$
\delta(x\,\cos\,\theta + y\,\sin\,\theta - s)dsd\theta
$$

$$
+ \beta[f(x,y) - g(x,y)] = 0. \tag{8.51}
$$

Equation (8.51) can be further rewritten as

$$
\int_{\hat{x}=-\infty}^{\infty} \int_{\hat{y}=-\infty}^{\infty} f(\hat{x},\hat{y}) \left[\int_{\theta=0}^{\pi} \int_{s=-\infty}^{\infty} \delta(\hat{x}\cos\theta+\hat{y}\sin\theta-s)\delta(x\cos\theta+y\sin\theta-s)dsd\theta \right] d\hat{y}d\hat{x} + \beta f(x,y)
$$

$$
= \int_{\theta=0}^{\pi} \int_{s=-\infty}^{\infty} p(s,\theta)\delta(x\cos\theta+y\sin\theta-s)dsd\theta + \beta g(x,y). \tag{8.52}
$$

Notice that

$$
\int_{\theta=0}^{\pi} \int_{s=-\infty}^{\infty} p(s,\theta)\delta(x\cos\theta+y\sin\theta-s)dsd\theta = \int_{\theta=0}^{\pi} p(s=\theta)|_{s=x\cos\theta+y\sin\theta}d\theta = b(x,y) \tag{8.53}
$$

is the backprojection of the projection data $p(s, \theta)$, and the backprojection is denoted as $b(x, y)$. It must be pointed out that $b(x, y)$ is not the same as $f(x, y)$ even when $p(s, \theta)$ is noiseless because no ramp filter has been applied to $p(s, \theta)$.

Also notice that

$$
\int_{\theta-0}^{\pi} \int_{s=-\infty}^{\infty} \delta(\hat{x}\cos\theta+\hat{y}\sin\theta-s)\delta(\cos\theta+y\sin\theta-s)dsd\theta
$$

$$
= \int_{\theta=0}^{\pi} \delta((x-\hat{x})\cos\theta + (y-\hat{y})\sin\theta)d\theta = \frac{1}{\sqrt{(x-\hat{x})^2 + (y-\hat{y})^2}} \tag{8.54}
$$

is the point spread function of the projection/backprojection operator at point (x, y) if the point source is at $(\hat{x}, \hat{y})$.

Using eqs. (8.53) and (8.54), eq. (8.52) becomes

$$\int_{\hat{x}=-\infty}^{\infty} \int_{\hat{y}=-\infty}^{\infty} f(\hat{x},\hat{y}) \left[\frac{1}{\sqrt{(x-\hat{x})^2+(y+\hat{y})^2}} + \beta \cdot \delta(x-\hat{x},y-\hat{y}) \right] d\hat{y} d\hat{x} = b(x,y) + \beta g(x,y).$$

(8.55)

The left-hand side of eq. (8.55) is a 2D convolution. Taking the 2D Fourier transform of eq. (8.55) yields

$$F(\omega_x,\omega_y) \times \left[\frac{1}{\sqrt{\omega_x^2+\omega_y^2}} + \beta \right] = B(\omega_x,\omega_y) + \beta G(\omega_x,\omega_y),$$

(8.56)

or

$$F(\omega_x,\omega_y) = \left[B(\omega_x,\omega_y) + \beta G(\omega_x,\omega_y) \right] / \left[\frac{1}{\sqrt{\omega_x^2+\omega_y^2}} + \beta \right].$$

(8.57)

Here the uppercase letters are used to represent the Fourier transform of their spatial domain counterparts, which are represented in lowercase letters; ω_x and ω_y are the frequency variables for x and y, respectively.

Equation (8.57) is in the form of "backprojection first, then 2D filtering" reconstruction approach and the Fourier domain 2D filter is $\sqrt{\omega_x^2+\omega_y^2}/\left(1+\beta\cdot\sqrt{\omega_x^2+\omega_y^2}\right)$. By using the central slice theorem, an FBP algorithm, which performs 1D filtering first and then backprojects, can be obtained. The Fourier domain 1D filter for this new FBP algorithm is the central slice of the 2D filter $\sqrt{\omega_x^2+\omega_y^2}/\left(1+\beta\cdot\sqrt{\omega_x^2+\omega_y^2}\right)$, which gives

$$\text{RAMP}(\omega) = |\omega|/[1+\beta\cdot|\omega|].$$

(8.58)

When $\beta = 0$, eq. (8.58) reduces to the ramp filter ω of the conventional FBP algorithm.

8.9 Noise-weighted FBP algorithm for uniformly attenuated SPECT projections

If the attenuator in SPECT imaging is uniform, Sections 4.5.1 and 4.5.2 of this book introduce two analytical image reconstruction algorithms: the Tretiak–Metz algorithm and Inouye's algorithm. The main difference between these two algorithms is that the Tretiak–Metz algorithm uses an exponential weighting function in the backprojector, while Inouye's algorithm uses a backprojector that is used as a regular FBP algorithm without the exponential weighting function. The backprojector plays an important role in noise amplification. The exponential weighting function in the backprojector worsens the image noise. Therefore, Inouye's algorithm is the preferred algorithm between these two algorithms.

As presented in Section 4.5.2, Inouye's algorithm converts the uniformly attenuated Radon transform into unattenuated Radon transform, as expressed by eq. (4.26). Therefore, the regular FBP algorithm can be used for image reconstruction.

In order to further control noise, the Landweber-FBP algorithm presented in Section 8.4.3 as well as the algorithms presented in Sections 8.5, 8.6, and 8.7 can be directly applied to the corrected, unattenuated projections provided by eq. (4.26).

8.10 Other applications of the Landwever-FBP algorithm

8.10.1 The Landweber-FBP version of the iterative FBP algorithm

In a regular iterative image reconstruction algorithm, the projection-domain data is directly backprojected into the image domain. Before backprojection, a noise-weighing function, w, is assigned to projection-domain data, which is the projection-domain discrepancy in an iterative Landweber algorithm.

In an iterative FBP algorithm, the projection-domain data is filtered with prespecified filter. This filter can be chosen as the ramp filter, a Laplacian filter, or a lowpass filter. In a Bayesian algorithm, the Bayesian terms in the objective function can also contain filters. As an example, an objective function can be expressed as follows:

$$\chi^2 = (AX - P)^T M(AX - P) + \beta(FX)^T \hat{M}(FX), \tag{8.59}$$

where M and $\hat{M}$ are two Toeplitz square matrices. We must point out that if the matrix M is a diagonal matrix, it acts as noise weighting instead of a filter. Noise weighting only scales a value up or down. On the other hand, a filter uses neighboring values as well.

A linear convolution can always be expressed as a Toeplitz matrix multiplication. A Landweber algorithm can be used to minimize this objective function. Using the similar method in Section 8.5, a Landweber Bayesian algorithm can be obtained as

$$X^{(k)} = (A^T MA + \beta F^T \hat{M}F)^{-1} \left[I - \left(I - \alpha A^T MA - \alpha\beta F^T \hat{M}F \right)^k \right] A^T MP$$
$$+ (I - \alpha A^T MA - \alpha\beta F^T \hat{M}F)^k X^{(0)}, \tag{8.60}$$

To illustrate how this can be translated into the Fourier domain, let's consider an example of $\beta = 0$, $X^{(0)} = 0$, and M is the Laplacian filter (that is, a second-order derivative filter). The transfer function of the 1D Laplacian filter is ω^2, and the transfer function of the 2D Laplacian filter is $\|\vec{\omega}\|^2$. Thus, eq. (8.60) has a 2D transfer function of

$$\frac{1-(1-\alpha\|\vec{\omega}\|)^k}{\|\vec{\omega}\|}\|\vec{\omega}\| = 1-(1-\alpha\|\vec{\omega}\|)^k. \tag{8.61}$$

The filter expressed in eq. (8.61) is applied to the pure backprojection of the sinogram P. The filtered image is the output of the kth iteration of the Landweber algorithm. After mapping eq. (8.61) into its 1D version as shown in eq. (8.62), an FBP algorithm is obtained:

$$1-(1-\alpha|\omega|)^k. \tag{8.62}$$

Notice that MP in eq. (8.60) is 1D filtering and $A^T MP$ backprojects this filtered result into the image domain. According to the central slice theorem, this is equivalent to filter the pure backprojected image with a 2D Laplacian filter $\|\vec{\omega}\|^2$.

This is an interesting image reconstruction algorithm, in the sense that high-frequency components converge first; the low frequency components converge later. This behavior is the opposite of what we have seen so far in iterative image reconstruction. In eq. (8.62), $|\omega|$ is a high-pass filter, $(1-\alpha|\omega|)^k$ is a lowpass filter, and $1-(1-\alpha|\omega|)^k$ is a high-pass filter. In the beginning, at $k=1$, the filter $1-(1-\alpha|\omega|)^k = \alpha|\omega|$ is a ramp filter. As $k \to \infty$, the filter tends to an all-pass filter except for a notch at DC. In an iterative FBP algorithm, you need to be very careful in selecting the filters. Not every filter leads to a useful reconstruction.

8.10.2 Estimation of the initial image's contributions to the iterative Landweber reconstruction

This section presents another application of the Landweber-FBP algorithm, which is an FBP algorithm acting like an iterative Landweber algorithm. All these three algorithms are linear, obeying the superposition principle. The conventional FBP algorithm has only one input: the sinogram. The other two algorithms have two inputs: the sinogram and the initial image. One can use an initial image to influence the final image.

Due to the linear nature of the algorithms, we can use the Landweber-FBP algorithm to efficiently study the impact of the initial image to an iterative algorithm without using any sinogram P. Equation (8.5) shows how the sinogram P and the initial image $X^{(0)}$ influence the output $X^{(k)}$. If we would like to see the impact only from the initial image $X^{(0)}$, we simple set the sinogram P as zero. Equations (8.4) and (8.5) are given here for reader's convenience:

$$H^{(k)} = \left(I - \alpha A^T WA\right)^{(k)}, \tag{8.4}$$

$$X^{(k)} = F^{(k)}P + H^{(k)}X^{(0)}. \tag{8.5}$$

Following the similar discussion in Section 8.4.2, the Fourier domain equivalent transfer function expression of eq. (8.4) is

$$(1 - \alpha w_m / \|\vec{\omega}\|)^{(k)},$$

(8.63)

which is a high-pass filter, applied directly to the initial image in the Fourier domain. As k increases, a high-pass-filtered version of the initial image will remain in the iterative reconstruction.

8.10.3 FBP implementation of the immediately after-backprojection filtering

First of all, what is the immediately after-backprojection filtering? Well, in an iterative image reconstruction algorithm, there is a backprojection step, and the backprojected image is somehow used to modify and improve the current image. No 2D filtering is necessary to process the backprojected image before the update of the reconstructed image. In the iterative FBP, a 1D filter is used *before* backprojection, as described in Section 8.10.1.

It has been demonstrated that the contrast-to-noise ratio in the final reconstruction can be improved if a lowpass filter is applied to the backprojection of the projection-domain discrepancy, and then this backprojection is used to update the reconstructed image from the previous iteration.

A regular iterative Landweber algorithm is given as

$$X^{(k)} = X^{(k-1)} + \alpha A^T \left(P - AX^{(k-1)} \right).$$

(8.64)

Let the immediately after-backprojection filter be expressed by a Toeplitz square matrix V. A revised iterative Landweber algorithm with an immediately after-backprojection filter V is given as

$$X^{(k)} = X^{(k-1)} + \alpha V A^T \left(P - AX^{(k-1)} \right).$$

(8.65)

If $X^{(0)}$ is a zero image, eq. (8.86) can be written as

$$X^{(k)} = (VA^T A)^{-1} \left[I - \left(I - \alpha V A^T A \right)^k \right] V A^T P.$$

(8.66)

This algorithm can be implemented as a Landweber-FBP algorithm, that is, an FBP algorithm with a 2D window function in the Fourier domain:

$$1 - \left(1 - \alpha \frac{v(\|\vec{\omega}\|)}{\|\vec{\omega}\|} \right)^{(k)},$$

(8.67)

where the lowpass filter $v(\|\vec{\omega}\|)$ is the Fourier-domain representation of the filter V and is freely chosen by the user. For example, this filter can be chosen as

$$v(\omega) = \left[\frac{1 + \cos(\omega)}{2}\right]^{10} \text{ with } -\frac{\pi}{2} < \omega \le \frac{\pi}{2}. \tag{8.68}$$

8.10.4 Real-time selection of iteration number

Determination of an optimal stopping point for an iterative image reconstruction algorithm has been an open problem for a long time. The stopping point is strongly dependent on the task that the radiologists are performing. For the lesion detection tasks, the stopping point is lesion size-dependent. There is no single iteration number that gives an optimal image for all lesion sizes. One image can be better for the detection of one lesion but is worse for the detection of another lesion.

Radiologists nowadays are provided with slices of a patient image volume to make a clinical diagnosis. The radiologists are only able to change the display grayscale window on the computer screen by using sliders on the screen. We can provide the radiologists another option: real-time adjustment of the iteration number, which is able to improve the lesion detectability for targeted lesion size. We observe that for any given lesion size, its contrast curve peaks at a certain iteration number. One distinguished advantage of the proposed method is that no special additional computational hardware is required. Thus it is applicable for tele- or mobile-radiology of the modern world.

As an alternative way to real-time adjustment of the emulated iteration number, the radiologists are able to click on a suspicious lesion and a background region. A gradient ascent algorithm is then searching for the optimal emulated iteration number to give the maximum contrast for this lesion or the maximum contrast-to-noise for this lesion.

8.11 Summary

- A linear iterative algorithm can be implemented as an FBP algorithm or a back-projection-then-filtering algorithm. The trick is the equivalence of the matrix multiplication with $A^T A$ and filtering with the transfer function $1/|\omega|$.
- When you actually implement the iterative Landweber algorithm and the windowed FBP algorithm, you will notice that the step size α in the iterative Landweber algorithm and the α in the FBP algorithm are different. This is because $A^T A$ and $1/|\omega|$ are not exactly the same. The α for $A^T A$ is determined by the maximum singular value of the matrix A, while the α for $1/|\omega|$ is determined by the sampling interval of the frequency ω.

- One should use a large iteration number for a Bayesian algorithm. This is a difficult and time-consuming task for the iterative algorithm. On the other hand, it is easy for the FBP algorithm, which we simply let $k \to \infty$.
- There are many ways to derive the windowed FBP algorithm. We presented two approaches in this chapter: the iterative Landweber approach and the calculus of variations approach.
- Implementation of the new FBP algorithms requires the discrete forms of the windowed ramp filter. We only have the continuous forms of these filters. Remember that if you directly sample the continuous forms, the resultant discrete forms may cause DC bias in the reconstruction. This situation is similar to Example 4 of Chapter 2.

Problems

Problem 8.1 Implement the iterative Landweber algorithm. Run an example for a pair of (a, k). Then run your code again with $(a/2, 2k)$. Then run your code again with $(a/3, 3k)$. Do you get almost the same results?

Problem 8.2 Implement the windowed FBP algorithm. Run an example for a pair of (a, k). Then run your code again with $(a/3, 2k)$. Then run your code again with $(a/3, 3k)$. Do you get almost the same results?

Problem 8.3 Implement the iterative Landweber algorithm that has view-based noise weighting $\{w_m\}$. Run an example for $(\{w_m\}, k)$. Then run your code again with $(\{w_m/2\}, 2k)$. Then run your code again with $(\{w_m/3\}, 3k)$. Do you get almost the same results?

Problem 8.4 Implement the FBP algorithm that has view-based noise weighting $\{w_m\}$. Run an example for $(\{w_m\}, k)$. Then run your code again with $(\{w_m/2\}, 2k)$. Then run your code again with $(\{w_m/3\}, 3k)$. Do you get almost the same results?

Bibliography

[1] Zeng GL (2012a) A filtered backprojection algorithm with characteristics of the iterative Landweber algorithm. Med Phys 39:603–607.

[2] Zeng GL (2012b) A filtered backprojection MAP algorithm with nonuniform sampling and noise modeling. Med Phys 39:2170–2178.

[3] Zeng GL (2012c) Filtered backprojection algorithm can outperform maximum likelihood EM algorithm. Int J Imaging Syst Technol 22:114–120.

[4] Zeng GL (2013) Comparison of a noise-weighted filtered backprojection algorithm with MLEM algorithm for Poisson noise. J Nucl Med Tech 41:283–288.

[5] Zeng GL, Zamyatin A (2013) A filtered backprojection algorithm with ray-by-ray noise weighting. Med Phys 40:031113, http://dx.doi.org/10.1118/1.

[6] Zeng GL, Li Y, DiBella ERV (2013a) Non-iterative reconstruction with a prior for undersampled radial MRI data. Int J Imaging Syst Technol 23:53–58.

[7] Zeng GL, Li Y, Zamyatin A (2013b) Iterative total-variation reconstruction vs. weighted filtered-backprojection reconstruction with edge-preserving filtering. Phys Med Biol 58:3413–3431.

[8] Zeng GL (2014a) Model-based filtered backprojection algorithm: A tutorial. Biomed Eng Lett 4:3–18.

[9] Zeng GL (2014b) Noise-weighted spatial domain FBP algorithm. Med Phys 41:051906, http://scitation.aip.org/content/aapm/journal/medphys/41/5/10.1118/1.4870989.

[10] Zeng GL (2015a) Comparison of FBP and iterative algorithms with non-uniform angular sampling. IEEE Trans Nucl Sci 62:120–130.

[11] Zeng GL (2015b) On few-view tomography and staircase artifacts. IEEE Trans Nucl Sci 62:851–858.

[12] Zeng GL, Divkovic Z (2016) An extended Bayesian-FBP algorithm. IEEE Trans Nucl Sci 63:151–156.

[13] Zeng GL (2016) Noise-weighted FBP algorithm for uniformly attenuated SPECT projections. IEEE Trans Nucl Sci 63:1435–1439.

[14] Zeng GL, Wang W (2016) Noise weighting with an exponent for transmission CT. Biomed Phys Eng Exp 2:045004.

[15] Zeng GL (2017) A fast method to emulate an iterative POCS image reconstruction algorithm. Med Phys 44:e353–e359.

[16] Zeng GL, Li Y (2017) Fourier-domain analysis of the iterative Landweber algorithm. IEEE Trans Radiat Plasma Med Sci 1:511–516.

[17] Zeng GL (2018) Estimation of the initial image's contributions to the iterative Landweber reconstruction. IEEE Trans Radiat Plasma Med Sci 2:27–32.

[18] Zeng GL (2018) Filtered backprojection implementation of the immediately-after-backprojection filtering. Biomed Phys Eng Express 4:047005.

[19] Zeng GL (2019) Estimation of the optimal iteration number for minimal image discrepancy. IEEE Trans Radiat Plasma Med Sci 3:572–578.

[20] Zeng GL (2019) Image noise covariance can be adjusted by a noise weighted filtered backprojection algorithm. IEEE Trans Radiat Plasma Med Sci 3:668–674.

[21] Zeng GL (2019) Real-time selection of iteration number. Biomed Phys Eng Express 5:047007.

9 Machine learning

Machine learning is currently the most active area in research and applications. It is data-driven and relied on a sufficient model or architecture. It can generate images better than those generated by analytic and iterative algorithms. This chapter introduces the basic principle of machine learning and presents its applications in medical image reconstruction.

9.1 Analytic algorithms versus iterative algorithms versus machine learning algorithms

The analytic image reconstruction algorithms are the most rigorous in mathematics and most challenging to develop. They rely on the closed-form solutions of an inverse problem. Due to the difficulty in solving an inverse problem, the researchers often make some strong assumptions such as the projections being noiseless line integrals. Once the analytic reconstruction algorithms have been developed, they can be applied to any objects. These types of algorithms are mathematics-driven.

The iterative image reconstruction algorithms are easier to develop than the analytic algorithms because a closed-form solution of an inverse problem is not needed. In the development of an iterative algorithm, only the forward data acquisition model is required. The discrepancy between the forward model outputs and the measurements is used to form an objective function. Then an optimization algorithm is used to minimize the objective function. Compared with the analytic algorithm development, the restrictions for a forward model are more relaxed. For example, noise model in the measurements can be incorporated. Some physics effects can also be incorporated. In this sense, the iterative algorithms are physics-driven.

The machine learning approaches require neither a closed-form solution to an inverse problem nor a forward model of data acquisition. They just need data, and a lot of it! The machine learning approaches are data-driven. A person without deep understanding of mathematics or physics can be very successful in producing excellent images by doing machine learning. The image quality obtained from machine learning methods are usually better than those obtained by analytical or iterative reconstruction algorithms. Figure 9.1 illustrates a general belief about these three categories of algorithms. The basic principles of machine learning are introduced in the next section.

https://doi.org/10.1515/9783111055404-009

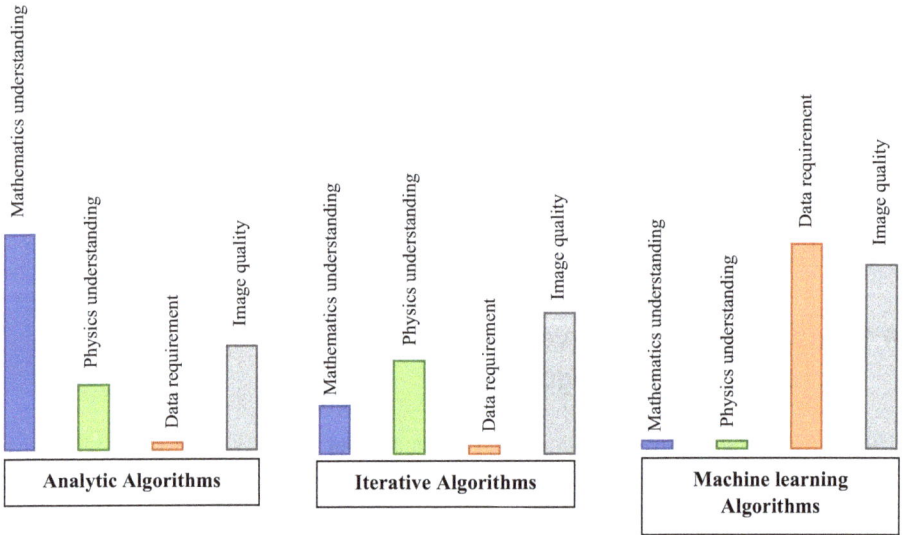

Fig. 9.1: A belief about the three categories of image reconstruction algorithms.

9.2 Basic principles of machine learning

The basic principle of machine learning is illustrated in Figure 9.2, where the black box is a general mapping, which transforms an input numerical string to an output numerical string. Here, a numerical string is a general term for a picture, a time series, a paragraph of text, and so on. Regardless the types of the input, they are first converted into numbers before fed into the black box. The output numerical string is sometimes referred to as the prediction. Therefore, the purpose of the black box is to make a prediction for a given input. This black box is fairly magical.

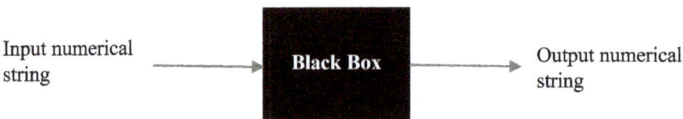

Fig. 9.2: A magical black box can transform an input numerical string to a desired output numerical string. This black box is a neural network.

Machine learning has two major types: classification and regression. Fundamentally, classification is about predicting a discrete value and regression is about predicting a continuous value. For a classification application, the label is always represented by a number, most commonly an integer. One example of a classifier is to have a black box to predict whether the input picture contains a dog or a cat. We can use 1 to label a dog and 2 to label a cat. As another important application, thanks to this black box, the photographs taken by license plate-reading cameras can be translated to digits and alphabets.

For image reconstruction, we are more interested in regression, which is a more difficult task than classification in machine learning. The input is a set of measurements, and the output is a reconstructed image.

Some closely related machine learning regression applications include image denoising and image deblurring. In image denoising, the input is a noisy image, and the output is a denoised version of it. In image deblurring, the input is a low-resolution image, and the output is a high-resolution version of it. In an image reconstruction task, image denoising and resolution recovery are usually included in machine learning.

The key component in machine learning is the black box, which is commonly referred to as the *neural network*. One needs to determine the architecture of the neural network. A neural network has many parameters, which are *weights* and *biases*. The parameters are calculated by training with a large amount of data pairs. Each data pair consists of an input and its desired output (referred to as *label*). After a neural network is trained, it is ready for image reconstruction in practical situations, where a desired output is unknown, and we want to predict it.

The performance of a neural network in image reconstruction boils down to the architecture of the neural network. How do we choose a particular neural network architecture for an application at hand? It is more an art than science. There is a theory telling us that a neural network exists for the task, but the theory does not tell us how to build it.

9.2.1 Adaptive linear neuron (ADALINE)

In as early as 1960, Prof. Bernard Widrow of Stanford University and his student Marcian Hoff proposed a simple neural network, which we now call the adaptive linear neuron (ADALINE) network. It only consists of weights, a bias, and a summation function, which can be expressed as

$$y = \sum_{j=0}^{n} x_j w_j, \tag{9.1}$$

with $x_0 = 1$. Thus, w_0 is the bias, and w_1, w_2, ..., w_n are the weights. A diagram of the ADALINE is shown in Figure 9.3. The error signal, e, is the difference between the model output, y, and the desired output, $\hat{y}$, as

$$e = \hat{y} - y. \tag{9.2}$$

A least mean squares (LMS) algorithm is used to determine the weights so that e^2 is minimized. This LMS algorithm is essentially the gradient descent algorithm. The theory behind this ADLINE neural network is the finite impulse response (FIR) filter, which is well known in the world of linear systems and signal processing. This FIR filter is essentially a convolution operator. Therefore, the LDLINE network works well if the input/output relationship is a convolution for a proper filter length n. For other input/output relationships, this network may not work well because this convolution model is not general enough.

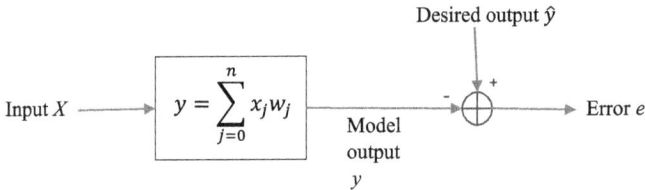

Fig. 9.3: A simple ADALINE neural network.

9.2.2 Universal approximation

A mathematical foundation is required to justify the use of neural networks. Over the years, many universal approximation theorems were proposed to support the use of neural networks. Here we use a 1D toy example to illustrate the meaning of a universal approximation theorem. Let us consider a continuous function $f(x)$ defined, without loss of generality, on $[0, 1]$. For prespecified accuracy, we can determine an integer n and divide the $[0, 1]$ into n subintervals such that $f(x)$ can be approximated by a piecewise constant function $\bar{f}(x)$. The function $\bar{f}(x)$ is a constant in $\left(\dfrac{k}{n}, \dfrac{k+1}{n}\right)$, with $k = 0, 1, 2, \ldots, n-1$ as shown in Figure 9.4. When n is large enough, the approximation is accurate enough. This approximation function $\bar{f}(x)$ is uniquely determined by n parameters, which are the n values of $f(x)$ at n evenly distributed points:

$$\frac{0.5}{n}, \frac{1.5}{n}, \frac{2.5}{n}, \ldots, \frac{n-0.5}{n} \tag{9.3}$$

in $(0, 1)$.

A piecewise constant function $\bar{f}(x)$ such as the one shown in Figure 9.4 can be expressed by a summation of weighted unit step functions:

$$\bar{f}(x) = \sum_{k=0}^{n-1} w_k u\left(x - \frac{k}{n}\right), \tag{9.4}$$

where the unit step function is defined as

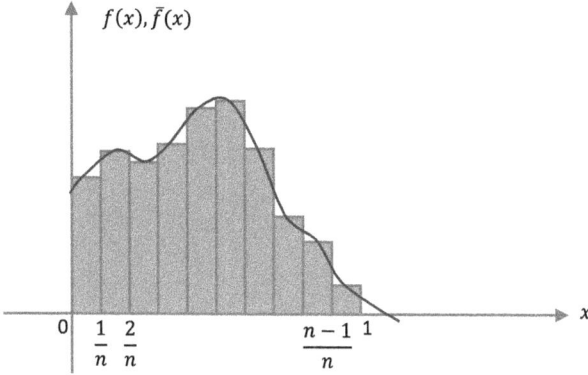

$f(x), \bar{f}(x)$

0 $\quad$ 1 $\quad$ 2 $\qquad\qquad$ $n-1$ $\quad$ 1

$\dfrac{1}{n}$ $\dfrac{2}{n}$ $\qquad\qquad$ $\dfrac{n-1}{n}$ $\qquad x$

Fig. 9.4: A continuous 1D function $f(x)$ and its piecewise constant approximation function $\bar{f}(x)$.

$$u(x) = \begin{cases} 1 \text{ if } x > 0, \\ 0 \text{ if } x < 0. \end{cases} \tag{9.5}$$

The weights w_k in eq. (9.4) are calculated as

$$w_0 = f(x_0), \tag{9.6}$$

$$w_k = f(x_k) - f(x_{k-1}) \text{ with } k = 1, \ 2, \ \ldots, n. \tag{9.7}$$

Expression (9.4) can readily lead to a neural network architecture as shown in Figure 9.5. This network has only one hidden layer. A hidden layer is a network layer between the input layer and the output layer. This hidden layer has n neurons. A neuron is computation unit. In a typical neural network, a neuron performs three operations: *inner product, bias*, and *activation* as shown in eq. (9.8):

$$y = u\left(\left(\sum_k w_k x_k\right) - b\right). \tag{9.8}$$

In eq. (9.8), $\sum_k w_k x_k$ computes the inner product between the weights $\{w_k\}$ and the input $\{x_k\}$; b is the bias; u is a nonlinear function. This nonlinear function in the field of machine learning is commonly referred to as an *activation function*. In our toy example of Figure 9.5, this activation function is the unit step function defined in eq. (9.7). We must point out that this activation function is hardly used because its derivative is zero almost everywhere it is not friendly for backpropagation in network training.

To overcome this drawback of the unit step function, the sigmoid function shown in eq. (9.9) and Figure 9.6 is a better substitution with its derivative as shown in eq. (9.10):

$$s(x) = \frac{1}{1 + e^{-x}}, \tag{9.9}$$

$$s'(x) = \frac{e^{-x}}{(1+e^{-x})^2} = s(x)[1-s(x)], \tag{9.10}$$

which implies that the derivative of the sigmoid function is the sigmoid function itself multiplied by 1 minus the sigmoid function. Equation (9.10) is useful in implementation of the backpropagation algorithm that involves sigmoid functions. It is clear to see that the sigmoid function $s(ax)$ tends to the unit step function $u(x)$ as $a \to \infty$. Three sigmoid functions with three different values of the parameter a are shown in Figure 9.6.

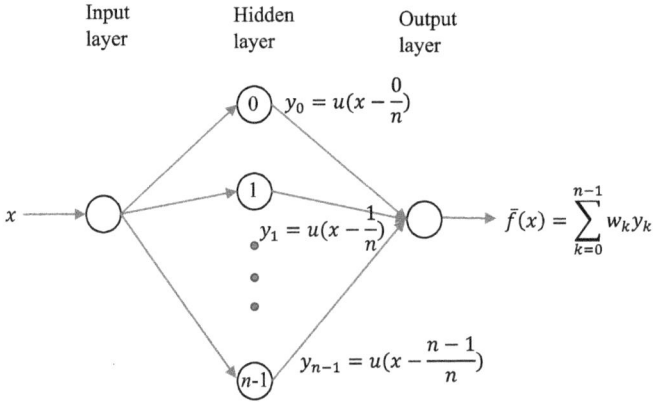

Fig. 9.5: This toy neural network has one hidden layer consisting of n neurons. The input is a scalar; the output is a scalar. This neural network represents a piecewise constant function.

In the toy example of Figure 9.5, the weight of neuron k in the hidden layer is 1 and the bias is $-k/n$. The activation function for each neuron is the unit step function. The output layer does not have an activation function and does not have a bias either. The weights for the output layer are calculated by eqs. (9.6) and (9.7).

A ReLU (rectified linear unit) activation, $\sigma(x)$, is defined by eq. (9.11)

$$\sigma(x) = \max(0, x) = \begin{cases} x \text{ if } x > 0, \\ 0 \text{ if } x \leq 0, \end{cases} \tag{9.11}$$

which is very popular in machine learning. Unlike the unit step function, the derivative of the ReLU function is nonzero when $x > 0$. We assume that we have the values of the true continuous function $f(x)$ at $n + 1$ evenly distributed points on $[0, 1]$:

$$\frac{0}{n}, \frac{1}{n}, \frac{2}{n}, \ldots, \frac{n}{n}. \tag{9.12}$$

As depicted in Figure 9.7, a piecewise linear function $\bar{f}(x)$ can be formed to approximate $f(x)$ as

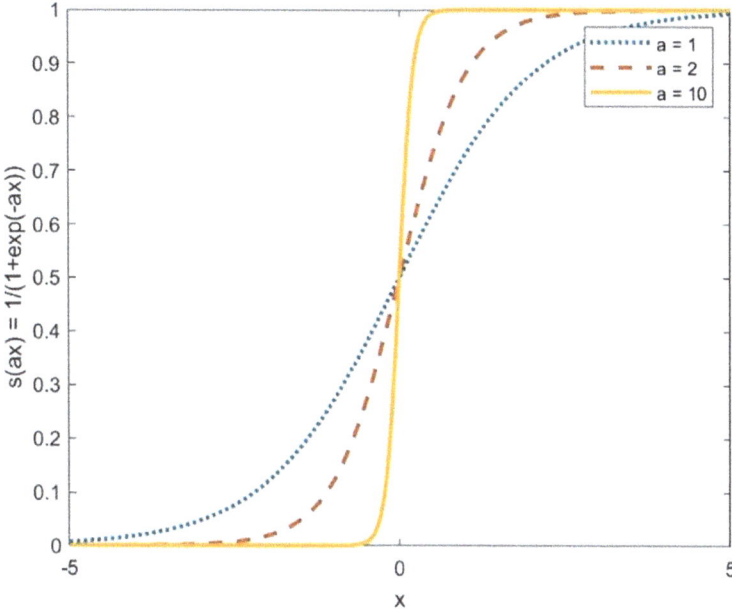

Fig. 9.6: Sigmoid function $s(ax)$ with $a = 1$, 2, and 10. As $a \to \infty$, the sigmoid function tends to the unit step function.

$$\bar{f}(x) = f(0) + \sum_{k=0}^{n-1} n\left[\sigma\left(x - \frac{k}{n}\right) - \sigma\left(x - \frac{k+1}{n}\right)\right]\left[f\left(\frac{k+1}{n}\right) - f\left(\frac{k}{n}\right)\right]$$

$$= f(0) + n\sigma(x)\left[f\left(\frac{1}{n}\right) - f(0)\right] + \sum_{k=1}^{n-1} n\left[f\left(\frac{k+1}{n}\right) - 2f\left(\frac{k}{n}\right) + f\left(\frac{k-1}{n}\right)\right]\sigma\left(x - \frac{k}{n}\right)$$

$$= b + \sum_{k=0}^{n-1} w_k \sigma\left(x - \frac{k}{n}\right),$$

$$(9.13)$$

where

$$b = f(0),\qquad(9.14)$$

$$w_0 = n\left[f\left(\tfrac{1}{n}\right) - f(0)\right],\qquad(9.15)$$

$$w_k = n\left[f\left(\tfrac{k+1}{n}\right) - 2f\left(\tfrac{k}{n}\right) + f\left(\tfrac{k-1}{n}\right)\right],\qquad(9.16)$$

with $k = 1, 2, \ldots, n-1$.

Similar to Figure 9.5, a simple neural network shown in Figure 9.8 with one hidden layer can be used to implement eq. (9.13). In the example of Figure 9.8, the weight of neuron k in the hidden layer is 1 and the bias is $-k/n$. The activation function for each neuron is the ReLU function. The output layer does not have an activation

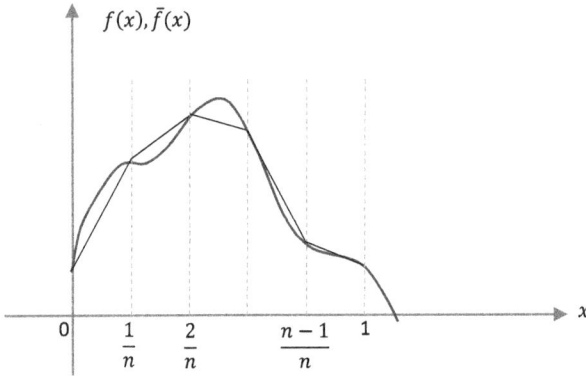

Fig. 9.7: A continuous 1D function $f(x)$ (illustrated by the thicker curve) and its piecewise linear approximation function $\bar{f}(x)$ (illustrated by the thinner linear segments).

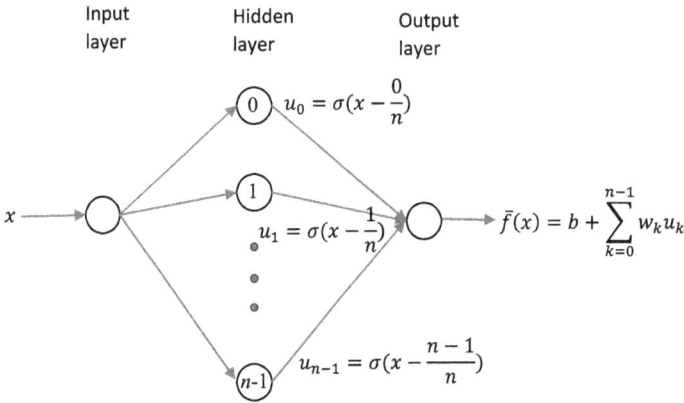

Fig. 9.8: This neural network has one hidden layer consisting of n neurons. The input is a scalar; the output is a scalar. The nonlinear activation function is the ReLU function. This neural network represents a piecewise linear function.

function. The bias in the neuron of output layer is given by eq. (9.14). The weights for the output layer are calculated by eqs. (9.15) and (9.16).

We can see that the common practice in machine learning is the use of piecewise polynomial approximation, which is also known as spline interpolation. The piecewise constant function approximation is the zeroth-order spline interpolation. The piecewise linear function approximation is the first-order spline interpolation. Of course, we can try quadratic or cubic splines if we want.

The piecewise constant function approximation uses the unit step function as the building block. As shown in Figure 9.9, from left to right, the amplitude of each sequential step function is the difference of the function amplitudes. The sampling points do not need to be evenly distributed. As a matter of fact, a better approximation can be obtained if finer samples are used in the regions where the function changes quickly. This is an advantage of machine learning that learns both the weights and delays during training with a large number of samples which are most likely not evenly distributed.

Likewise, the piecewise linear function approximation uses the ReLU function as the building block. As shown in Figure 9.10, from left to right, the slope of each sequential ReLU function is determined by the difference of the function slopes. Also, a better approximation can be achieved if the function sampling points are wisely positioned. Machine learning methods are able to automatically find the optimal function sampling points.

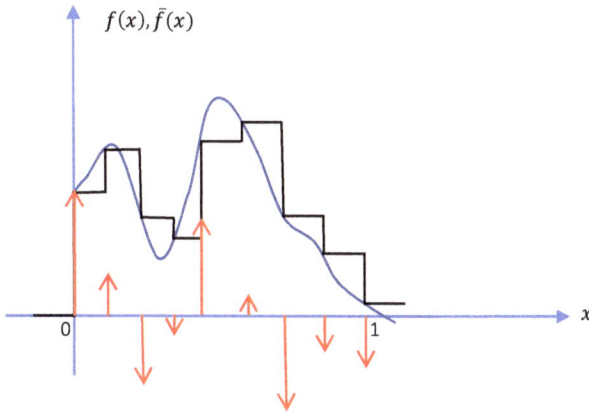

Fig. 9.9: A continuous 1D function $f(x)$ and its piecewise constant approximation function $\bar{f}(x)$. The red arrows indicate the function value changes between the sampling points.

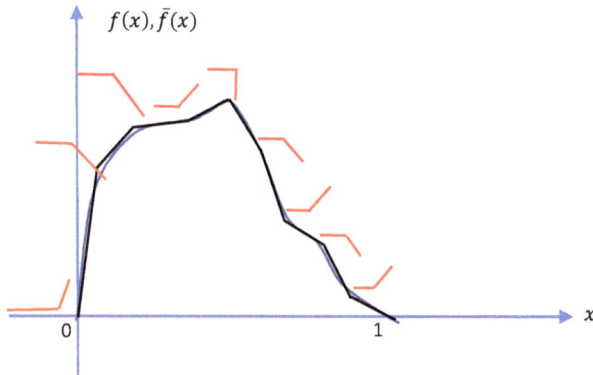

Fig. 9.10: A continuous 1D function $f(x)$ and its piecewise linear approximation function $\bar{f}(x)$. The red polylines indicate the function slope changes between the sampling points.

Besides the unit step function $u(x)$, the sigmoid function $s(x)$, and the ReLU function $\sigma(x)$, the fundamental building block to approximate a 1D continuous function $f(x)$ can use other building blocks such as a ramped step function as shown in Figure 9.11, which is a composition of two ReLU functions or a hat function as shown in Figure 9.12, which is a composition of three ReLU functions. In fact, the first line of eq. (9.13) represents the piecewise linear function $\bar{f}(x)$ via the ramped step function $\sigma\left(x - \dfrac{k}{n}\right) - \sigma\left(x - \dfrac{k+1}{n}\right)$; the second line of eq. (9.13) represents the piecewise linear function $\bar{f}(x)$ via the hat function $\sigma\left(x - \dfrac{k-1}{n}\right) - 2\sigma\left(x - \dfrac{k}{n}\right) + \sigma\left(x - \dfrac{k+1}{n}\right)$.

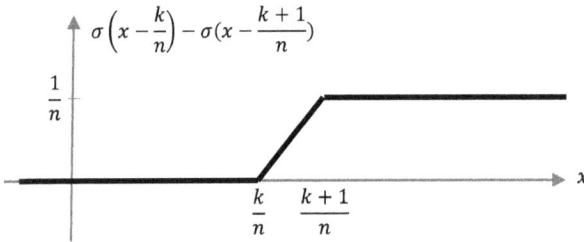

Fig. 9.11: A ramped step function can be used as a building block to approximate a continuous function.

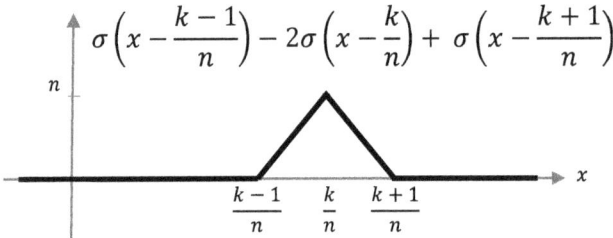

Fig. 9.12: A hat function can also be used as a building block to approximate a continuous function.

9.2.3 A deep model for a continuous univariate function

Section 9.2.2 showed that a shallow and wide network is able to approximate a continuous univariate function. Increasing the width of the network reduces the approximation errors. Closed-form formulas exist to calculate the network weights. In fact, deep networks can also perform this task and have closed-form formulas. One deep network model that approximates a continuous univariate function is shown in Figure 9.13.

In Figure 9.13, the functions $v_0(x)$ and $v_1(x)$ are nonlinear activation functions, and they are defined as

$$v_0(x) = \begin{cases} 2x & \text{if} & 0 \le x < 0.5 \\ 0 & \text{otherwise} \end{cases} \tag{9.17}$$

and

$$v_1(x) = \begin{cases} -2(x-1) & \text{if} & 0.5 \le x < 1 \\ 0 & \text{otherwise} \end{cases}, \tag{9.18}$$

respectively. These two functions are depicted in Figure 9.14. The sum $v_0(x) + v_1(x)$ forms the well-known *hat* function or the *triangle* function, commonly used as the kernel function for linear interpolation as shown in Figure 9.15.

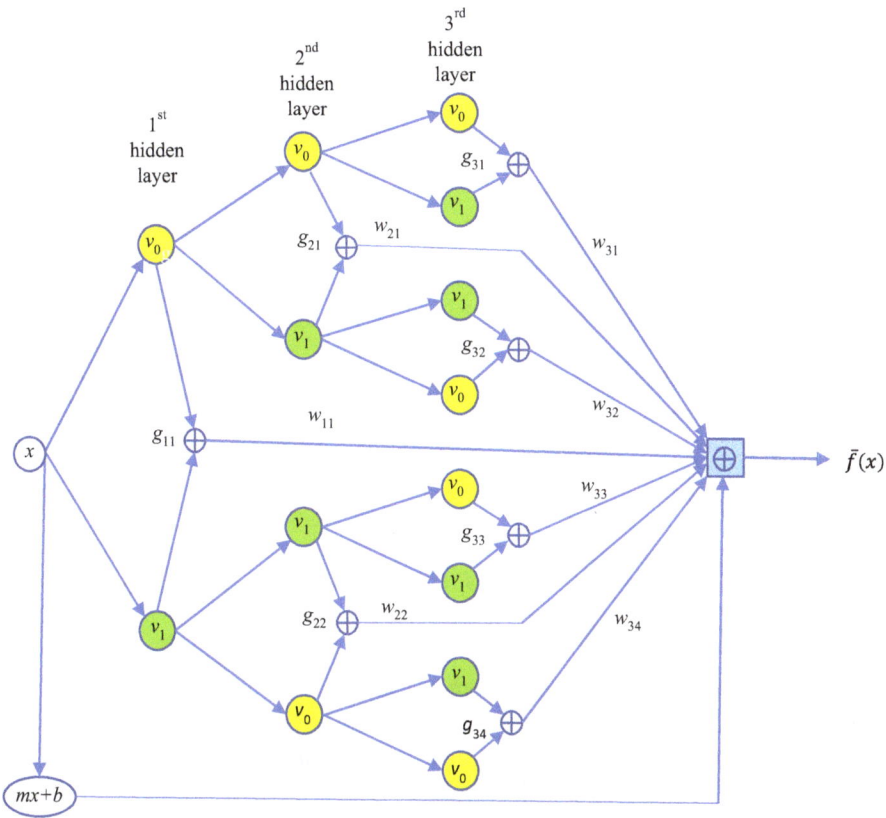

Fig. 9.13: A deep network uses a piecewise linear function to approximate a general target function $f(x)$. Three hidden layers are used in this toy example.

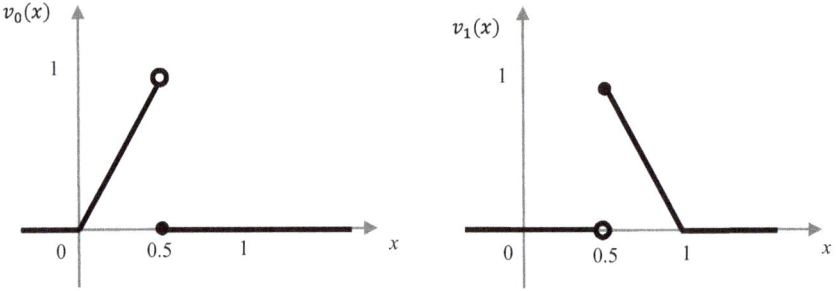

Fig. 9.14: The left neuron $v_0(x)$ and the right neuron $v_1(x)$ are two nonlinear activation functions.

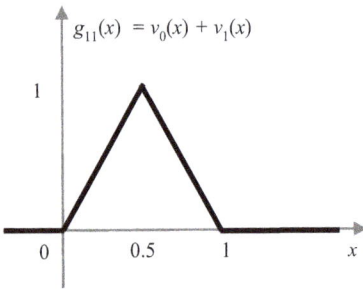

Fig. 9.15: The hat function $g_{11}(x) = v_0(x) + v_1(x)$.

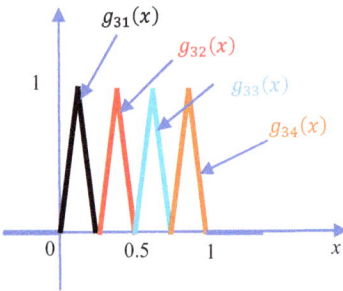

Fig. 9.16: The four hat functions in the third hidden layer: $y_{31}(x)$ in black, $y_{32}(x)$ in red, $y_{33}(x)$ in blue, and $y_{34}(x)$ in orange.

Similar to the triangle function shown in Figure 9.15, all other functions $g_{ij}(x)$ are triangle functions with various base widths and locations (see also eq. (9.21)). For example, the four triangle functions at the third hidden layer are depicted in Figure 9.16. The general formula for the weights at the ith hidden layer is given as

$$w_{ij} = f\left(\frac{2j-1}{2^i}\right) - \frac{f\left(\frac{2j-2}{2^i}\right) + f\left(\frac{2j}{2^i}\right)}{2} \tag{9.19}$$

for $j = 1,\ 2,\ 3,\ \ldots,\ 2^{i-1}$, and $i = 1,\ 2,\ 3,\ \ldots, N$. The expression of the approximation function $\bar{f}(x)$ is given by

$$\bar{f}(x) = mx + b + \sum_{i=1}^{N} \sum_{j=1}^{2^j} w_{ij} g_{ij}(x)$$

$$= [f(1) - f(0)]x + f(0) + \sum_{i=1}^{N} \sum_{j=1}^{2^j} a_{ij} g_{ij}(x)$$

(9.20)

where $g_{ij}(x)$ is the hat function defined by

$$g_{ij}(x) = 2^i \left[\sigma \left(x - \frac{2j-2}{2^i} \right) - 2\sigma \left(x - \frac{2j-1}{2^i} \right) + \sigma \left(x - \frac{2j}{2^i} \right) \right]$$

(9.21)

with the ReLU function $\sigma(x) = \max(0,x)$, for $j = 1, 2, 3, \ldots, 2^{i-1}$, and $i = 1, 2, 3, \ldots,$ N. Of course, there are many other deep neural network models to approximate a continuous function.

9.2.4 When the input is an *n*-tuple

We now consider the case when the input is a 2-tuple. In other words, the input is a vector of two components x and y, where each component is confined on [0, 1]. A function of (x, y) can be treated as a real-valued image $f(x,y)$, which can have positive and negative values. We will have a neural network model to approximate $f(x,y)$. The difficult part in doing so is to convert a two-dimensional function into some one-dimensional functions because the activation functions in a neural network are univariate (i.e., functions of a single variable). This difficulty can be overcome by using our knowledge of tomography.

Since $f(x,y)$ has a finite support, it can be reconstructed by a filtered backprojection algorithm from its Radon transform as discussed in Chapter 2 of this book. Let us assume that the parallel line-integral projections of object $f(x,y)$ are measured over M views evenly over 180°. After the application of a ramp filter, the filtered sinogram is $q_m(t)$ with $m = 1, 2, \ldots, M$. The variable t is along the 1D detector. The approximated function is the backprojection of $q_m(t)$ as

$$\bar{f}(x,y) = \sum_{m=1}^{M} q_m \left(x \cos \frac{m\pi}{M} + y \sin \frac{m\pi}{M} \right).$$

(9.22)

In eq. (9.22), a 2D function is expressed as a summation of univariate functions. This can be illustrated in Figure 9.17 as a neural network structure. Each box in Figure 9.17 contains a univariate function $q_m(t)$, which is evaluated at a value determined by the inner product between a point in the image $\vec{x} = [x,y]^T$ and a unit directional vector $\vec{\theta}_m = \left[\cos \frac{m\pi}{M}, \sin \frac{m\pi}{M} \right]^T$. For a 2D continuous function $f(x,y)$ and the imaging geometry, the univariate functions $q_m(t)$ are well-defined. The functions $q_m(t)$ are readily obtained by applying the ramp filter to the sinogram in the detector direction. Using the results in Section 9.2.2, each univariate function can be approximated by a shallow neural network with a nonlinear activation faction, for example, the ReLU function. Therefore, a 2D continuous function $f(x,y)$ can be approximated by a neural network shown in Figure 9.17.

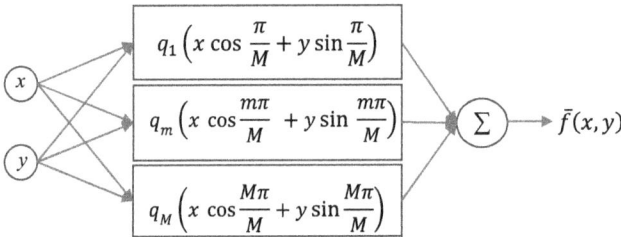

Fig. 9.17: A backprojector is illustrated as a neural network structure that sums the filtered sinogram functions. The filtered sinogram functions are the univariate functions, which in turn can be approximated by neural networks discussed in Section 9.2.2.

Similarly, when the input is a 3-tuple (x,y,z), where each component is confined on [0, 1], we consider the 3D Radon transform of the function $f(x,y,z)$. Section 5.2 of this book discusses the 3D Radon transform, which is parallel plane integral of a 3D object. The Radon inversion formula is a 1D second-order derivative followed by a Radon backprojector. If we denote the second-order differentiated Rodon data as $q_{ij}(t)$, the 3D function $f(x,y,z)$ can be reconstructed by a neural network shown in Figure 9.18.

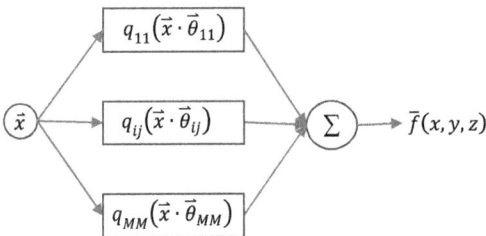

Fig. 9.18: A backprojector is illustrated as a neural network structure, where the filtered sinogram functions are univariate functions, which in turn can be approximated by neural networks discussed in Section 9.2.2.

9.2.5 An *n*-tuple to *m*-tuple mapping

In Section 9.2.4, we explained how a neural network with univariate activation functions can be used to approximate an n-D continuous function $f(\vec{x})$. This network is a mapping from an n-tuple to a scalar. For any n-tuple $\vec{x}$ within a specified range, a real number $\bar{f}(\vec{x})$ is calculated using two types of calculations. The first type is the inner product between two vectors; one of the vectors depends on the input $\vec{x}$; the other vector depends on the underlying function f. The second type is the evaluation of a nonlinear univariate activation function. The univariate activation function usually contains a bias term. Such a neural network unit is commonly referred to as a perceptron. A general expression of a perceptron is

$$y = \sigma(\vec{x} \cdot \vec{w} + b), \tag{9.23}$$

where $\vec{x}$ is the input vector, $\vec{w}$ is the weight vector, b is the bias, σ is the univariate activation function, and y is the output scalar. A perceptron is the fundamental building element of the neural network.

A perceptron can be used for function regression. When we are given many $(\vec{x}, y)$ pairs, the function regression procedure is to find a function $\bar{f}(\vec{x})$ with a specific model (for example, a linear function, a cubic function, an exponential function, and so on) such that $\bar{f}(\vec{x}) \approx y$ for the given pairs. This function $\bar{f}(\vec{x})$ is the best fit to the training pairs $(\vec{x}, y)$. If the training pairs have an underlying $f(\vec{x})$, then this model $\bar{f}(\vec{x})$ is an approximate of the underlying function $f(\vec{x})$, and for any unseen $\vec{x}$ in the specified range $\bar{f}(\vec{x})$ can be used to approximate $f(\vec{x})$. Clearly, a perceptron neural network is a good candidate for a function regression task.

Another clear application of a perceptron neural network is in classification. If the n-tuples $\vec{x}$ in different regions belong to different classes, a continuous function $f(\vec{x})$ can be formed such that a dedicated integer value is associated with a class. Since the function $f(\vec{x})$ is continuous, the boundaries of the classes may not be clearly defined. A neural network similar to that shown in Figure 9.18 can be implemented to approximate the function $f(\vec{x})$. All we have to do is to add an output layer contains different threshold values. The quantized integer values indicate the classes that unseen points $\vec{x}$ should belong to.

Figure 9.19 shows a 2D classification problem, where the points in the first and third quadrants belong to one class (say, class 1) and the points in the second and fourth quadrants belong to the other class (say, class −1). This is the famous exclusive-or problem. A continuous function f can be created for this classification purpose as shown in Figure 9.20. This function f can be approximated by a neural network that generates a function $\bar{f}$. A final output layer is appended to this network. The final output layer contains a threshold value of 0. If $\bar{f} > 0$, the input $\vec{x}$ is classified as in the class 1; otherwise it is classified as in the class −1.

Now we consider a neural network configuration, whose input is an n-tuple and output is an m-tuple. If $m < n$, this is the dimension reduction situation. Dimensionality reduction refers to techniques that reduce the number of features in a data set. Dimensionality reduction is useful in data compression, data visualization, and feature extraction.

The n-tuple-input/m-tuple-output neural network is fairly easy to build. All we have to do is to construct m sub networks, each of which has an n-tuple input and a scalar output. Roughly speaking, the size of the n-input/m-output neural network is m times larger than an n-input/one-output network. Figure 9.21 illustrates a 4-input/3-

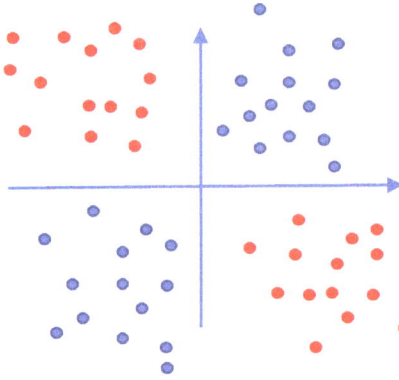

Fig. 9.19: An example of exclusive-or (XOR) classification problem. The blue dots belong to one class and the red dots belong to another class.

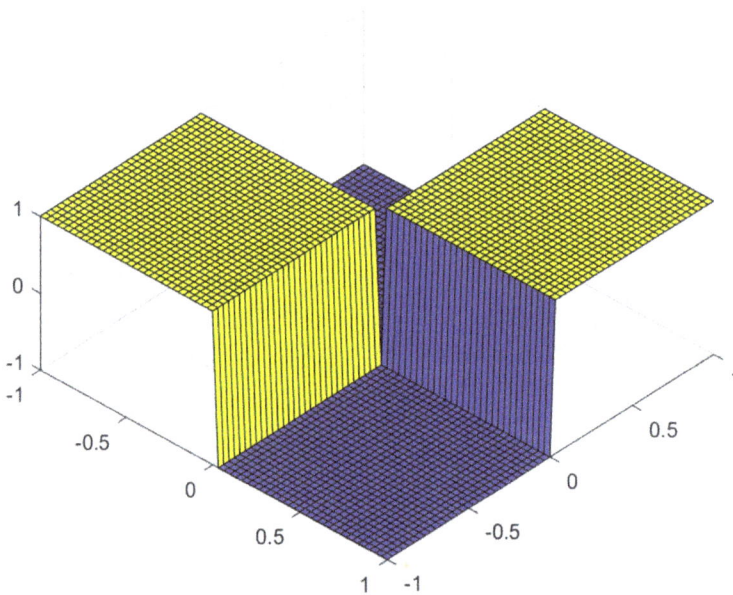

Fig. 9.20: A continuous 2D function is created for the XOR classification problem. This function can be approximated by a neural network using perceptions. The classification decision is made by using a threshold value of zero.

output neural network that is built by combining three 4-input/1-output networks. In reality, an n-input/m-output neural network may not have this structure. It may have multiple layers as in a deep network; it may have convolutional layers; it may have a fully connected output layer, and so on. A fully connected layer is also known as a dense layer. There are so many possible architectures.

Dimensionality reduction networks are easy to construct but may not be easy to train. The difficulty in training is the lack of availability of the target data. One clever way to train a dimensionality reduction network is by an autoencoder, which uses the

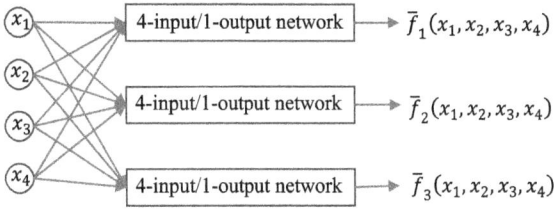

Fig. 9.21: A 4-input/3-output neural network can be constructed by combining three 4-input/1-output networks.

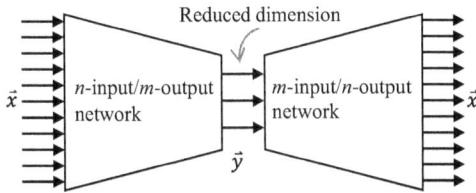

Fig. 9.22: An autoencoder uses the same input and output to train this network. The dimension of $\vec{x}$ is greater than the dimension of $\vec{y}$. Here $\vec{y}$ is the extracted principal feature vector from the original vector $\vec{x}$.

same data as both the input and the output as shown in Figure 9.22. Then the availability of the training data becomes relatively less challenging.

9.3 Image reconstruction as an n-tuple to m-tuple mapping

In Section 9.2, we learned that a (shallow or deep) neural network can accurately approximate an n-tuple to m-tuple mapping. If we treat measurements as the input n-tuple and a reconstructed image as the output m-tuple, there exist neural networks that can perform this mapping. Here, the input n-tuple is the measurements represented in the vector format, and the output m-tuple is the reconstructed image represented in the vector format. These neural networks can be obtained by training with many measurements/reconstruction pairs. A drawback of this naïve approach is that a fully connected layer (i.e., a dense layer) is necessary to map the measurement-domain images into the reconstruction-domain images (see Figure 9.23). Due to the large sizes of the measurements and the images, the number of the parameters in the fully connected layer can be too large, especially in 3D imaging, for a regular computer. This approach currently may not be the most practical choice.

A more practical neural network image reconstruction approach is to use the FBP algorithm or the iterative ML-EM algorithm to perform the mapping between the sinogram domain and the image domain and to use the neural network to process the images in the sinogram domain or in the image domain. The image processing tasks include denoising, deblurring, and super-resolution. The image processing neural networks usually contain much fewer parameters than those required for fully connected layers. For example, the image processing can use the convolutional neural networks (CNNs), which we will discuss next.

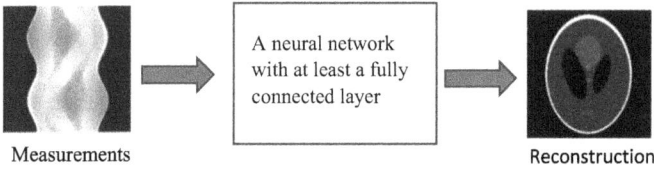

Measurements

A neural network with at least a fully connected layer

Reconstruction

Fig. 9.23: A neural network with a fully connected layer is able to map the measurements to the reconstruction.

9.4 Convolutional neural network (CNN) as a matched filter for identification

The main purpose of a neural network according to Section 9.2 is to build a universal model that maps an n-tuple to an m-tuple. Each element of the m outputs is a function of the n input variables. This mapping is continuous with respect to the n input variables. The CNNs do not try to fulfill the same purpose. CNNs are good at pattern recognition.

In the machine learning community, the "convolution" is, in fact, implemented as the cross-correlation. The 1D convolution and cross-correlation are expressed in eqs. (9.24) and (9.25), respectively.

$$\text{Convolution: } y(t) = \int x(\tau)h(t-\tau)d\tau, \tag{9.24}$$

$$\text{Cross} - \text{correlation: } y(t) = \int x(\tau)h(t+\tau)d\tau. \tag{9.25}$$

Let us consider a task, in which you are given a long string (an n-tuple $\vec{x}$) and a short pattern (a k-tuple $\vec{h}$). The size n of the input data $\vec{x}$ is much larger than the size k of the pattern $\vec{h}$. The cross-correlation operation is a procedure that moves $\vec{h}$ over $\vec{x}$ and calculates the inner products to find a match. When a match is found, the cross-correction will produce a spike, which is a larger value than others. The effectiveness of the cross-correlation in pattern recognition is illustrated by the following 1D toy example.

Here we have a long binary (using 1 and −1) string:

[1, −1, 1, 1, −1, −1, 1, 1, 1, 1, 1, −1, 1, −1, 1, −1, 1, −1, 1, 1, 1, −1, 1, −1].

We want to know if this long string contains a short string

[−1, 1, 1, −1].

The cross-correlation of these two strings is calculated as

$$[-2, 4, 0, -4, 0, 2, 0, 0, 2, -2, 0, 0, 0, 0, 0, 0, -2, 2, 2, -2, 0].$$

A peak of value 4 in the cross-correlation result indicates that the string $[-1, 1, 1, -1]$ is found in the long string and is located at the second place (see Figure 9.24). If the binary strings use 1 and 0, the cross-correlation results will have a bias. The purpose the cross-correlation operation is to find the similarity of the two strings or of two functions, segment by segment.

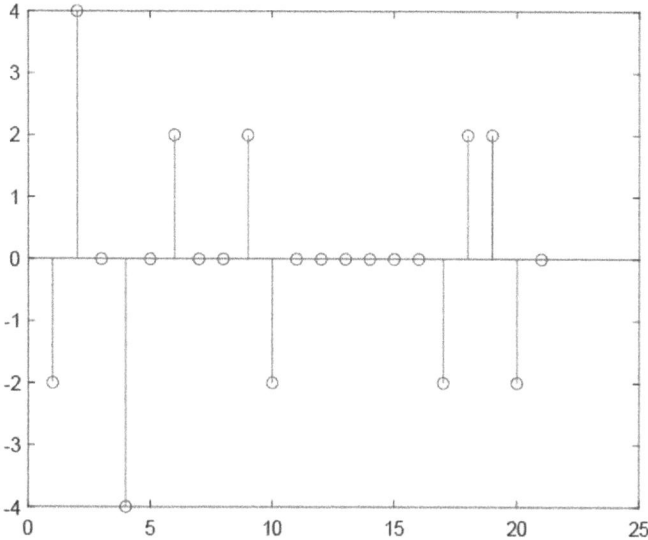

Fig. 9.24: The cross-correlation plot has a peak of value 4 at position 2.

We now encounter another problem. The target pattern string and the input string may have similar patterns, but different in size. For example, we consider $[-1, 1, -1]$ and $[-1, -1, 1, 1, -1, -1]$ as a match. One solution is to vary the resolution of the input, as done in the U-net. A simplified U-net is illustrated in Figure 9.25. The steps marketed as "Down sample" and "Up sample" change the image resolution.

As a side note, there are some *skip* links in the U-net. Without a *skip* link, the CNN tries to model the labels provided during training. With a *skip* link, the CNN tries to model the "difference" between the input and its label. Here, a *label* is a desired output.

Let us explain what this "difference" is. In a denoising task, the input of the network is a noisy image. A regular network would use a noiseless image to train. If the difference of the input and output is used to train the network, the network is a *residual* network. If we assume the noise to be additive, the label of the network is the noise (see Figures 9.26 and 9.27). The network output is the estimated noise, which may be positive and negative.

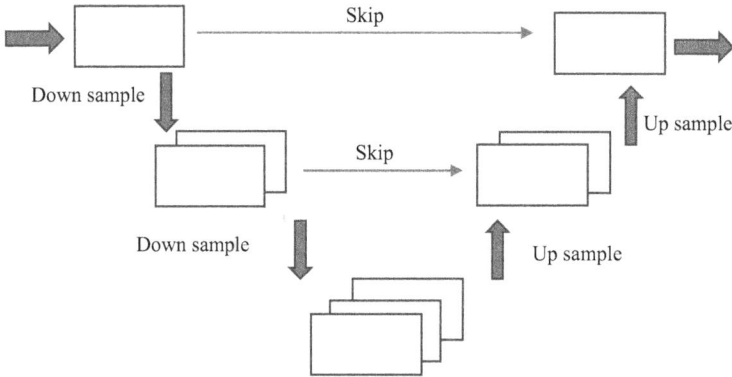

Fig. 9.25: A simplified U-net.

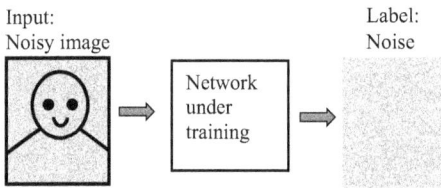

Fig. 9.26: Training of a residual network. The residual is the difference between the noisy image and the true noiseless image.

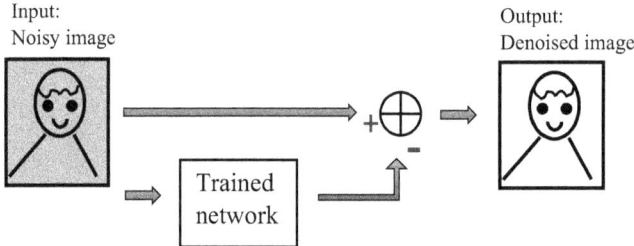

Fig. 9.27: A residual network in action. The denoised image is the difference between the noisy image and the estimated noise.

The idea of using cross-correlation to identify a pattern can be extended into 2D and 3D images. For example, if we want to know whether a big 2D picture contains a bicycle, we could perform the 2D cross-correlation between the big picture and a bicycle template. We could also break down the bicycle template as explained below.

The modern machine learning uses the CNN (or ConvNet) approach with multiple layers. In fact, the U-net that we saw earlier is a CNN. The strategy of CNN is to break down the pattern template into smaller multiple templates. Let us first divide the bicycle into wheels and the frame. At the next level, a wheel is subdivided to arcs, and the frame is further decomposed into smaller segments (see Figure 9.28). We break down the bicycle template step by step. Eventually, the smallest templates can be as small as 3×3.

When we break down the pattern template to smaller templates, no fine details are lost. The earlier layers of the CNN are to catch the building components, and the later layers are to piece up the lower-level components to build a bicycle.

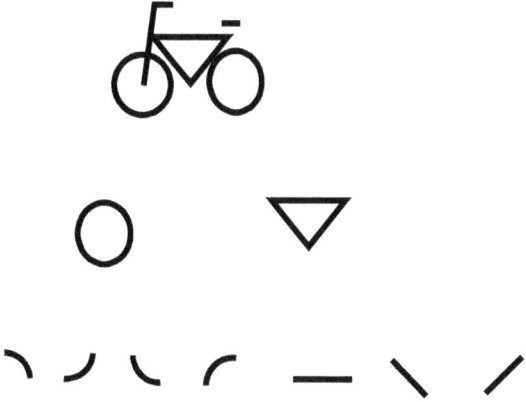

Fig. 9.28: A small template of a bicycle is further broken down into much smaller templates of segments.

Here, we assume that the CNN has already been trained for bicycle detection and is ready to use. In a CNN, the procedure of identifying a bicycle is as follows. First, we find the cross-correlation images between the given big picture and all the lowest-level (with the finest details) templates. For any lowest-level template, a bright dot will show up in its associated layer output image when a match is found. The smaller cross-correlation values show up as black dots with a value 0 if there is no match, thanks to the nonlinear ReLU function and the bias at each cross-correlation unit (see Figure 9.29).

Some illustrative partial results are shown in Figures 9.29–9.31. The gray-color objects in these figures are not parts of the images or templates but are for location reference only.

After the first layer (see Figure 9.29), we know if the given picture contains the elements of arc segments.

After the second layer, one combination of the inputs (which are the outputs of the first layer units) matches the small template of the second layer. Hence, a circle is identified, and a bright dot shows up at the circle location in the associated output image of second layer. If there are two circles, two bright dots will appear at an output image of the second layer. Thus, a pair of side-by-side circles (i.e., the wheels) is identified. If no circle is identified, this cross-correlation will produce a black dot at the location, thanks to bias term and the nonlinear ReLU function attached to the end of the cross-correlation units.

Of course, each layer also contains small templates for the bicycle frame. The bicycle frame is identified similarly. Layer by layer, the scope of attention is zooming out, and larger regions are considered. Finally, at the end, the last templates check if all the components are at the right location relative to each other. If all the components are present and are at the correct locations relative to each other, a bicycle is recognized (see Figure 9.31).

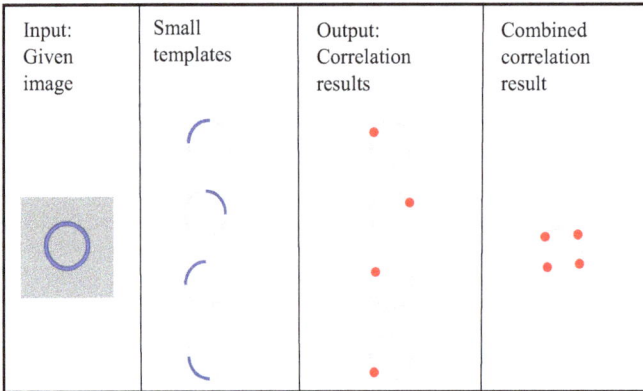

Fig. 9.29: The happenings in the first layer.

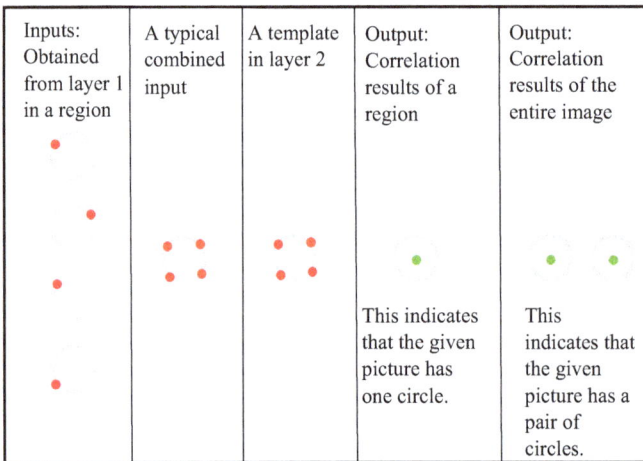

Fig. 9.30: The happenings in the second layer.

We can use a small (e.g., 3 × 3) convolutional kernel in every layer. However, the *attention* region size is different at each layer. For example, at the first layer, a convolution kernel can only pay attention to a small portion of a circle (i.e., a wheel). At the second layer, a convolution kernel of the same size pays attention to the entire circle (i.e., the

Inputs: the dots	A typical combined input	A template in this layer	Output: Correlation result
			This indicates that the given picture has a bicycle.

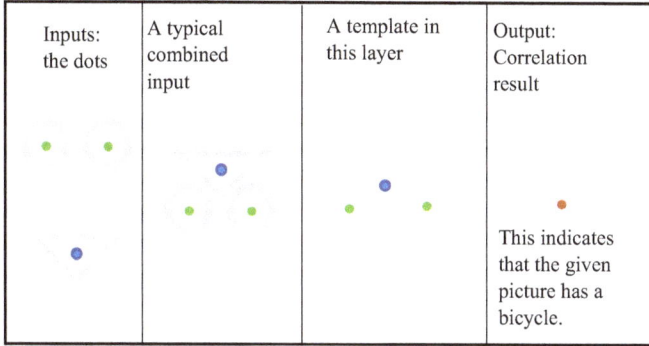

Fig. 9.31: The happenings in the last layer.

entire wheel). At the third layer, a convolution kernel of the same small size can pay attention to the entire bicycle. This changing of the attention scope is achieved by changing the image resolution from one layer to the next layer. As a result, the images at different layer have different spatial resolutions.

There are many ways to change the image resolution. The most common down-sampling method is the *max pooling* method. For example, in a 2×2 neighborhood, the maximum value is selected, and the other three values are discarded because the maximum value represents a matched feature. This *max pooling* method assumes that no features can occur too close to each other.

Not all CNNs use this strategy. Some CNNs do not vary the image resolution at all. In fact, we do not have to down sample the images and the templates; we do not have to make all templates the same small size.

9.5 Convolutional neural network (CNN) as a nonlinear denoise filter

In many neural network image reconstruction algorithms, the neural network's duty is to denoise. The mapping from the measurement domain to the image domain is performed by a regular backprojector. The overall image reconstruction algorithm is a hybrid algorithm, which may involve the FBP or an iterative reconstruction algorithm. The machine-learned denoising filter may also include the tomographic ramp filter.

One advantage of using a CNN to denoise is that it uses much fewer parameters than a fully connected network. In a CNN, a small convolutional kernel is applied to the entire image by assuming the shift-invariant property of the image. The small convolutional kernel is the pattern feature template that was discussed above. The convolution output is a feature map. For example, a high-pass convolutional kernel will produce a feature map that emphasizes the edges and high-frequency noise.

An almost equivalent larger template can be achieved by using multiple convolutional layers with smaller templates. A larger template can capture a more complicated image pattern. In a CNN, we do not treat the 2D image input as a vector (or an n-tuple) because the 2D cross-correlation operation is sensitive to neighboring image pixels. The CNN pays attention to a pixel and its nearby pixels. On the other hand, a fully connected layer treats every input as a vector (i.e., an n-tuple).

A CNN layer has multiple channels. Each channel is associated with its own convolution kernel and produces its own feature map. We now consider a CNN with three convolution layers. The first CNN layer has three channels; this layer will produce three images each corresponding its own convolution kernel. The second CNN layer has two channels; this second layer will generate two images, associated with two convolution kernels. The third CNN layer has one channel; this third layer will produce one output image, as illustrated in Figure 9.32. On the other hand, a fully connected layer only produces one image as the output.

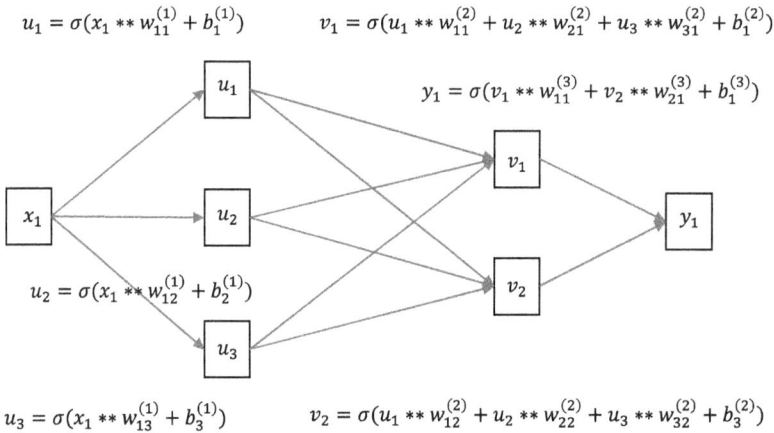

$$u_1 = \sigma(x_1 ** w_{11}^{(1)} + b_1^{(1)}) \qquad v_1 = \sigma(u_1 ** w_{11}^{(2)} + u_2 ** w_{21}^{(2)} + u_3 ** w_{31}^{(2)} + b_1^{(2)})$$

$$y_1 = \sigma(v_1 ** w_{11}^{(3)} + v_2 ** w_{21}^{(3)} + b_1^{(3)})$$

$$u_2 = \sigma(x_1 ** w_{12}^{(1)} + b_2^{(1)})$$

$$u_3 = \sigma(x_1 ** w_{13}^{(1)} + b_3^{(1)}) \qquad v_2 = \sigma(u_1 ** w_{12}^{(2)} + u_2 ** w_{22}^{(2)} + u_3 ** w_{32}^{(2)} + b_3^{(2)})$$

Fig. 9.32: A CNN with three convolutional layers. The first convolutional layer has three channels, the second convolution layer has two channels, and the third convolution layer has one layer.

In Figure 9.32, x_1, u_1, u_2, u_3, v_1, v_2, and y_1 are 2D images; $w_{ij}^{(k)}$ are 2D convolution kernels for the kth convolution layer associated the ith input image and the jth output image; $b_j^{(k)}$ are bias scalars for the kth convolution layer associated the jth output image; σ is a nonlinear element-by-element activation function, e.g., the ReLU function; $**$ denotes the 2D cross-correlation operator (also known as convolution operator in machine learning community). By comparing Figure 9.32 with Figure 9.8, we can see the clear differences between a CNN and a dense network. In a dense network, a neuron is a scalar output, calculated as

$$\sigma\left(\sum_{i=1}^{n} x_i w_i + b\right),\tag{9.26}$$

where x_i is the ith scalar element of the input n-tuple, w_i is the ith scalar element of the n-tuple weights, and b is the bias scalar. Here, n is the number of neurons in the previous layer. In a CNN network, a generalize neuron is an image output, calculated as

$$\sigma\left(\sum_{i=1}^{n} x_i \ast\ast w_i + b\right),\tag{9.27}$$

where x_i is the ith image of the input images, w_i is the ith 2D convolution kernel of the n kernels in this layer, and b is the bias scalar. Here, n is the number of channels. In a CNN, a common optional trick is to change the image size from one layer to the next layer.

The universal approximation theory discussed in Section 9.2.2 only applies to dense networks; it does not apply to CNNs. Why do we believe that a CNN can give us better results than some conventional methods, for example, in denoising? At least, we know that the machine learned denoising filter should outperform the conventional filters. This is simply because the conventional filters are convolution-based and are the special cases of the machined learning models. A conventional filter can be considered as a one-layer, one-channel filter without a nonlinear activation function.

Let us look at a simple two-layer denoising CNN in action. The input of the network is a sinogram. The first convolution layer has two channels, and the second convolution layer is the output layer that has only one channel. Each convolution kernel is 3×3 and the activation function is the ReLU. The outputs of the channels are shown in Figure 9.33. The first layer produces two filtered sinogram images. One image contains many bright dots and looks like a result from a high-pass filter, and the other image looks like a result from a low-pass filter. The second layer further filters the two images from the first layer and combines them in a nonlinear way to produce the final denoised result. The dot-like pattern is removed by the ReLU function. On the other hand, the conventional filter is linear. A linear combination of two linear filters is a third linear filter, which is not equivalent to a nonlinear filter. Nonlinear filters have better performance than linear filters in noise reduction while keeping the sharp edges.

A machine learned filter may or may not be shift-invariant. A shift-invariant filter applies the same action in every region in the input image, regardless the location of the region. The convolution (or cross correlation) action is shift-invariant. The max pulling action makes the filter shift-variant. If you want a shift-invariant denoising filter, you need to choose a shift-invariant action to change the image dimension, for example, average pooling.

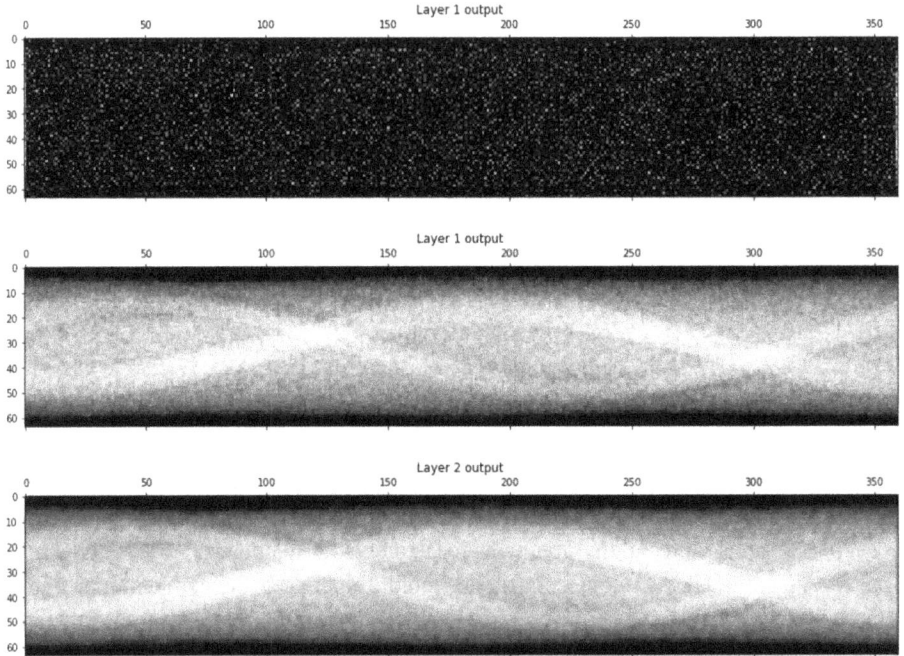

Fig. 9.33: Channel outputs. Top: From channel 1 of the first layer. Middle: From channel 2 of the first layer. Bottom: From the second layer.

9.6 Super-resolution

In medical imaging, sometimes the measurements are undersampled. To estimate the unmeasured measurements is closely related to super-resolution image estimation. We always have doubts whether it is even possible to obtain a high-resolution image from a low-resolution image. We believe that image processing is unable to create unmeasured information. Successful evidence suggests that under some circumstances super-resolution is possible. Super-resolution images may be visually appealing, but they may not be identical to the true original high-resolution images.

The traditional linear space expansion theory may not be able to explain the machine learning models due to the nonlinearities in the models. The traditional approach is to engineer a procedure to enhance the high frequency components so that the blurred low-resolution image becomes a sharp high-resolution image. This traditional approach does not work well because noise and Gibbs oscillation artifacts will be amplified.

One explanation suggests that the neural network somehow remembers the sharp edges of the training labels and find the closest sharp edges for the new input low-resolution images. This explanation may not be what actually happens in reality.

Another hypothetical explanation of the magic of achieving super-resolution is through image mapping. A mapping is essentially a table lookup approach as illustrated in Figure 9.34. The blurred curve on the left of Figure 9.34 is the input n-tuple of the machine learning model. The sharp curve on the right of Figure 9.34 is the output n-tuple of the model. After enough training pairs are used to train the model parameters, the model is expected to output a sharp curve for an unseen blurred curve input. How satisfactory of the output will depend on how similar the unseen input is to the training inputs.

Still another explanation is that the neural networks are nonlinear combinations of edge detection filters. As a result, the high-frequency edges are enhanced. To reduce the pixelization effects in the low-resolution images, a depth-to-space transform by shifting the images from different channel outputs to form a higher-resolution image is illustrated in Figure 9.35, where the phrase "feature extraction" is another way to say "filtration." The filtered images are sometimes referred to as feature maps. In Figure 9.35, the input is a 4×4 low-resolution image, and the output of the convolutional layers consists of four channels with a 4×4 at each channel. The depth-to-space transformation merges these four channels into one 8×8 high-resolution image.

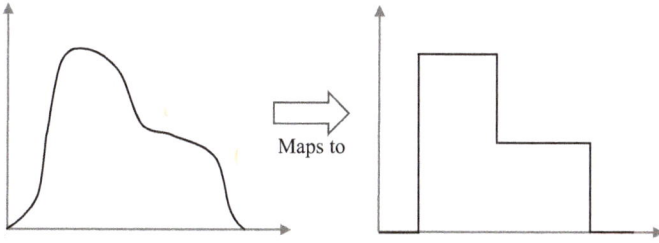

Fig. 9.34: Super-resolution is achieved by mapping or "table lookup" .

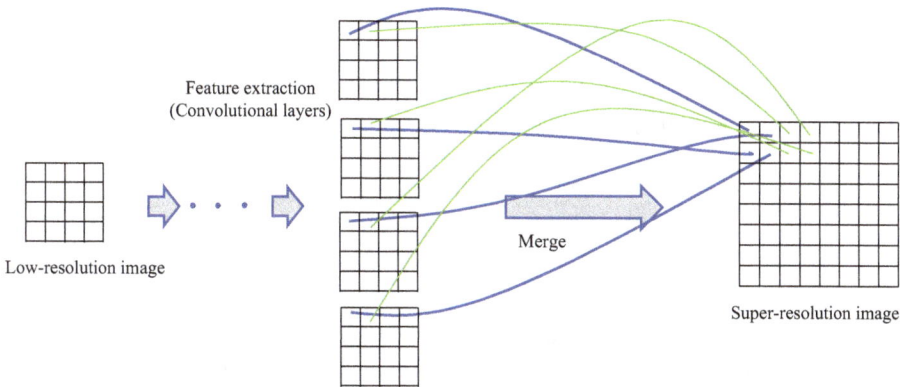

Fig. 9.35: Merging four channels of lower-resolution images into one higher-resolution image using the depth-to-space transform.

9.7 Training

Model training is similar to iterative image reconstruction in the sense that it is an optimization procedure. The objective function (also known as the cost function or loss function) in model training is the distance between the model output and the *label*. In machine learning terminology, a label is a desired output. In denoising, a label is a noiseless image, and an input is a noisy image. In super-resolution, a label is a high-resolution image, and an input is a low-resolution image. In direct (end-to-end) image reconstruction, a label is a true image, and an input is a sinogram. A general training setup is shown in Figure 9.36. Training is to find the model parameters W by minimizing the objective function with a lot of input/label pairs. Training is an iterative procedure that goes through all input/label pairs multiple times. Each iteration is called an *epoch* in machine learning community. After training, the trained model is ready to accept any unseen input and to produce an output in a fast and one-pass way without any iterations.

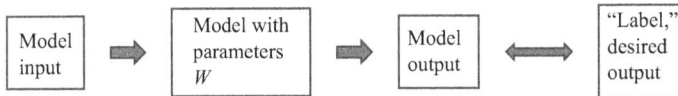

Fig. 9.36: A general setup for parameter training.

In machine learning, the training algorithms are mostly gradient descent-based. The gradient of the objective function with respect to the parameters is evaluated according to the chain rule, which is referred to as backpropagation in machine learning. As a toy example, let us consider a neural network shown in Figure 9.37 and find the gradient $\partial J / \partial w_1$, where J is the objective function

$$J = \frac{1}{2} \sum_{i=1}^{N} (y_i - \hat{y}_i)^2,$$

(9.28)

where $\{x_i, \hat{y}_i\}$ is an input/label pair, there are N input/label pairs, and y_i is the model output. The gradient $\partial J / \partial w_1$ is calculated using the chain rule as follows:

$$\frac{\partial J}{\partial w_1} = \sum_{i=1}^{N} (y_i - \hat{y}_i) \frac{dy_i}{d\sigma} \left(\frac{dv_1}{d\sigma} x_i w_3 + \frac{dv_2}{d\sigma} (0) w_4 \right)$$
$$= \sum_{i=1}^{N} (y_i - \hat{y}_i) u(y_i) u(v_1) x_i w_3,$$

(9.29)

where $u(\cdot)$ is the unit step function defined in eq. (9.5). Once the gradient of the objective function with respect to each parameter is known, a gradient-based iterative algorithm is ready to be implemented for training the network.

$$v_1 = \sigma(xw_1 + b_1)$$

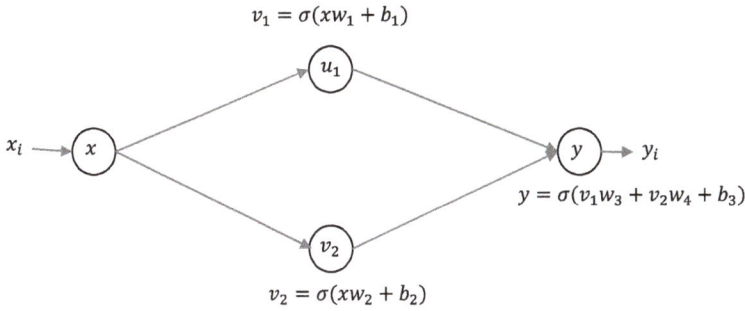

$$y = \sigma(v_1w_3 + v_2w_4 + b_3)$$

$$v_2 = \sigma(xw_2 + b_2)$$

Fig. 9.37: A small dense network.

9.8 Attention

There is a strategy in machine learning called *attention*. In a network layer, a pixel in the output is strongly influenced by some certain pixels in the input image. The network will pay more attention to those input pixels. In image denoising and super-resolution applications, the neighboring pixels have the most influence. The CNN architecture has a built-in attention scheme to focus only the nearby input image pixels covered by the convolution kernel. The uncovered pixels are ignored, as illustrated by Figure 9.38.

For a direct end-to-end reconstruction network, a pixel in the reconstructed image is strongly influenced by the sinogram data along a particular sinewave, as illustrated by Figure 9.39. The reconstruction network should pay more attention to those sinogram pixels along this particular sinewave. The sinewave attention region in the input varies with the output image pixel location.

A general attention architecture is illustrated in Figure 9.40, which looks like the residual network shown in Figure 9.27 excepting that the addition operator $\oplus$ is replaced by an element-by-element multiplication operator $\otimes$. Attention can help the network pay more attention to the high-frequency features. The high-frequency components usually are edges, texture, and other details. The output of the assisting network is called the *attention map* (also known as the *attention mask*). The values in an attention map range from 0 to 1. A value close to 1 in the attention map indicates that

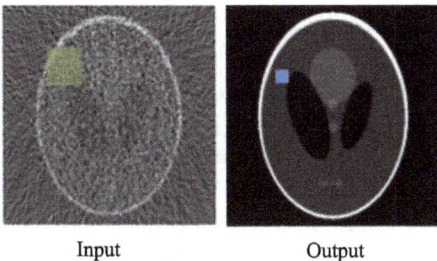

Input   Output

Fig. 9.38: The attention region marked in the left input image strongly influences the marked pixel in output image on the right.

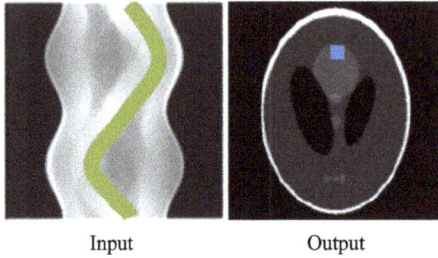

Input   Output

Fig. 9.39: The attention region marked as a sinewave in the left input sinogram strongly influences the marked pixel in output reconstruction image on the right.

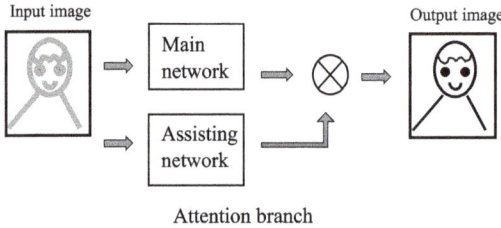

Fig. 9.40: Architecture of an attention branch that generates an attention map.

the associated pixel value in the feature map (i.e., the output of the main network) at the same location has more contribution to the main task in mind.

9.9 Perspectives

The original theoretical foundation of neural networks is to treat everything as an n-tuple to m-tuple mapping. The universal approximation theory is the backbone to support the field of neural network research. A neural network with sufficient parameters is able to approximate this mapping. Machine learning is to approximate this underlying function (mapping) through a large amount of data.

In practice, however, the universal approximation theory may not be needed. To approximate an n-D function on $[0,1]^n$ using a coarse (sparse) sampling of 10 samples in each dimension, a model may need as many as 10^n parameters. For a small image of size 32×32, we need as many as 10^{1024} parameters. This number of 10^{1024} is astronomical!

The success of neural networks in classification tasks may be credited to the inner product operation. In Section 9.4, we have used the matched filter perspective to explain how a CNN can be used to recognize a pattern (for example, a bicycle). The application of a matched filter is the same as the inner product operation, where one operand is an input image and the other operand is a desired image pattern. The main difference between a fully connected (also known as dense) neural network layer and a CNN layer is that a CNN kernel (i.e., the desired pattern) traverses all over the input image and looks for a match. In a dense layer, nothing moves. We now use the MNIST handwritten digit dataset to illustrate this point. This dataset is as basic and as famous as phrase "Hello, World!" as a test message for a beginner in programming.

For example, the digit "4" can be written in different ways, some examples are illustrated in Figure 9.40. It is difficult to find a universal pattern W such that if the input image X is "4" then the inner product $W \cdot X$ is greater than a threshold b, and if the input image X is not "4" then the inner product $W \cdot X$ is less than the threshold b.

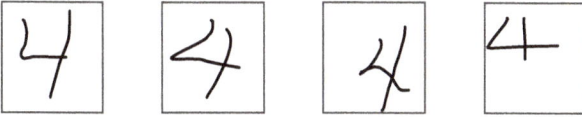

Fig. 9.41: Handwritten samples of digit "4".

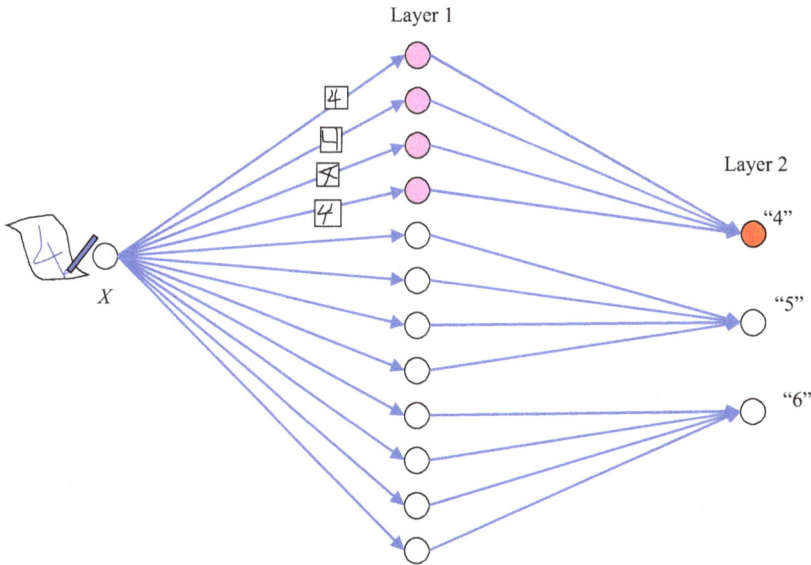

Fig. 9.42: The first dense layer has multiple patterns for each digit; each subcategory of a digit belongs to one neuron. In the second dense layer, each neuron corresponds to a unique digit that collects all its subcategories in the first layer.

One solution to resolve this difficulty is to have multiple subcategories of the digit "4." Each subcategory represents similar styles of "4" and uses a pattern W_i and a threshold b_i. We can build a dense layer neuron, $\text{ReLU}(W_i \cdot X - b_i)$, for the ith subcategory of "4." The next dense layer collects all the subcategories of "4" into one neuron; all subcategories of other digits are clumped up in the similar manner, as shown in Figure 9.42.

The toy example of a dense neural network shown in Figure 9.42 tries to classify three handwritten digits: "4," "5," and "6." Each digit is subdivided into four subcategories according to different handwriting styles. The output of each neuron in the first layer is a scalar, calculated by ReLU($W_i \cdot X - b_i$), where W_i is the pattern for the subcategory and b_i is a threshold value. If the input X is a handwritten "4," some of the pink neurons are activated with positive values.

Each layer is fully connected. An image X or its corresponding pattern image W_i can be represented in the vector format (in any order without any effect on training) so that the inner product $W_i \cdot X$ makes sense.

The second layer is, in fact, a dense layer. The links with zero weights are not shown in Figure 9.42. Each neuron in the second layer essentially sums up the output from all its corresponding subcategories from the first layer. This gives the one-hot result in the second layer. Here by one-hot, we imply that the neuron with the maximum output value is referred to as hot with a binary value 1 and other neurons are assigned a binary value 0. Optionally, a "softmax" activation function is used in the second layer. The softmax function normalizes the neuron output values into a probability-like distribution.

The actual MNIST handwritten digit dataset has 60,000 training image/label pairs. It contains handwritten digits from "0" to "9." If a fully connected neural network (similar to that shown in Figure 9.42) is used for the classification task, the first hidden layer may contain as many as 2,000 neurons and the second contains 10 neurons.

For the end-to-end image reconstruction, the input is a raw measurement-domain sinogram, and the output is a tomographic image. The image reconstruction is a regression (instead of classification) task. The image can be implemented as using a dense layer to perform the backprojection and then using a series of convolutional layers to perform image processing, which includes the ramp filtering. This dense layer can be considered as a "hardwired" backprojector on the pixel-by-pixel base. A backprojector for one image pixel can be ready represented as an inner product between the input sinogram and a sinewave pattern. There is no need for a threshold or for a nonlinear activation function for this dense backprojection layer. In fact, this dense layer is not dense at all. Almost all the weights are zeros except for the sinewave pattern.

For an MRI end-to-end image reconstruction task with a general k-space sampling scheme, the same neural network architecture as above can be used. The dense layer performs the k-space re-gridding and the 2D inverse Fourier transform. An inner product expression can be readily obtained for the re-gridding and inverse Fourier transform. Notice that the k-space measurements are complex with real parts and imaginary parts. Our neural network commonly performs real number calculations. We can simply store a complex number as two real numbers, one for the real part and the other for the imaginary part, in any order, taking the advantage of the fully connections.

A dense layer is able to perform image processing in the sense of traditional convolution, which is essential in denoising and super-resolution tasks because convolution is a special case of inner product and every common neuron contains an inner product operation.

Both of convolutional layers and dense layers require multiple kernels for a single input image. The fully connected architecture implements these multiple kernels by using multiple neurons in each layer. The CNN architecture implements these multiple convolution kernels by using multiple channels in each layer. The CNN architecture further assumes that the convolutional kernels are shift-invariant, while the kernels for the fully connected network generally do not have the shift-invariant property. Therefore, it is not proper to call the inner products in a fully connected network as convolutions.

9.10 Summary

– The power of machine learning is in its ability of creating a general mapping between two numerical strings or two images. The key characteristic of the mapping is nonlinearity, which is more general and more powerful than the linear mapping. When given a group of n-tuple/m-tuple pairs, the network tries its best to fit those pairs if the network is large enough. A nonlinear m-dimensional "label" space is created.
– After the network is trained, when an unseen n-tuple is fed to the network, an m-tuple in the "label" space is selected as the output. This selected m-tuple has the common features of those training labels.
– From our own perspectives, the real power of machine learning in imaging is through the versatility of the inner product operation. It can be used for matched filtering, image backprojection, re-gridding, inverse Fourier transform, and so on.
– The training of a network is to determine the model parameters through optimizing an objective function, which is formed by the training pairs and the distance between the model outputs and the labels. The training algorithms are mostly gradient-based. The gradients are calculated by the chain rule, which is referred to as the *backpropagation*.
– The machine learning techniques can be applied to image reconstruction in many ways:
    – End-to-end direct reconstruction: The network input is a sinogram (or k-space data in MRI). The network output is the reconstructed image.
    – Analytic reconstruction: The network is used as nonlinear denoising filter, which can be applied in the measurement domain or in the image domain. The nonlinear network filter can also replace the ramp filter.

- Iterative reconstruction: The network can be used as a denoising filter which can be applied to the images at every iteration. The network output can also be used in the objective function of an image reconstruction algorithm.
  - The network model can be used in other areas such as extension of the under-sampled measurements, motion corrections, artifact reduction, and so on.
- Machine learning models contain nonlinearities. Even a quadratic least-squares cost function is used for the training, though the cost function has many local minima. It is most likely that our trained neural network is suboptimal, regardless the optimization algorithms. Some cost functions and some optimization algorithms perform better than others.
- There is no substitute for the measurements of high-quality data. Image denoising and super-resolution do not generate reliable results especially in the medical fields because image denoising or super-resolution neural networks may remove vital details.
- Finally, we should always be alert that the machine learning generated model may not give correct results in some cases. For example, if the sinogram is under-sampled and the object contains something that is never found in the training pairs, it is unlikely that this thing can be recovered by the machine-learned model. If the image is too noisy, the output of a denoising model is still noisy. Therefore, it is unrealistic to expect the network model to produce a perfect result.

Problem

Problem 9.1 Consider one layer of a dense layer, whose input is an image consisting of 512^2 pixels and the output is an image of the same size. Calculate the total number of parameters in this dense layer. Then consider only layer of a CNN with one channel and the convolution kernel is 3×3. Calculate the total number of parameters in this convolutional layer.

Problem 9.2 Consider a U-net. Calculate the total number of pixels in all channels at each layer.

Problem 9.3 For the small dense network shown in Figure 9.34, find the gradient $\partial J / \partial b_1$.

Problem 9.4 Write an objective function for iterative Bayesian image reconstruction for the following situation. The image may have some artifacts, which can be effectively detected by a neural network model m. When an image x is fed to m, the model produces a positive value symbolically denoted as $m\, x$. A large value $m\, x$ indicates severe artifacts in the image x. No artifacts are found if $m\, x$ is zero.

Bibliography

[1] Widrow B, Samuel D (1985) Stearns: Adaptive Signal Processing, Prentice Hall, ISBN 0-13-004029-0.

[2] Hornik K, Stinchcombe M, White H (1989) Multilayer feedforward networks are universal approximators (PDF). Neural Networks 2:359–366, Pergamon Press.

[3] Lu Z, Pu H, Wang F, Hu Z, Wang L (2017) The expressive power of neural networks: A view from the width. Adv Neural Inf Process Syst 30:6231–6239, Curran Associates, arXiv:1709.02540.

[4] Zhu B, Liu JZ, Cauley SF, Rosen BR, Rosen MS (Mar 2018) Image reconstruction by domain-transform manifold learning. Nature 555(7697):487–492.

[5] Reader AJ, Corda G, Mehranian A, D. Costa-luis C, Ellis S, Schnabel JA (Jan 2021) Deep learning for PET image reconstruction. IEEE Trans Radiat Plasma Med Sci 5(1):1–25, doi: 10.1109/TRPMS.2020.3014786.

[6] Wang G (2016) A perspective on deep imaging. IEEE Access 4:8914–8924, doi: 10.1109/ACCESS.2016.2624938.

[7] Zeng GL (2021) A deep-network piecewise linear approximation formula. IEEE Access 9:120665–120674, doi: 10.1109/ACCESS.2021.3109173, PMCID: PMC8442618, NIHMSID: NIHMS1738616, PMID: 34532202, https://ieeexplore.ieee.org/document/9526604?source=authoralert.

[8] Lv L, Zeng GL, Zan Y, Hong X, Guo M, Chen G, Tao W, Ding W, Huang Q (2022) A back-projection-and-filtering-like (BPF-like) reconstruction method with the deep learning filtration from listmode data in TOF-PET. Med Phys 49(4):2531–2544, https://doi.org/10.1002/mp.15520.

[9] Zeng GL (2022) Examples of machine learning models from classic to modern. 2022 Intermountain Engineering, Technology and Computing (IETC), 1–6, doi: 10.1109/IETC54973.2022.9796796. https://ieeexplore.ieee.org/stamp/stamp.jsp?tp=&arnumber=9796796.

Index

https://doi.org/10.1515/9783111055404-010